UNIVERSITY OF STRATHCLYDE
30125 0054720
KV-119-720

ANDERSONIAN LIBRARY
WITHDRAWN
FROM
LIBRARY
STOCK
UNIVERSITY OF STRATHCLYDE

POLYMER PROCESS ENGINEERING 97

POLYMER PROCESS ENGINEERING 97

Edited by
P. D. Coates
Professor of Polymer Engineering and
Associate Director of the IRC in
Polymer Science and Technology
Mechanical & Manufacturing Engineering
University of Bradford

The Institute of Materials

The University of Bradford

Book 681
First published in 1997 by
The Institute of Materials
1 Carlton House Terrace
London SW1Y 5DB

ISBN 1-86125-044-4

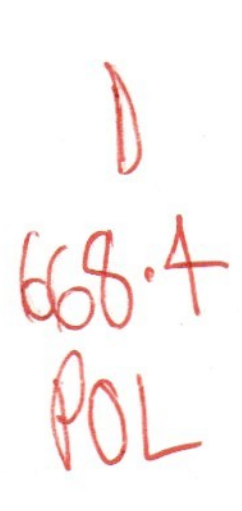

Reproduced from authors' camera-copy
Printed and bound at
The University Press, Cambridge

CONTENTS

Foreword *vii*

Seven Drivers of Highly Effective Plastics Suppliers 1
E. Natalini, Ford Motor Co, Germany

1: INJECTION MOULDING DEVELOPMENTS

Innovative Injection Moulding Processes for Better Products 2
H. Eckhardt, Battenfeld GmhH, Germany

Selective Foaming Technology 17
B. Penney and P. Clarke, Coralfoam Ltd UK

Appraisal of 2K Moulding 24
J. Tinson, Sifam Ltd, UK

Air Intake Manifolds – Lost Core Moulding versus Vibration Welded 32
M. Hazel, DuPont, UK

Surface Layer Technology 39
S. Kilim, Addmix Ltd, UK

Structure Development in Injection Moulding 40
E. M. Gil, G. Kalay, P. S. Allen & M. J. Bevis, Brunel University, UK

2: TRENDS IN PROCESS MONITORING & CONTROL

Process Control for Rotational Moulding 55
R. J. Crawford, The Queen's University of Belfast, UK

Illuminating the Manufacturing Process – On-line UV Spectroscopic Analysis 65
H Herman, Actinic Ltd, and P S Hope, BP Chemicals, UK

Dual Thermocouple Loops for Extruder Temperature Control 78
D. Ferris, Davis Standard, HES Division, USA

In-Line Measurement of Copolymer Composition and Melt Index 89
M. G. Hansen and S. Vedula, University of Tennessee, USA

Quality Regulation Techniques for Injection Moulding 103
D. Kazmer, University of Massachussetts, Amherst USA
R. Thomas and G. Sherbelis, Moldflow Pty, Australia

Injection Process Monitoring and Process SPC 115
A. Dawson, A. Key, M. Kamala, R. M. Rose and P. D. Coates
Bradford University/IRC in Polymer Science & Technology, UK

Neural Computing: Problem Solving and Applications 125
J Pilkington Scienflfic Computers,
R Amin & T Hyde, ERA Technology, UK

3: EXTRUSION DEVELOPMENTS

Microlayer Coextrusion Processing and Applications 137
C. Mueller, J. Kerns, T. Ebeling, S. Nazarenko, A. Hiltner and E. Baer, Case Western Reserve University, USA

Technical Blow Mouldings from Engineering Polymers, Including Two Component Mouldings 158
P. Deane, DuPont, UK

Multi-Layer Blow Moulding of Fuel Tanks 173
G. Pickwell & N. Thacker, BPTA, UK

Intermittent Encapsulation of Plastics and Other Techniques for Discontinuous Extrusion 178
S. Dominey, Killion Inc, USA

Scale up of Extruders, Practicalities and Pitfalls 187
J. A. Colbert, Betol Machinery Ltd, UK

Chemical Assist Foaming and the Role of Supercritical Fluids in Extrusion 197
M. Reedy, Reedy International Corporation, USA

Integrated Compounding Technology for the Preparation of Polymer Composites Containing Waste Materials 202
C. E. Bream, E. Hinrichsen, P. R. Hornsby, K. Tarverdi and K. S. Williams, Brunel University, UK

4: PROCESS RHEOLOGY AND FLOW MODELLING

Simulation of the Injection Moulding Process - A Practical Validation 214
P. A. Glendenning, PERA Technology, UK

Gas-Assisted Injection Moulding Modelling 227
P. Gorman, Plastic Moulding Consultants, UK

Extensional Viscoelastic Measurements of Polymer Melts 240
M. Rides, National Physical Laboratory, UK

Flow visualisation for Extensional Viscosity Assessment 264
C. Nakason, M. Kamala, M. T. Martyn, P. Olley and P. D. Coates, Bradford University/IRC in Polymer Science & Technology, UK

Biaxial Extensional Rheometry of PET Relevant to Process Conditions 275
C. Gerlach and P. Buckley Oxford University, UK
D. P. Jones, ICI Films, UK

In Line and On Line Rheometry of Filled Polyolefins 289
D. Barnwell and K. Martin, Raychem Ltd, UK
A. L. Kelly, M. Woodhead and P. D. Coates, Bradford University/IRC in Polymer Science & Technology, UK

Foreword

Professor P. D. Coates, FEng

Polymer processing is an exciting, fast moving area of industry, which has also attracted a strong, lively academic presence worldwide. The growth in the use of plastics, driven by considerations of property, styling, mass production capabilities and cost, ensures that the processing sector will continue to develop, in the quest to ensure enhanced quality and efficiency. The scope of this volume is to address the issues: *where is polymer processing going? and what are the key trends in technology at the end of the 20th century in this vital international industry?* Each of the papers in the book was presented at the Polymer Process Engineering International Conference, July 1997. In total they cover engineering excellence in polymer processing, with leading edge developments in polymer processing technology, in-process measurements and process flow modelling, and control

Key developments in the major generic processing areas include important variants on the standard injection moulding and extrusion processes in terms of machine and material control, and novel processing techniques. These are complemented by presentation of trends in process monitoring and control, where developments in at-process or in-process sensing are opening up new possibilities for quality control in extrusion and injection moulding processes – these include the move of spectroscopic techniques nearer to the process, philosophies of control approaches, and also strategies for dealing with large amounts of process data. The underlying process rheology and computer modelling of process flows is well represented, ranging from basic extensional and shear rheology studies – off line and in-process – related to processing, through to evaluations and applications of commercial flow modelling software. Throughout there is considerable interest in sensors and their application for enhanced monitoring of polymer, machinery and process, in process engineering and the application of first class engineering science to the manufacturing process.

I believe that this book provides a timely and most interesting spectrum of Polymer Process Engineering, and I am grateful to all the authors and colleagues who have made this volume possible.

The PPE conference, a biennial international conference, is organised by:

- The Polymer Processing and Engineering Committee of the Institute of Materials,
- The UK Interdisciplinary Research Centre in Polymer Science and Technology, and
- The University of Bradford.

Professor P. D. Coates FEng
Conference Organising Committee Chairman;
Associate Director,
Interdisciplinary Research Centre in Polymer Engineering and Science
Dept of Mechanical & Manufacturing Engineering, University of Bradford,
Bradford BD7 1DP

The Institute of Materials
Polymer Processing & Engineering Committee

Seven drivers of highly effective plastics suppliers

E Natlani
European Supplier Technical Assistance Manager
Ford Werke AG,
Köln, Germany

ABSTRACT

The next generation in plastics is accompanied by a broadening of scope. The plastics industry is undergoing change to keep pace with OEMs. Ford 2000 and similar restructuring at other automotive companies and the industry at large are forcing a shift in the operating principles of the plastics suppliers.

There are seven systems drivers fuelling this:

- Globalization
- Full service suppliers
- Modules
- New Technology
- Recycling
- Total Cost Management, and
- Quality

The plastics supplier of the future, whether raw material or component supplier, will be required to think on a systems level for development, manage cost and improve quality while embracing the changes these seven drivers require. This presentation outlines these seven drivers and the resulting paradigm shift that will have to take place in the industry.

The Supplier 2000 will have 'fit for use' products that are on time, meet cost targets and are defect free.

Innovative Injection Moulding Processes for better products

Dipl.-Ing. Helmut Eckardt
Process and Engineering Dept., Battenfeld GmbH, Postfach 1164/65
D-58540 Meinerzhagen, Federal Republic of Germany
Polymer Process Engineering 97,
University of Bradford, UK

Injection moulding has been known for many decades and has been developed as a process for moulding the full range from very small to very large injection moulding. Modern machine and mould technology in combination with the raw material enable the production of high quality mouldings. The technology has been further developed, modern C-technologies (CAD/CAE) are being used more and more.

Moulding with thick wall sections create sink marks and tend to warp. Mouldings with flexible and rigid sections made in conventional injection moulding need to be produced in two injection steps and assembled afterwards. Extremely large size mouldings create problems by internal stresses. To overcome the limits of standard injection moulding innovative processes have been developed, not to compete with but to support standard injection moulding.

The following explanations and ideas show, that by means of these processes not only new ideas can be realised but also quality can be improved and costs reduced.

1. Innovative Injection Moulding Processes

1.1 Structural foam injection moulding

The process has been known for more than 25 years. In the seventies and early eighties there was a large interest in this process because of the possibility to produce large size moulding with thick walls without any sink marks and no warping.

The process works in such a way, that a plastic material containing a blowing agent is injected into a cavity. During the rapid injection the blowing agent starts to foam and fill the cavity. As a result of the filling method the parts have regular foam structure inside without any sinking and warping. Splashes resulting from the stretched blowing bubbles on the surface are the disadvantage of this process.

There have been developed a number of modified structural foam processes to overcome the disadvantage of the surface defect. The most important one is the gas counter pressure process. In this process cavity before injecting plastic containing a blowing agent is between 10 and 25 bar have to be used in the cavity.

The advantage is a good surface quality in combination with a regular fine foam structure inside.(Fig. 1) The disadvantage is that the moulds have to be gas sealed. In the late eighties there was less interest in these processes mainly because the co-injection process and also the gas assisted injection moulding became popular.

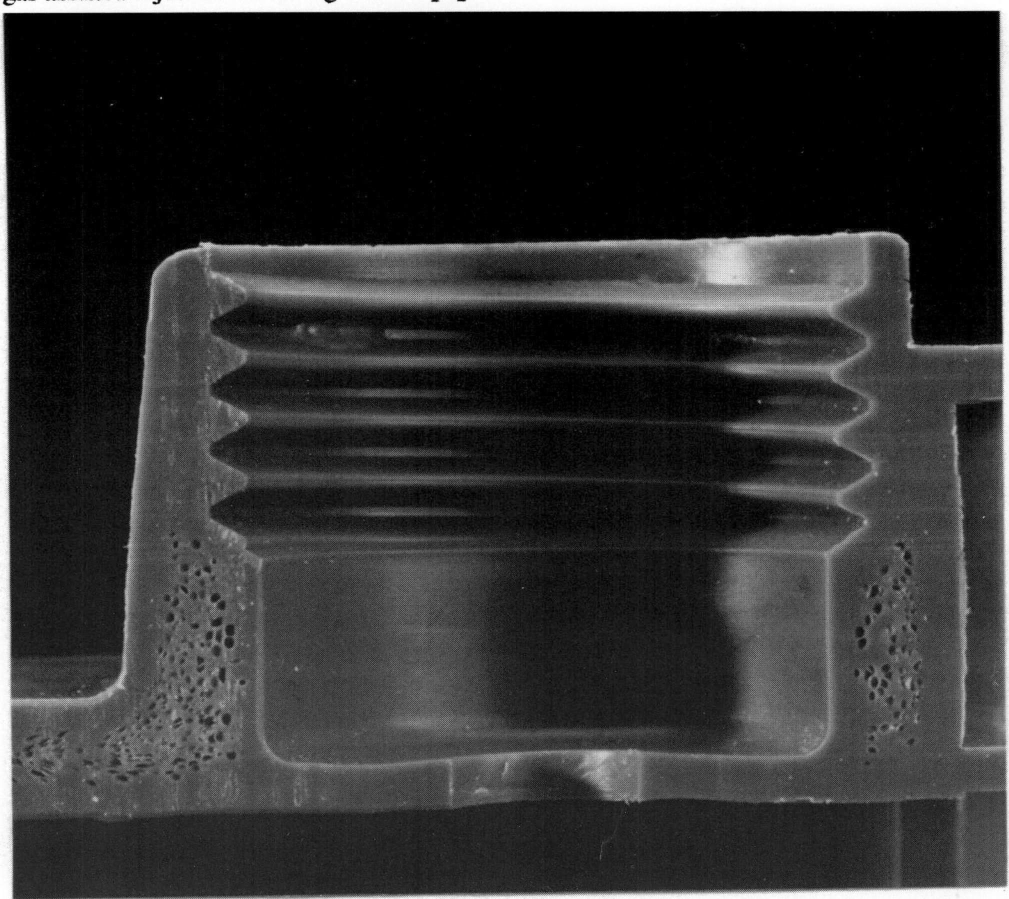

2. Multi-Polymer Injection Moulding

2.1 Definition

Although we are talking in general about the Multi-polymer injection moulding process in most cases we use two types of material only.

Because of the diversity of different injection moulding processes and also the rapid developments of the mould technology, new mould technologies arose which offer now product characteristics. Unfortunately, a confusion occurred as there is not yet a standard naming of the different process so that at first a clarification of the definitions of the different processes used today has to be found.

2.1.2 Two-Polymer Injection Moulding Processes

This process has already been used for many years, the start was to use polymers of different colours, that the process was called two-colour moulding. Today the process is better named two-material injection moulding, as two polymers of different grades can be injected.

Parts consisting of two materials which are different in colour or type are injected in a way that one polymer is injected in the first cavity.(Fig. 2) Then after opening of the mould the moulded product comes into action with a second cavity for injection of the second material. One uses rotary moulds or injection moulding machines with a rotary table or sliding table.

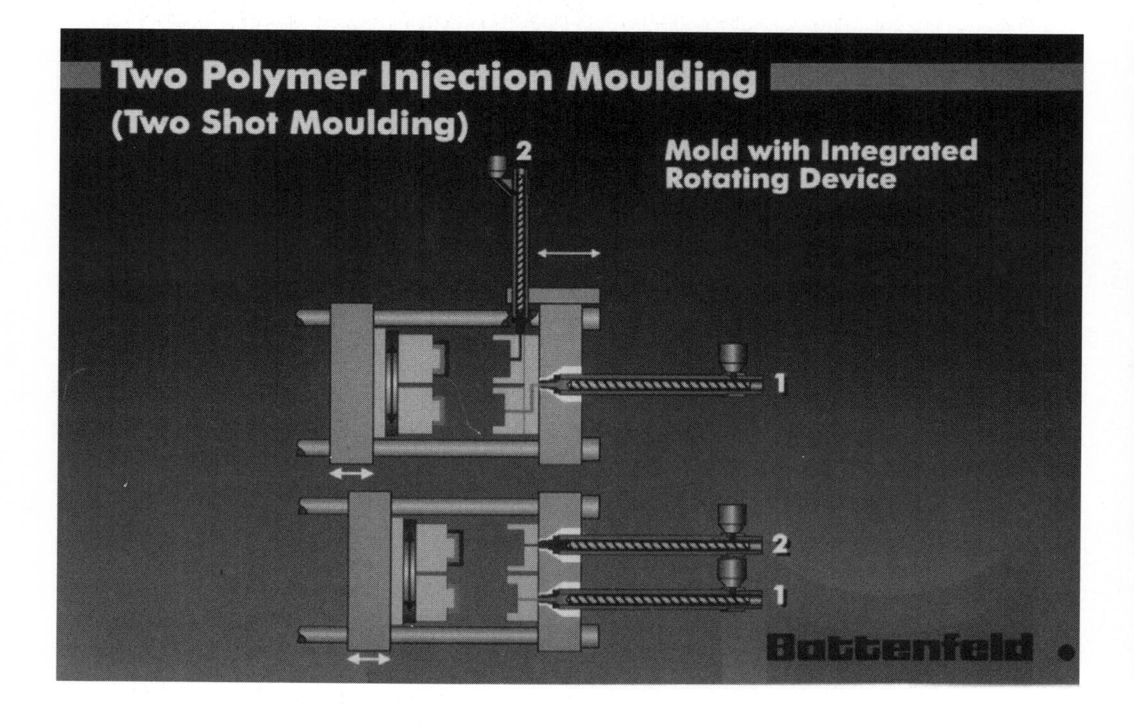

If more than two polymers are used, e. g. three or four, in some cases one has to employ as many moulds as types of materials. Applications of this process areas multi-colour keys and automotive rear lights.

2.2.2 Composite Injection Moulding (Combiform)

Another technique for the injection of two or more materials is by means of movable segments or cores in a closed mould.(Fig. 3) The cavities for the second material are initially closed and, after injection of the first material, they are opened for the second one by means of core movements.

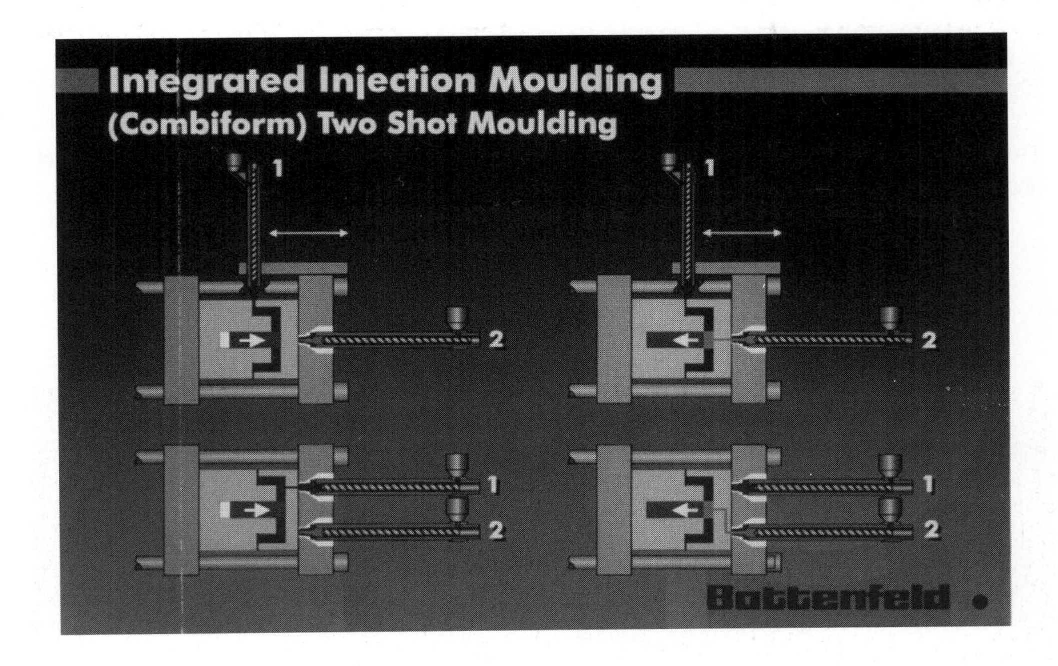

The process offers similar possibilities like the two-polymer injection moulding process but there is only one mould needed which is more expensive. A mould opening between the injection of the first and second material is not necessary. This process was introduced as the "Combiform process" by Battenfeld in 1979.

This process in general is suitable for all mouldings where by means of a core movement in the mould the cavity for the second material to be injected can be closed.

For most applications we need good adhesion between material one and two, but there are also applications where good adhesion among materials is not desired. The moulding with movable moulded sections can be produced in a single operations.
The Two-polymer and the structural injection moulding are Two Shot Processes, that means injection of first and second polymer are sequential.

2.2.3 Multi-Component Injection Moulding Process

The Multi-component injection moulding process already expresses in its name the possibility to employ more than two materials. As most of the time only two materials are used this process is also called two-component injection moulding.

Both expressions, Multi-component injection moulding process and Two-component injection moulding process are today equally titled side by side, or even co-injection or Sandwich Moulding.

This process will be described in the following:

Two materials are injected through a specially designed nozzle (Fig. 4) into one mould so that one material encapsulates the other completely.

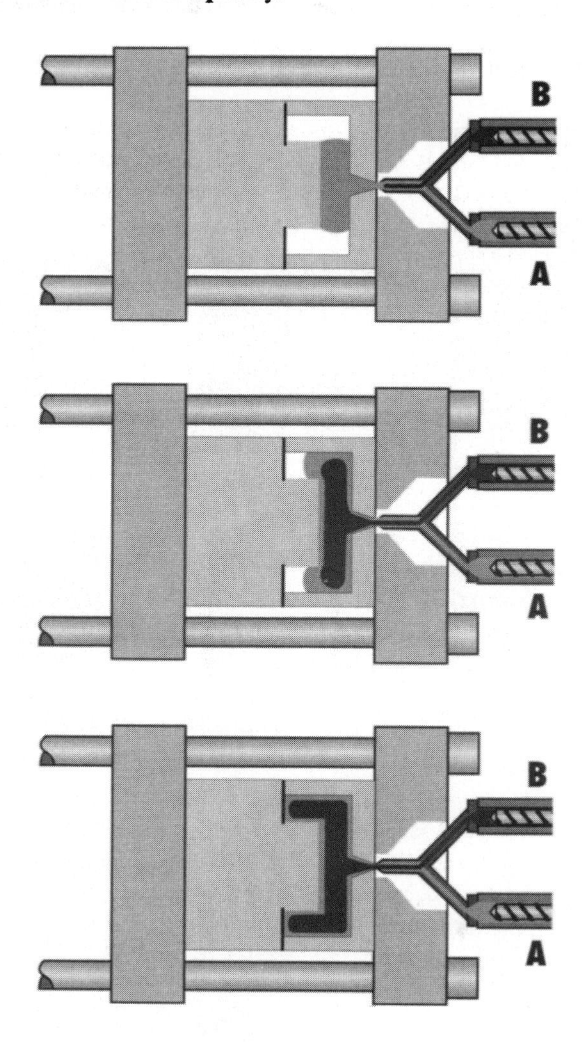

For moulding with wall thickness of 4 and more mm the core material contains of a blowing agent, for thinner walls the core is injected compact.

3. Innovative Applications

3.1 Structural foam

There was the idea to make a telephone cabin in one shot only to reduce assembly cost and time. When thinking about the technical possibilities by means of the different injection moulding processes, the only process offering the design complexity and enabling the production of housings of that large shot weight was structural foam.

In a team consisting of the customer, the mould maker, the raw material supplier and the machine maker there were discussed all possibilities and also potential problems to make sure that these did not occur. First problem to overcome was the demoulding of this very complex part.
The next was the filling because of the extremely large size.

The result is a telephone cabin (Fig. 5) moulded in fibre glass reinforced polycarbonate structural foam with a shot weight of 60 kg. The injection moulding machine consists of a vertical clamp having 2.500 t clamping tonnage and 3 injection units.
The mould size is also enormous. The total mould weight is 185 t.

Every time the mould opens a complete telephone cabin can be moulded. The telephone cabin then will be painted and assembled.

3.2 Two-polymer injection moulding

The wet room socket

The traditional production method was injection moulding of the socket and the seal moulded in a flexible thermoplastic material separately. These two mouldings were assembled before mounting of the wet room socket.

To reduce costs and to improve quality there was the idea to do the injection moulding of the socket and the seal in one moulding operation. It was discussed how to modify the design of the sealing and also the socket to enable the production by means of this innovative process.

Cost studies covering machine and mould costs, polymer costs, cycle time etc. showed that at the end of this development the moulding could be made at even lower costs compared to the former method not calculating the reduction in logistics and assembling costs. (Fig. 6)

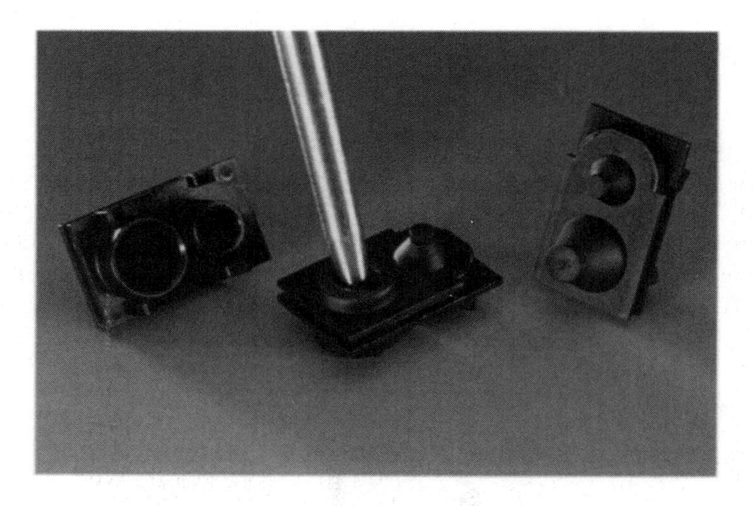

Cable guide in PP with TPE seal*
*Moulder: Stewing

Rear light

Rear lights for cars today are moulded in the two-polymer injection moulding process. The rear light (left and right) was to be moulded in three materials, red, yellow and transparent. As the red and the yellow colour are not getting in touch to each other, there was the possibility to inject red and yellow in one production step and then after opening and moving the premouldings to a next station, overmould the total product by means of transparent PMMA. (Fig. 7)

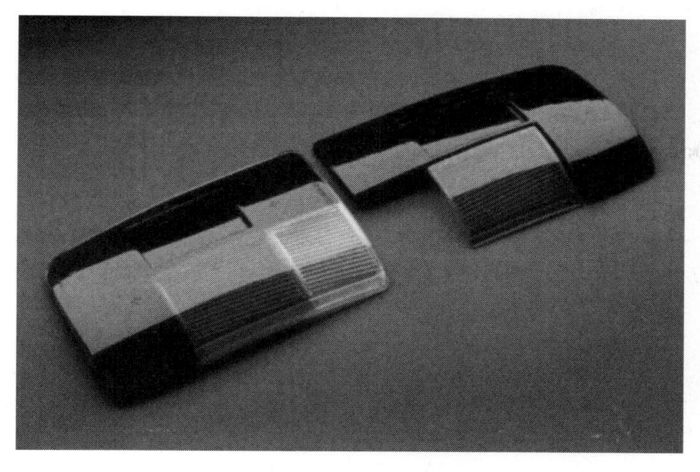

Section through bumper. The red part is injected before the black.

The mould is quite complex in shape to make sure that in the first step of the production the cores are in the forward position and then without opening the mould and after pulling the cores injecting of the second material.

Automotive cover with seal lip

The development of automotives leads to more use of plastics and closer connections of the plastic parts to each other and smaller gaps in between all different parts. That means, there is a great interest to get mouldings which have an integrated seal lip. To enable the production of mouldings with a seal lip the Combiform process is a very good solution. The moving cores in the initial stage are blocking off the cavity for the second material. After having injected the first material the cores are retracted to allow the injection of the second material. (Fig. 10)

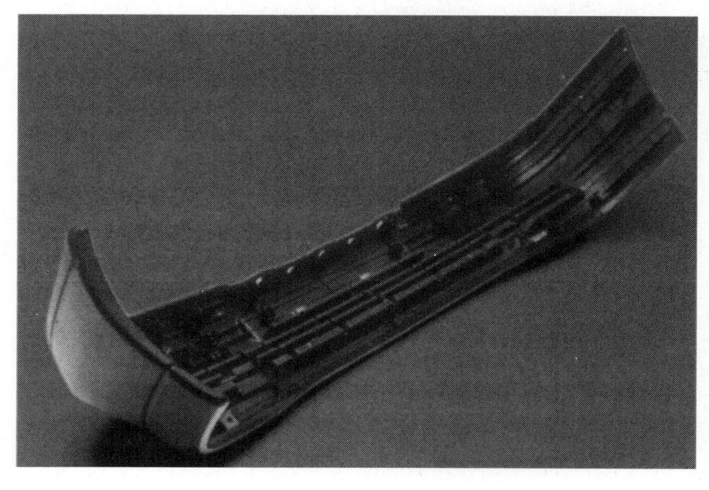

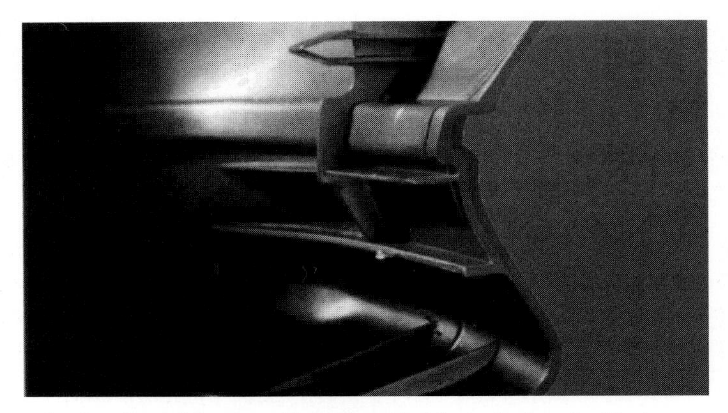

Section through bumper. The red part is injected before the black.

The mould is quite complex in shape to make sure that in the first step of the production the cores are in the forward position and then without opening the mould and after pulling the cores injecting of the second material.

Automotive cover with seal lip

The development of automotives leads to more use of plastics and closer connections of the plastic parts to each other and smaller gaps in between all different parts. That means, there is a great interest to get mouldings which have an integrated seal lip. To enable the production of mouldings with a seal lip the Combiform process is a very good solution. The moving cores in the initial stage are blocking off the cavity for the second material. After having injected the first material the cores are retracted to allow the injection of the second material. (Fig. 10)

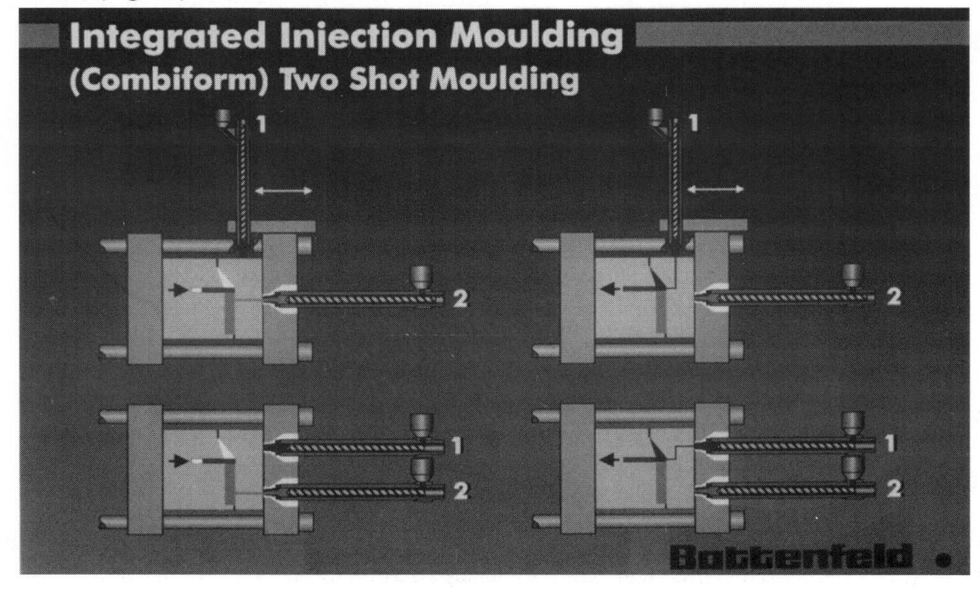

That means the contour of the core must be designed in such a way to block off the second cavity in the initial stage and for the finished moulding to give the contour of that section of the moulding.

For the production of rigid mouldings with soft seals the shot weight between rigid and soft material is quite different. The ratio can be 7:1 up to 10:1.

The cover for the automotive was moulded on a machine having a clamping tonnage of 650 t, the second injection unit is about 1/10 in shot weight compared to the first one and it is mounted hucky pack above the first injection unit. By means of this production quality could be improved and costs reduced, as there is no need for any assembling or secondary operation.

Washing machine cover

The cover for a washing machine was to be moulded with a seal. The former models were equipped with flexible seals which had to be mounted in a second step to the injection moulded rigid product.

To reduce the costs and to avoid any potential problems or faults in assembling the flexible seal to the cover it was created the idea to use the Combiform method for the production.

A cover was moulded in talcum filled polypropylene, the seal is TPE. (Fig. 11) The machine is equipped with a clamping tonnage of 400 t, the second injection unit is located above the basic injection unit.

Cover in PP with TPE seal*

3.4 Multi-Component Injection Moulding

Paper sorter

For a copy machine there was to be designed a suitable paper sorter. Because of the size of the component in the design it was clear that normal injection moulding was not suitable.

To make the moulding strong and rigid enough there was a need to use wall thickness' of around 6 mm. These wall thickness' in normal injection moulding cannot be managed without sinking and warping of the moulding. The answer was the use of the co-injection method which enables the production of thick walled mouldings above 4 mm with a foamed core material, for thinner walls skin and core are injected compact.

By using a foamed core material in combination with an unfoamed skin material it could be guaranteed that the moulding would get a perfect surface finish but without sinking and warping.

To get a uniform distribution of the core all over the moulding the gating position and the wall thickness have to be selected and designed properly. In this moulding the part had to be gated centrally, because of the given design there was no alternative.

By means of the processing conditions and wall thickness adaptation it could be assured that the core material was distributed over the moulding.

The material is a PC/ABS alloy with the same core material but foamed. The shot weight is 7,5 kg.

The moulding is made on a co-injection machine having a clamping tonnage of 1.050 t being equipped with two injection units of size 2 x 10.000 nozzle.

The multi-component injection moulding machines of series BM are very flexible in use. They can be used for

- co-injection with a solid skin and a foamed core
- co-injection with a solid skin and a solid core
- compact injection moulding with one or the other injection units or both
- structural foam
- gas counter pressure process with the relevant moulds and the gas counter pressure kit
- gas assisted injection moulding (Airmould) with the relevant kit

The system allows raw material combinations such as

- skin compact - core foamed
- skin compact - core compact
- skin reinforced - core not reinforced
- skin not reinforced - core not reinforced
- skin flexible - core rigid
- skin rigid - core flexible
- skin shielded - core not shielded
- skin not shielded - core shielded
- skin compact - core barrier properties
- skin virgin - core recycled material

By means of this process tailor made properties of the mouldings can easily be achieved.

Cover

The moulding of PC/ABS alloy was designed for normal injection moulding. Because of the u-shaped design there was the problem of warping which could not be controlled by means of the moulding conditions.

By using the co-injection process and adding a little of blowing agent to the core material the internal stresses could be reduced. That is why the mouldings could be made in a much better quality without any further warping.

3.5 Airmould

The Airmould process is a gas assisted injection moulding process. The pressure controlled process enables the production of high quality mouldings with hollow sections. In general this process is suitable for four different kinds of products

- Rod shaped mouldings

The advantages are

weight reduction
reduction in cycle time

- Panel shaped mouldings

Panel shaped mouldings need gas channels to control the distribution of the gas. By means of the gas assisted injection moulding process these products can be made with

no sink marks
high rigidity because of rigid rips
reduction of clamping tonnage

- Panel shaped mouldings with locally increased heavy sections

These are mouldings which have some areas with increased sections. By means of the Airmould process these thick sections can be moulded without any sink marks. Advantages are

no sink marks
easier mould design when undercuts can be eliminated

- Problematic mouldings

These are all mouldings which cause problems in normal injection moulding.

Advantages are

no sink marks
low internal stresses
low warping
lower clamping tonnage

The process offers a wide range of applications.

The Airmould Modular Concept consists of three basic modules

- pressure generating system
- pressure control systems
- gas injection

<u>Door handle</u>

For a car an exterior door handle had to be developed. The material is a polyamide 6.6 with 35% fibre glass reinforcement. By creating a hollow section in the centre of the handle the weight should be reduced and cycle time as well. It has to be guaranteed that no gas is going into the side section to assure the stiffness and the impact of the moulding. The solution was the side injection of the gas directly by means of the gas injection module. These modules are extremely small (3 mm) that the integration even into existing moulds is no problem, the gas injection modules fit in between the cooling channels.

The mould was designed as a two cavity mould where each gas injection was achieved by means of one gas injection module (Fig. 12) being controlled by a pressure control module.

The pressure control modules are designed in such a way that because of their small size they can be mounted very close to the mould. This ensures a very accurate pressure control and also saves energy because of the low losses of the gas.

The handles are filled in a short shot method, the complete filling is made by means of the gas. Because of the very exact close loop control of the pressure control systems there is no hesitation mark to be seen on the moulding.

Cover for copy machine

The housing has quite large dimensions. The former generation was designed for structural foam because of the rigidity.

As the surface quality of structural foam was not sufficient without painting in structural foam there was the interest in making the mouldings with a perfect surface finish. Because of the size and the thin wall normal injection moulding could not be used, the part would have warped and would not have been rigid enough.

The solution was designed for Airmould. There were designed three basic channel systems connecting all the important bosses and rips to the channel system. The gas is injected directly into each channel system.

The polymer is a polycarbonate ABS alloy.

The moulding machine is an 800 t press.

A basic advantage of the Airmould Modular System is that the control cabinet can be easily connected to each machine by means of an interface. One control cabinet can feed up to four pressure control modules.

3.6 Co-Injection and Airmould

The combination of co-injection and Airmould offers the possibility to make thick walled mouldings with different materials for skin and core and to core out the inner section by means of gas.

The production of a high quality arm rest for a business chair was designed for this process. For the skin there was used a nonreinforced and soft polyamide, the core material is a highly fibre glass reinforced polyamide to give rigidity. The gas is injected into the centre, the filling was achieved from one end.

The result is a high quality soft touch arm rest with integrated stiffness by means of the reinforced polyamide whereby the reduction in weight by means of the gas compared to the compact injection moulded product is around 50%.

3.7 Film insert moulding and Airmould

Insert moulding of plastic films and sheets has been known for a long time.

The in-mould decoration is a process which enables the production of finished products a decor in different colours in a one shot operation. The film with the decor is transported through the open mould. The plastic injected is activating the hot temperature glue of the decor, thus the decor remains on the moulding when the mould is opened. A finished product can be achieved by means of this method.

It is also possible to use thicker films and foils which e. g. have been painted. These films with wall thickness' up to several 10/ of a mm can be either laid into the cavity and formed by means of the injected plastic or they can be even vacuum formed before inserting into the cavity.

Side body panel

Side body panels with relatively thick walls are injected in this method. A painted ABS film is used. The film is inserted into the open cavity and overmoulded by means of polyurethane. To avoid any sinking or warping of the moulding gas by means of the Airmould injection moulding process is injected afterwards.

Front grill

For a car a new front grill was to be designed. This front grill has several thick sections, the material is a polycarbonate PBTP alloy. The target was to make the front grill with a finished surface. A painted ABS film first is vacuum formed and then inserted into the cavity. Then the PC/PBTP alloy is injected and the gas is injected into those areas where the thick sections are.

By means of the combination of the processes a perfect surface finish in combination with a structural strength of the moulding can be achieved.

4. Conclusions

Innovative injection moulding processes offer a wide range of possibilities to make innovative and structural injection mouldings.

The injection moulding processes are available and a large know how of how to design the products is also known.

It is important to not only think of how to use a specific process for a specific application but to look at the properties of the specific moulding and think about suitable processes. As shown in the previous examples sometimes a combination of more than one injection moulding process is the right solution.

Innovative injection moulding process will become more and more important for the future to improve quality, reduce costs and enable the production of innovative and integral injection mouldings meeting the specific requirements of the moulding.

Dipl.-Ing. Helmut Eckardt * Battenfeld GmbH * D-58540 Meinerzhagen

SELECTIVE FOAMING TECHNOLOGY

Barrie Penney & Peter Clarke

Coralfoam Ltd, Unit 12, Hamper Common Industrial Estate
Petworth, West Sussex, GU28 9NR

ABSTRACT
Coralfoam is the new process technology that gives moulders' and designers the possibility to produce injection moulded parts that have thick foamed sections, but that mould in remarkably short cycle times. With Coralfoam technology it is possible to produce mouldings with higher stiffness, lighter weight and in a shorter cycle time. The process also brings other benefits. These include lower stress, reduced clamp, improved flow, easy mould release and complex shapes from simple moulds.

Coralfoam is a patented technology and Coralfoam Limited has been established to develop and licence the technology. At present there are 8 patents applied for and development work is continuing to further increase the cost savings that are possible.

The Coralfoam process is based on selective foaming certain areas of the moulding to expand to a thickness that is greater than the section in the closed mould. If we look at each of the characteristics of Coralfoam you will appreciate how this process technology works and how it can be applied to a large number of parts.

CORALFOAM SELECTIVE FOAMING TECHNOLOGY

The Coralfoam process is based on selective foaming certain areas of the moulding to expand to a thickness that is greater than the section in the closed mould. If we look at each of the characteristics of Coralfoam you will appreciate how this process technology works and how it can be applied to a large number of parts. Compared to conventional injection moulding the characteristics of Coralfoam do present some contradictions.

CORALFOAM CHARACTERISTIC 1

10MM THICK RIBS - 6 SECOND CYCLE TIME.

Using conventional moulding techniques stiffening ribs provide stiffness but are also a source of stresses and give problems due to sink marks. To avoid problems of this type the normal design requirement is to keep the rib to two thirds of the wall section and provide a generous radius to avoid a concentration of stresses. The resultant mass of material controls the cycle time and the amount of hold on pressure that is needed. With Coralfoam technology such limitations do not apply. Reinforcement ribs can be placed onto surfaces where they are needed. The ribs can be 5mm to 50mm wide even when placed on thin walled mouldings of 0.5mm thickness. The shape of the rib is then controlled by the expansion which is predetermined by the process conditions.
Compared to conventional designs Coralfoam technology produces closed section ribs which have greater stiffness, and that in particular, provide great torsional stiffness.

A simple plaque mould was developed to prove the theory on thick ribs. The plaque has an average thickness of 1.2mm.

To create the reinforcement rib with a section of 10mm x 7mm the wall section of the plaque is thickened locally. The material is injected and the tool is opened within 2 seconds. The release of the hold on pressure allows the thicker section to expand. The expansion is controlled by the release of carbon dioxide gas from within the melt.
By adding an endothermic blowing agent as a masterbatch at the hopper it is possible to create a solution of carbon dioxide throughout the melt.
Especially with polypropylene materials there is a high solubility and a perfect distribution of gas can be achieved. This means that at each point in the cavity, or in each cavity of a multi cavity tool there is an equal gas pressure and concentration.
As soon as the skin has formed in the thinner sections of the moulding the mould can be opened. In the thin areas the carbon dioxide cannot deform the skin so the carbon dioxide goes to atmosphere and the transition from a liquid to a gas provides additional cooling in the material. In the sections where expansion is to taker place the gas creates a foam structure or a void and deforms or moves the skin to the required position. In the case of our plaque tool a rib having a width of 10mm and a height of 7mm is moulded in 6 seconds. Normally a compact moulding of a similar dimension would require at least 30 seconds of cooling time. The plaque is a simple mould but the principal translates to more complicated parts.

Experience so far shows that the ribs can be placed at different distances from the gate and material can flow through different section thickness without affecting the ability for a predictable foaming and expansion to take place. Significantly a large rib can be placed at the end of the flow path and a rib will be produced of the anticipated dimension. In this respect we have the second contradiction of conventional mould design, that thick sections can be fed through thin sections, and they can be at the ends of flow paths.

CORALFOAM CHARACTERISTIC 2

SAME PART WEIGHT - 300% STIFFER.

Light weighting exercises in packaging, automotive and other applications areas have left many products with a poorer quality image. In containers the parts are too flexible. In engineering applications they no longer have the physical presence or the feel that helped communicate quality.

Once a lightweighting programme has been started Coralfoam technology can make a major contribution by bringing additional stiffness to the part and adding sections that appear to be generous with the amount of material used. In fact it is an illusion that comes from the perceived quality of a generously radiussed closed rib, or solid piece of material that has high stiffness. As an example, Coralfoam took a horticultural flower pot which lacked stiffness and which had an average wall section of 0.45mm. The wall section was constant to achieve a fast cycle time and the surface of the mould was highly polished. To eject the polypropylene flower pot a stripper ring was necessary. We removed metal from the tool around the lip of the flower pot increasing the wall section to 0.6mm.
The stripper ring was disengaged and ejection was modified to only use an air blast circuit. The mould was placed back on the machine and although metal had been taken away, the shot weight was reduced by 3% due to foaming.

An endothermic blowing agent was added as a 50% active masterbatch at the hopper in a concentration of 2%. The mould was opened as soon as the general side wall had frozen in the 0.45mm area and the 0.6mm area was allowed to expand to form a reinforcing ring of 3.5mm cross section.
The compressive hoop strength of the flower pot was now 3 times the hoop strength of the base moulding. The top load strength remained the same.
The impact strength in the thin section remained the same. Using Coralfoam it had been possible to increase the hoop strength 3 times with the same volume of polymer and without loss of properties.
In this case the benefits of Coralfoam technology, are in the physical propeties of the flower pot which has a higher stiffness and appears to be of a higher quality, without a cost penalty being incurred. Although the material cost, which now included the blowing agent, had increased, the reduction in cycle time to 3.7 seconds balanced the equation.

CORALFOAM CHARACTERISTIC 3

FOAM STRUCTURES - HIGH GLOSS SURFACES

Most peoples experience with blowing agents and foams is that there is a trade off with regard to surface finish. A grained surface or splash marks tend to appear. This has not been the experience with Coralfoam technology. With Coralfoam technology mouldings with a matt, semi gloss or gloss surface can all be produced. The material will take up the contour of the mould during injection, even though a full shot is not made. The carbon dioxide within the melt provides the final expansion that gives the overall product form. The carbon dioxide is a low pressure gas and providing the correct conditions are maintained (fast injection, correct tool temperature, low clamp, poor venting) the gas will not break through the melt front and good appearance parts can be produced.

This low pressure expansion ensures that mould surface details are accurately replicated, but at the same time the lack of packing eliminates flashing or poor quality edges. Surface distortion due to stress is not a problem. The mould is then opened to allow the expansion of the Coralfoam selectively foamed sections. After expansion the surface remains practically the same as the mould, although textured matt surfaces will have a tendency to become more glossy.

In the thin sections where no expansion takes place, the carbon dioxide goes to atmosphere and leaves no visible change on the surface of the moulding. In the development of our drinking cup we were able to show the possibilities. The cup has a grit blasted surface to give a light texture, but we have also polished in two small windows. No break through of the foam occurs, and even though the windows are positioned after a foamed section, a high gloss and perfect surface quality are achieved.

CORALFOAM CHARACTERISTIC 4

50% LIGHTER - 40% FASTER

Now that we are beginning to understand the characteristics of Coralfoam it becomes possible to take a different approach to the design of moulded articles. In some cases this will result in substantially thinner sections that have the same stiffness as thicker non Coralfoamed articles.

First let us look at the drinking cup that we developed. The design was based on providing thermal insulation for hot drink usage.
Unfortunately, even by increasing the thickness of the material in a foamed or non foamed structure, within the mould , it was not possible to achieve the necessary insulation.

Using Coralfoam technology the thick section was allowed to expand to 8 x the starting thickness. This provided the insulation needed and also increased the compressive hoop strength by 5 times.
A thinner section of Coralfoam selective foam satisfied the insulation requirement, and it was possible to achieve the stiffness requirement from 50% of the new raw material. The cup has now been redesigned to meet the application requirement with a 50% thinner wall reducing the designed weight. The cup was moulded using Coralfoam technology and cycled in 6 seconds compared to 8 seconds non foamed.
After making the materials savings, cycle time of less than 4 seconds was possible. This represents a major cost saving and doubles the utilisation of the moulding machine.

The same design principal translates to other parts but in different proportions. We took a 2 litre container having a part weight of 54 grams. The objective was to reduce the weight to 40 grams and improve the stiffness. The existing lidding feature was required to work. The overall wall section was reduced by 25% and thicker sections were then added back on a localised basis to provide stiffening ribs and flow aids. The container had previously run with a polypropylene melt flow 12 material. The material was changed to a medium low temperature impact grade. Previously the part was moulded in a 300 ton clamp machine. This was now reduced to a 175 ton clamp machine. The cycle time was reduced from 6.0 seconds to 4.5 seconds. The mould was opened as soon as the thin wall had stabilised and the thick sections expanded to provide the selectively foamed areas. Even with the lower modulus of the easy flow grade of polypropylene the stiffness from the foamed sections resulted in a part that had much higher stiffness at a lower overall partweight.

In our next case we redesigned a noodle pot used for packaging hot fill dried noodles as well as for serving the prepared food. Most of the designs for noodle pots are flanged containers with external ribs that provide a degree of insulation from the hot contents. Our redesign looked at the possibility of making an improved pot from an insulation point of view, and with a shorter cycle time.

The design we finalised on was submitted as an entry in the Awards for Excellence at the recent Interplas exhibition in the UK. The Coralfoam noodle pot was awarded first prize.

Within the design we allowed foaming to take place in 3 areas. Firstly in the lip. The purpose was to expand the lip to provide the necessary stiffness when filled with hot contents. The rigidity from the lip also helps prevent panelling when the container is hot filled with dried noodles.

The radiussed lip replaces the moulded flange with several benefits. The foamed lip is a 'soft' shape. When users drink from it the feel to the mouth is comfortable and pleasing. The 'hard'

flange has non of these attractive features. As the radiussed lip expands the tool split line is absorbed. There are no sharp edges. The radiussed foamed lip represents a much shorter flow length than the moulded flange.
With thinner wall sections being moulded, the shorter flow length is a real benefit. To create the top hat shape from the raduissed lip there is no change in direction of the melt with the resultant pressure drops. The whole shape is made with a shorter flow length. Moulded flanges may include two 90 degree changes in flow.

Noodle pots need to be lidded. To obtain a good lidding performance the flange is normally designed with energy directors. These are not needed with Coralfoam technology.
The foamed section is radiussed and consists of a thin skin with a small mass of material behind it. This matches very well the material on the foil, and hot foil lidding can be carried out quickly and repeatability. The foil will also peel evenly. The foamed lip also provides excellent thermal insulation. The insulation value of a foamed lip is far greater than a vertical rib as far less material is being used to create the volume, and a significant gap is created between the hot contents and the plastic.

The second area to be foamed was in small ellipsoids around the side of the container. This shape was chosen for two reasons. Firstly the ellipsoid can be allowed to expand both inwards and outwards. This gives a greater air gap and increases the insulation effect. Generally, if the design was a continuous band, then any inward expansion would cause the band to compress itself and this would cause a tyre effect. By breaking the band into ellipsoids the compression of the ring is eliminated. By breaking the section into ellipsoids more possibilities for different styles of decoration are also created.

The third area of selective foaming was in the base where the feet are provided with thermal insulation.

The process will work in multi cavity tooling. Each cavity can be fed with a hot runner. With the noodle pot a weight saving of 30% is possible for the same stiffness and thermal insulation is increased. A cycle time of less than 6 seconds is achievable.

CORALFOAM CHARACTERISTIC 5

SELECTIVE FOAM - TRANSPARENT SECTIONS

Transparent foams sounds like a complete contradiction. With Coralfoam it is possible to produce clear polypropylene mouldings with areas that have been selectively foamed. For this a very good control is needed over the injection pressure, and the best results to date have been achieved on the Netstal Synergy machine using the Graphtrak control.

In the thin sections the pressure is maintained preventing the formation of discrete bubbles. Once the skin has hardened the tool is opened and the thicker sections expand whilst the thin sections remain clear. In products such as drinking glasses this makes it possible to place a foamed lip at the top to give great stiffness and an attractive shape to drink from. It also provides an additional possibility to decorate the side wall of the glass or container with a logo or decoration. The decoration will stand out as a white 3D effect in a clear container.

CORALFOAM CHARACTERISTIC 6

10MM UNDERCUTS - NO SPLIT CAVITIES

Because the foaming takes place in a open mould the Coralfoam effect can be used to produce shapes or features that would only have been possible with much more complicated tooling.

Using this feature it is possible to produce mouldings that can lock together. A drum moulding can have a reinforcing ring on the internal bore.

A door pocket on a car can be integrated with simple tooling. Conventional ribs can have foamed sections at the end to provide reinforcing beams of foam.

All of these are shapes that would require expensive tooling features using conventional injection mould design.

CORALFOAM CHARACTERISTIC 7

CONCAVE TOOL - CONVEX PART

We have recently started work on the next generation of shapes. We are now able to predictability produce mouldings from concave moulds that have convex double walled forms. You can see in the example that the mould which starts as a concave form generates a moulding that is a mirror image of the starting mould. In this way sections of enormous stiffness can be produced from relatively small quantities of material. In developing rod shaped mouldings we have produced mould forms with a cross section 'w' out of which we produce mouldings with a closed figure 8 section. Cycle times for sections 3.2mm thick have been less than 3 seconds.

THE CORALFOAM STRATEGY

Coralfoam is a licencable technology and as you can see from the presentation we have done most of our work so far in the food packaging sector. We expect to grant further options for Coralfoam licences in the food packaging sector during 1997. Other option have already been granted in food service and horticultural market sectors. We are also talking to a number of companies about automotive, TV, and toy applications.

We expect in the automotive sector to be able to make considerable savings on interior trim articles, and we also feel that Coralfoam will contribute to safety through the possibility to produce mouldings with radiussed edges and lower stress levels.

We expect Coralfoam to gain widespread acceptance as a new process technology as it works well on existing plant and equipment and due to the fact that no special skills are needed with regard to mould-making or processing.

After a few days training most moulders will feel comfortable applying the principals. The mouldings produced can be moulded to precise tolerances that will satisfy the needs of either packaging or automotive customers. Once conditions are established the processing window is generous and operator friendly.

Products that would benefit from Coralfoam technology include:- pallets, crates, boxes, ice cream containers, yoghurt pots; snack food containers; razor handles; safety helmets; car bumpers; pens; closures; lids; mushroom punnets; garden furniture; cutlery; plates and bowls; toilet seats; car door trays; airline disposables and reusables; dual ovenable dishes; meat trays; toys; housewares; take away food containers; planters; body armour for sports, etc

It will be interesting to see where Coralfoam offers the greatest savings.

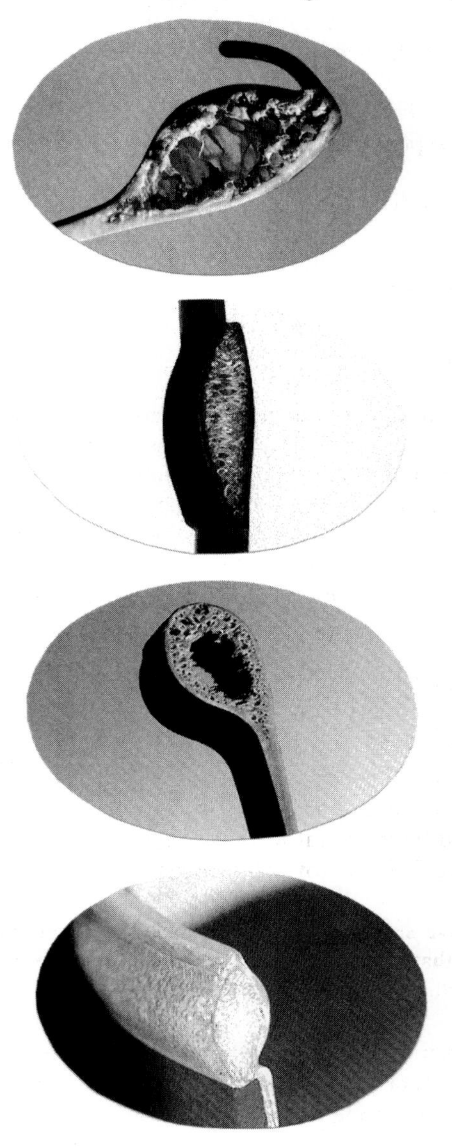

Sections in some products using Coralfoam Selective Foaming technology

Appraisal of 2K Moulding

JOHN TINSON
SALES & MARKETING DIRECTOR OF SIFAM LTD

Multi-shot injection moulding is a process whereby a number of thermoplastic materials are injected simultaneously, or sequentially, into a mould on a single machine during one machine cycle - the most common form being a two material process, perhaps one material being a soft elastomer. Machines are now available that can inject up to six different materials, a two material injection process would be called 2K.

In this paper, the author explores the different injection methods, concentrating mainly on a rotating platen 2 or 3K process whereby two or three materials/colours are injected to build up a part. The benefits in product cost and point of sale advantage are discussed with case studies in the field of power tools, mobile phones, domestic appliances, cars and household products.

The advantages of such a process in designing-in gaskets, soft touch areas, eliminating of printing operations, product colour enhancement and moulded in windows are shown with recent case studies from the UK's leading specialists in the field of 2 & 3K moulding and tooling, SIFAM Ltd of Torquay.

INTRODUCTION

SIFAM, the Torquay based plastics and measurement technology group, has been involved in multi-colour/material 2K moulding for its own in-house use for over ten years. Recent investments in toolroom and mould shop machinery have allowed SIFAM to expand into custom injection 2 & 3K moulding for the general industrial and consumer markets.

Multi-colour or multi-material moulding has been around for over twenty five years in such applications as computer key pads or car dashboard controls. Many companies are now combining the use of multi-colour/material moulding with the advent of new soft touch materials to gain an ergonomic or visual advantage over the competition. This paper will look at some of the reasons behind this, as well as some of the advantages to be gained from this technique.

DEFINITION

Two colours or materials can be combined in the moulding process in a number of ways. The innermost part can be taken out of one moulding machine and placed into a second moulding machine for over moulding; a common example of this has been two material screw driver handles. A robot arm would lift the inner element made in a standard single shot tool and place it into a mould on another machine. A second material would be moulded around it to form a two material part.

Another method would be to inject two colours or materials simultaneously from different sides of the tool. If they were allowed to meet without interference, then a very strong bond between the materials could be achieved. This principle has been used to make workmen's boots with a high glass fill nylon toe cap rather than a steel toe cap. Where the material is forced to meet along a dividing wall in the tool, a straight line would occur as is found in tail lights on cars. These can be three or four shot injection moulded.

Perhaps the most common method today is to revolve the tool on a two shot moulding machine through 180 degrees on an index plate, such that the second injection shot takes place within one machine cycle. It is this method that will be discussed. SIFAM uses this revolving index plate method to mould up to three different colours or materials all within the same machine cycle. The moulding machines used at SIFAM are either Arburgs or Engels.

Illustration 1 - Two shot 175 tonne machine

Illustration 2 shows a typical general assembly of a two shot injection mould tool designed to run on a two shot machine with one horizontal and one vertical barrel. This particular tool illustrates a *concept* two shot telephone with the first shot being a window and the second shot the main body.

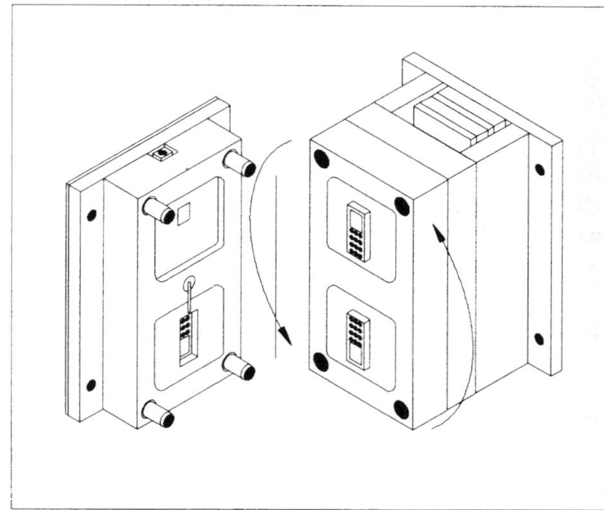

Illustration 2 - Concept two-shot telephone tool

The tooling is constructed on the principle of a constant core, hence on the moving half there are duplicate cores top and bottom - the principle being that the first shot is moulded in one cavity, the tool opens with the first shot moulding retained on the core, the tool then rotates 90, 180, 120 degrees (dependant on tooling layout) and is introduced to the second shot cavity still retained on the original core, where the second shot material is injected. As the two shot press is equipped with two barrels, injection is simultaneous; hence each time the press opens, a completed moulding is ejected. Further developments in technology mean that it is possible to robotically insert an inmould appliqué into the first shot cavity prior to the cycle commencing, this would offer the ability to introduce not only a printed mask but a scratch/static resistant finish to the final product. This is technology currently being explored in the telecommunications field.

Material selection is an important part of the two shot process. The best results can be achieved by using materials that give a chemical bond when overmoulded. An enclosure such as the one illustrated could be moulded using polycarbonate for the window and an ABS/PC blend for the case. This material selection offers very good adhesion as well as good strength and durability properties. The technique has three main attractions: the first is to add an area of colour or graphic to a product, the second is to add a window or soft material area for ergonomic reasons, and the third is to save money by eliminating subsequent assembly or print operations on the part. In truth, it is the latter which is most attractive to many volume manufacturing companies; the former two advantages come as a very handy by product of the process.

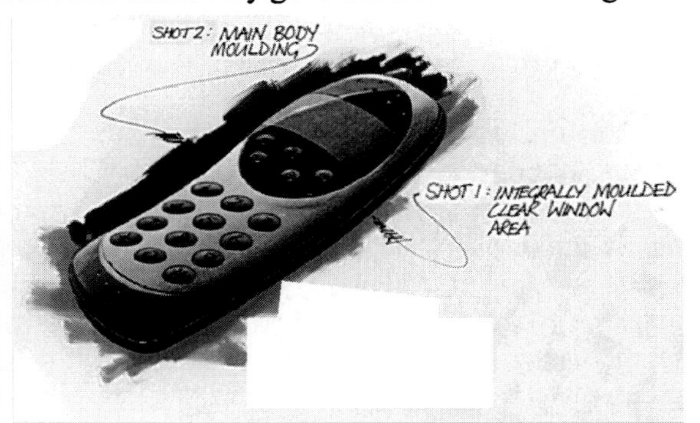

Illustration 3 - Concept telephone

The Importance Of Design

In the 1980's, products were differentiated at point of sale, let's say a high street shop or buyer's office, by their technical features. Items such as a camera would come with a long list of features made possible by the latest microprocessor technology. Consumers may have made the purchase decision using a matrix of features against price as shown in the "Which" consumer guides of the time.

Towards the end of the 80's came the realisation that technology alone was not the complete answer. Most people did not use fifteen wash cycles on their washing machine and did not need to record five different programmes simultaneously on their video recorder. Products started to become simpler and a greater emphasis was placed on their external form, texture, colour and weight to gain point of sale advantage.

The shape of products also took a more natural form with soft curves replacing straight lines. Shapes in nature and, in fact the human form itself are rarely straight lines, a designer drawing freehand will never draw in straight lines. Once products such as cars and electronic equipment followed these natural curves, they began to take on the more organic and friendlier shapes that appealed, not just because of the product's function, but as a design in its own right. The trend in the 90's is to aim the product at a very specific section of the total market for that item by using design stimuli that will appeal to this section. We have seen this in the advertising that is aspirational, but we now also see it in the design form itself.

In the car industry, cars like the Mazda MX5 were designed to appeal to certain sub-sections of the market (in this case women between the age of 25-35), and were very successful because of it. The ultimate in use of design stimuli is perhaps the Toyota Lexus. The note of the engine, the clunk of the door and the specially enhanced smell of the leather seats have all been tuned for maximum appeal within the car's target executive market. The keys were even weighted to give that extra quality feel when handed over for the test drive.

Research indicates that over 60% of the decisions to purchase are made at the actual point of sale where the customer in the shop or buyer in the office handles the product. At this point the customer has invariably boiled the decision down to two or three choices that will all technically do the job. At the point of picking the product up, the look, feel, and texture of the product will make a big difference in the final purchase decision. It is the job of the designer in the 90's to use the latest manufacturing techniques and materials to maximise sales of the product. The use of visual response, i.e. colour, as well as tactile response is critical to this.

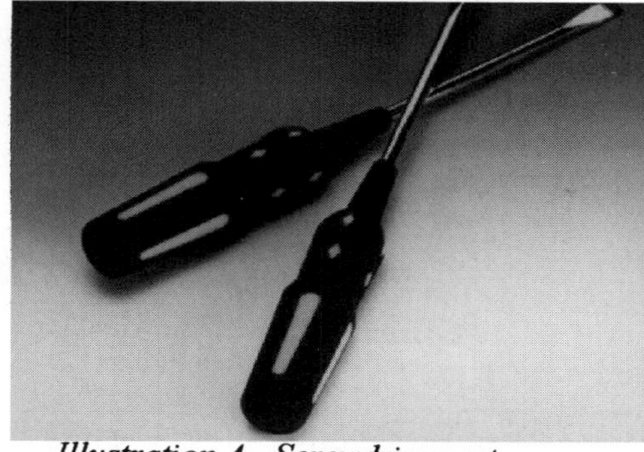

Illustration 4 - Screwdriver set

As an example, a person in a DIY store deciding between a set of screw drivers might be swayed by the fact that Stanley have used a soft touch material, coupled to strong colours, to give a warm ergonomic feel. The customer knows that this is a product that will not give him blisters. It is also a bright, cheerful product that stands out on the shelf. A classic example in fact of good multi-colour and material moulding to gain point of sale advantage.

Another example would be that of the simple toothbrush. One might have thought that over the past fifty years, designers would have run the gamut of design possibilities. A visit to any supermarket will show that, with the addition of colour and flexibility offered by different materials, the combinations for design are almost unlimited.

A very interesting aside to the toothbrush story is the sales price itself. Before the advent of two-shot toothbrushes, the market in the UK was flat and the price stuck below £1. A quick look at your supermarket shelf today will show that most brushes are now multi-shot and the price has gone up to £1.75. The single shot brushes are still stuck below £1. The additional works cost in moving from single to two shot? A matter of one or two pennies!

Illustration 5 - Toothbrushes

APPLICATIONS FOR MULTI COLOUR/MATERIAL MOULDING

Elimination of Assembly

In an effort to compete with far eastern production costs, every possible saving must be made in the works cost. Any human intervention in product build is looked at with the aim of designing it out. In the past, Philips would have glued or over-moulded the window area that surrounds the neon bulb on their standard kettle range. Using SIFAM's 2K machines, the part is now moulded in one cycle.

Elimination Of Printing

As has already been mentioned, this technique has been around for some time. Using a two shot tool, the print work on a component is moulded as the first shot and then the body of the part is moulded around it. In this way the area that carries the legend stands flush, or proud of the body surface. The legend is then wear-resistant since it is moulded into the part, and a secondary operation, i.e. printing, has been avoided.

Illustration 6 - Philips switch

Addition Of Colour

The example shown below is that of the three colour Trio knob from SIFAM. This is a standard part sold by SIFAM for use on audio mixing consoles. On these consoles, the different channels tend to be represented by different cap colours for the control knobs. Typically, desks can have over 1000 control knobs on them, so differentiation and visual appeal is important. Cost is another key factor with so many parts per desk. SIFAM offers from its standard catalogue 1.25 million combinations of colour, shaft, and graphic, and supplies over 50 million parts per year.

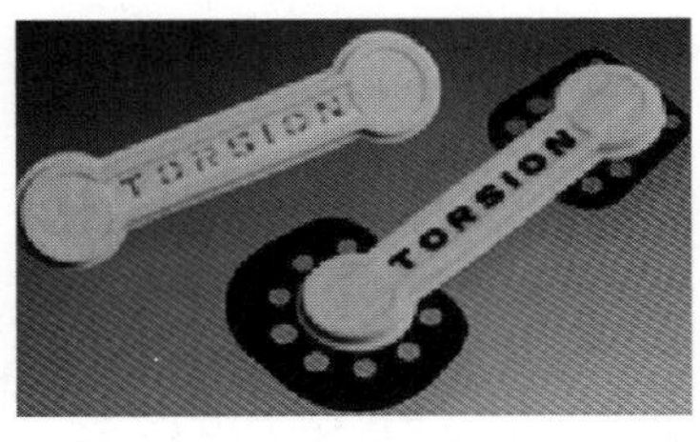

Illustration 7 - Torsion bars

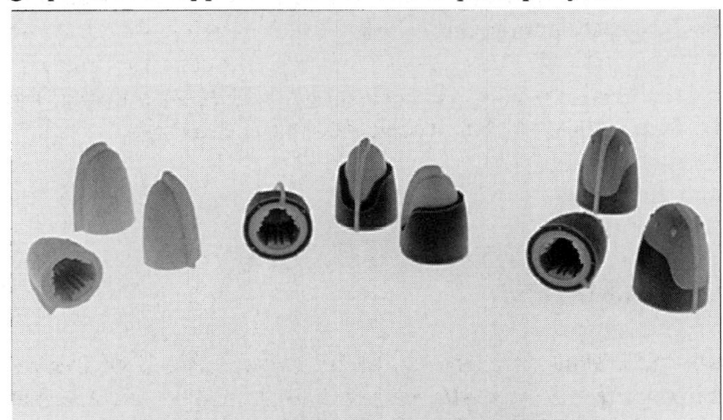

Illustration 8 - Three stages of Trio moulding

The Trio knob allows customers to work with three different areas of colour and material to achieve the required channel differentiation and visual impact. These plastic parts are shipped around the world, including Japan, Korea and Taiwan. SIFAM must be one of few European companies who can ship pure plastic components to Taiwan and compete with local moulders.

Three Shot Key Fob

As a further example of multi colour/material moulding, SIFAM commissioned the design of a car key fob. The key could be electronic to lock the car by infra red, or it could be a simple key with built-in light.

The three elements of the process are a clear, but coloured area for transmission of the light signal, a hard body area which is of a special plastic material chosen & designed to look and feel like stone, and a soft touch elastomeric area which is used for the push button, gasket and grip area of the key.

This key fob is moulded in one machine cycle but contains three different materials and colours. The tooling for such a key fob would actually be cheaper than the traditional single shot tooling, due to the elimination of interlock features in the different mouldings. The parts cost is far lower than a traditional product due to the complete elimination of any assembly. The latest BMW keys, though single colour only use a two shot process. SIFAM has used colour and its three shot machine capability to show what could be achieved in the automotive field.

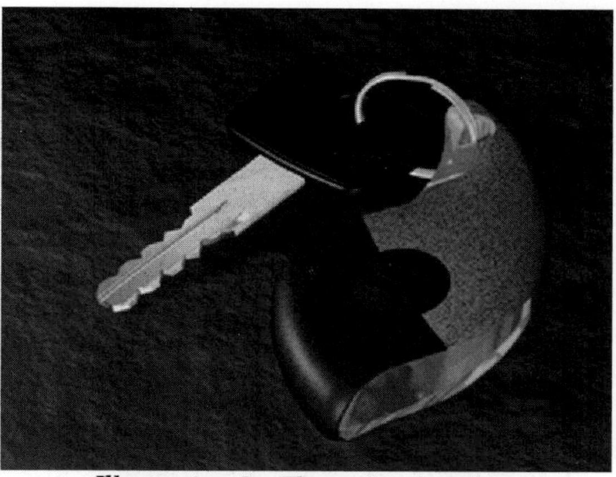

Illustration 9 - Three shot key fob

Illustration 10 - Alessi kettle

The Alessi Kettle

Each year Philips commission the top Italian designer, Alessi, to look at a particular product from the Philips range. In the picture you will see their Alessi kettle. Now this is certainly a product that will differentiate itself at the point of sale. Many of you may have seen this in the shops already over the past two years. The use of exciting colours and materials, as well as the form itself give this product not only a distinctive look, but feel as well. SIFAM moulded the handle using a two shot process and soft elastomer.

Other Ideas

2K parts do not have to bond during the moulding process to be useful. Black & Decker employ a switch on their power drills using a 2K acetyl/acetyl combination with talc fill that allows one part to slide over the other. The switch mechanism is enclosed but still allowed to move and comes off the machine in one cycle.

Material bonding

Despite the above example material selection, bonding is a key element in 2K moulding. Illustrations 11 & 12 provide a summary chart of material bonding strengths when using the 2K process.

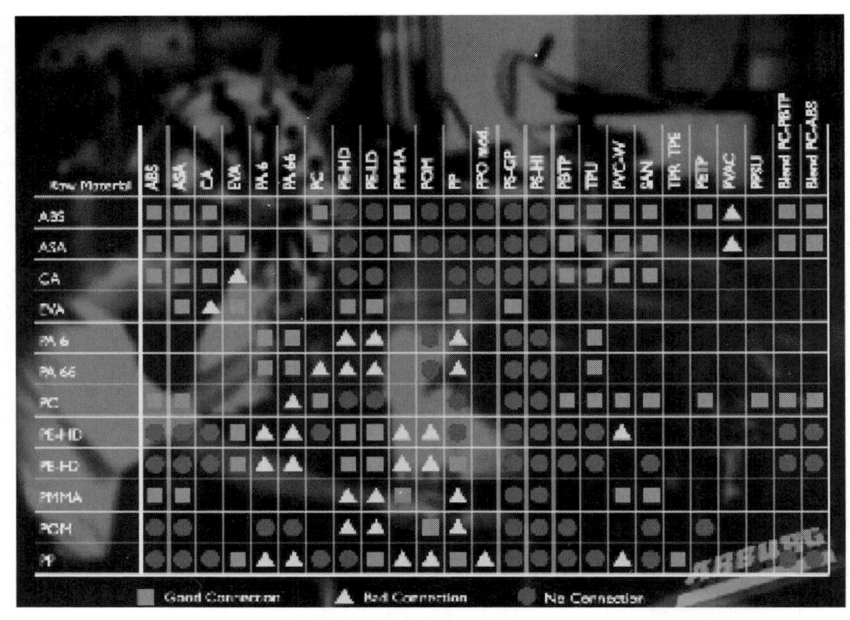

Illustration 11 - Compatible materials

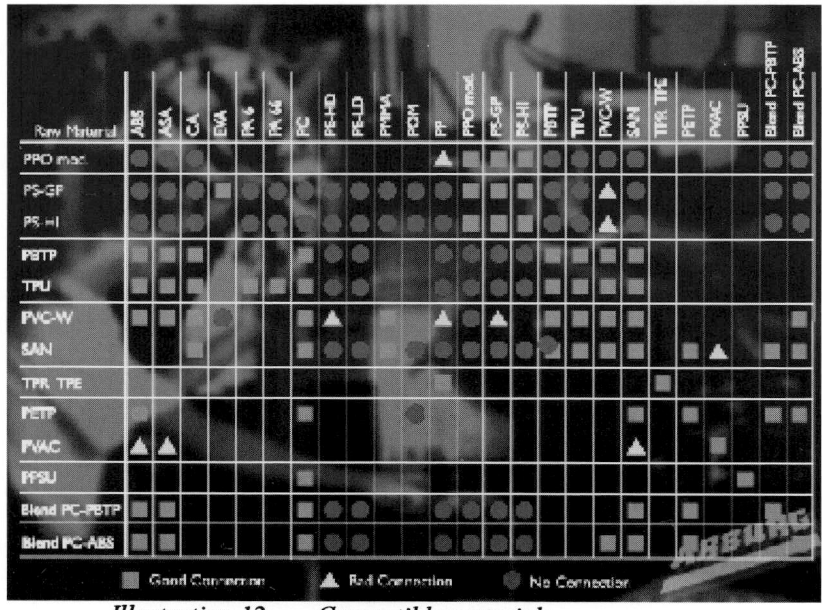

Illustration 12 - ....Compatible materials

Conclusions

Applications for multi-shot moulded parts are now widespread, SIFAM being one of the top UK suppliers. SIFAM's multi-shot mouldings are used on the grip areas of Black & Decker drills, the on/off switches of Philips kettles, the control knobs on Electrolux cookers, control areas of Heatrae Sadia showers, control panels in cars, front covers of mobile phones, and even in baby care accessories, as well as forming elements such as gasketry and windows in a variety of hand-held electronic enclosures.

The applications go well beyond the addition of colour or texture to a product. Non-bonding parts can be moulded together in one cycle to form such items as valves within one machine cycle; gaskets and buttons can now form part of the main body, as can windows. The applications for multi-shot moulding techniques are limited only by the imagination of the designer and the skill of the toolmaker.

Air Intake Manifolds – Lost Core vs Vibration Welded

MARK HAZEL

Dupont (UK) Ltd, Maylands Avenue, Hemel Hempstead HP2 7DP

Over the last 8 years Automotive underbonnet applications have been the major growth market for glass reinforced nylon, and of these the fastest growing is Air Intake Manifolds.

There are 2 primary production methods for manifolds:-

A) Lost Core - where the plastic is moulded over a Tin/Bismith core which is melted out afterwards

B) Vibration Welded - where 2 or possibly more shells are vibration welded together.

This paper will examine the 2 main production methods and the reasons for choosing which method to use. The choice of material. which follows on from the choice of production method will be reviewed.

MARKET GROWTH

In Europe the first glass reinforced nylon intake manifold was commercialised at BMW in 1990, in 1996 approximately 35% of intake manifolds were produced in plastic and it is estimated by 2000 almost 65% will be glass reinforced nylon.

The reasons for such a fast conversion from aluminium to plastic were primarily a cost and weight benefit. The plastic part is about 50% lighter than aluminium and offers a significant cost saving, the actual amount depending on the volume, production method and level of component integration.

Secondary benefits discovered during the development cycle were a 2% improvement in performance, due to the plastic surface structure and better hot starting; since the plastic does not take up as much heat as the aluminium, the fuel injector remains at a lower temperature and so does not vaporise the fuel in the injector.

The first plastic manifolds were produced by the lost core method in glass reinforced nylon 6.6 with the first welded manifold being commercialised about 4 years later.

In 1991 virtually 100% were lost core, in 1996 about 60% were lost core and by 1998 this will reverse with 60% being vibration welded.

DETAILS OF PRODUCTION METHOD

1. *Lost Core Production Cell*

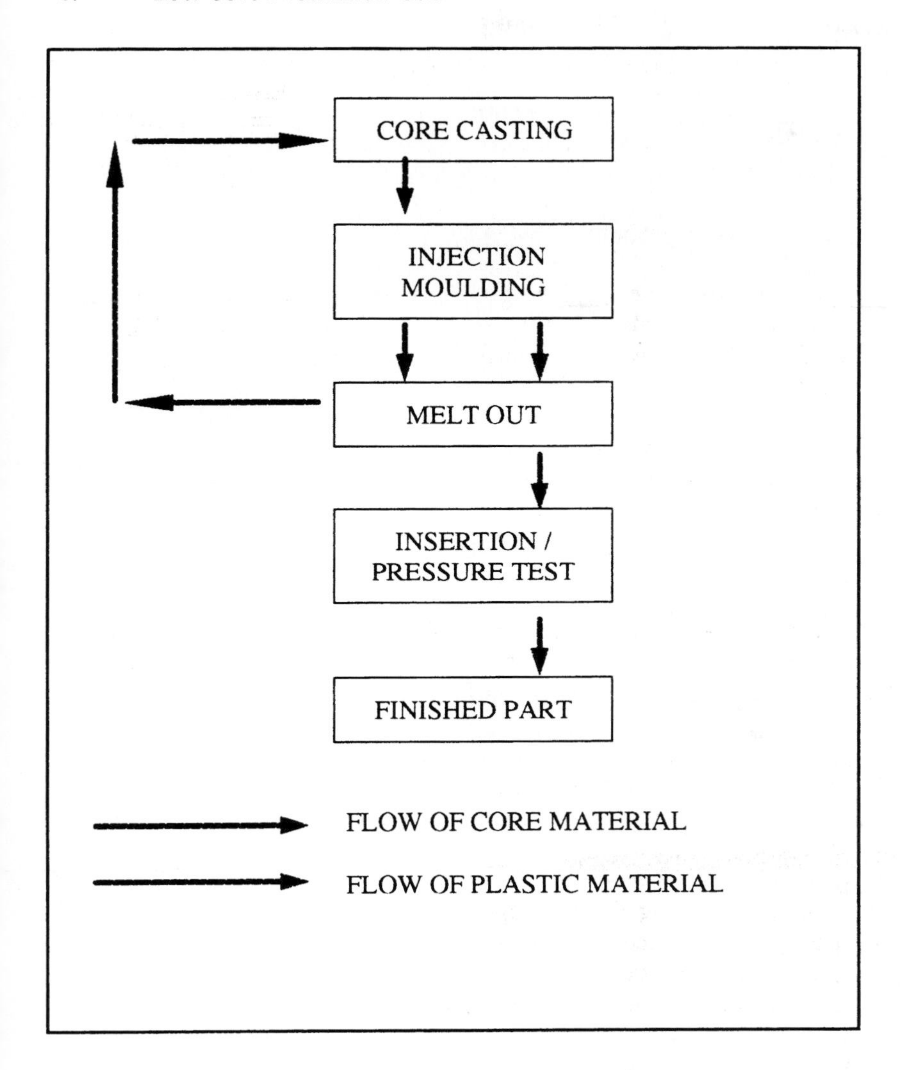

The core and finished moulding of a Peugeot Ltd manifold are shown in the photograph below:

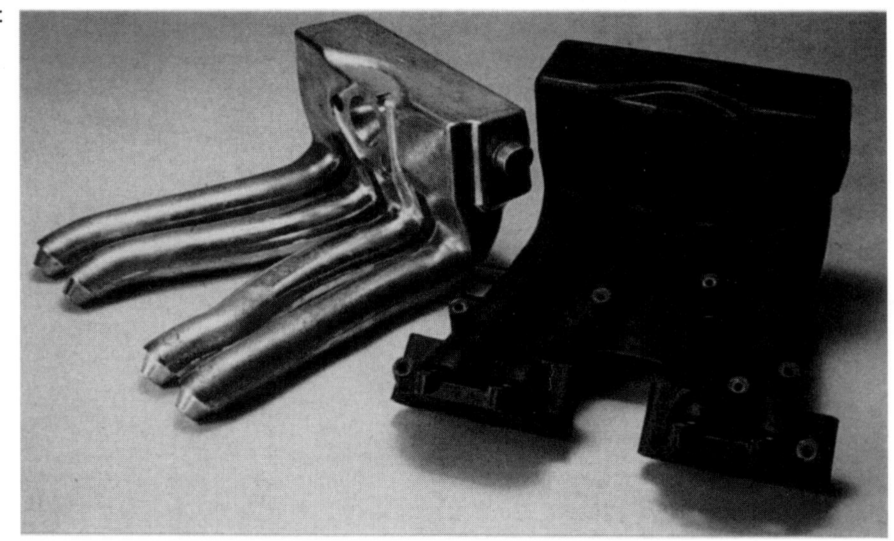

The first part of the process is casting the cores in Tin/Bismuth. This is a pressure diecasting process. The cores are solid (not hollow) and for a 4 cylinder manifold weigh ≅ 30Kg. The Tin/Bismuth is a eutectic alloy, with a melting point of 137° c and a density of 8.58 g/cc. It is used because it is dimensionally stable, from casting to melt-out, and therefore does not expand to crack the moulding during melt-out.

The rule when designing the core tool is to fill from the lowest point of the core and be continually pushing the material upwards, otherwise the material runs downhill freely and it forms a wrinkled surface as the tool is filled so giving a poor finish.

The core is transferred cold to the injection moulding machine where the glass reinforced nylon is moulded around it - with typical mould close to mould open times being approximately 1 minute.

During injection moulding the pressure of the molten plastic if applied to one side of the core only can displace it (phenomenon known as core shift) causing a thin wall situation, so DuPont have developed an analysis method using FEA and Moldflow to predict this core shift and optimise the gate position, wall thickness and core support footprints to give the best pressure balance around the core.

The Tin/Bismuth core is melted out at in an oil bath 170°C for approximately 100 minutes and there will usually be 100 parts passing through meltout bath at any one time. The process for melt out is either oil bath or eddy current (under oil). The straight oil bath at 170°C is the normal route for melt out.

The production cell is usually either a conventional horizontal moulding machine with a single tool, giving a maximum daily production rate of 600, or alternatively a vertical press with a rotary table, having 2 bottom tools and one top tool, giving a maximum production rate of 1200 per day.

Vibration Welding Production Cell

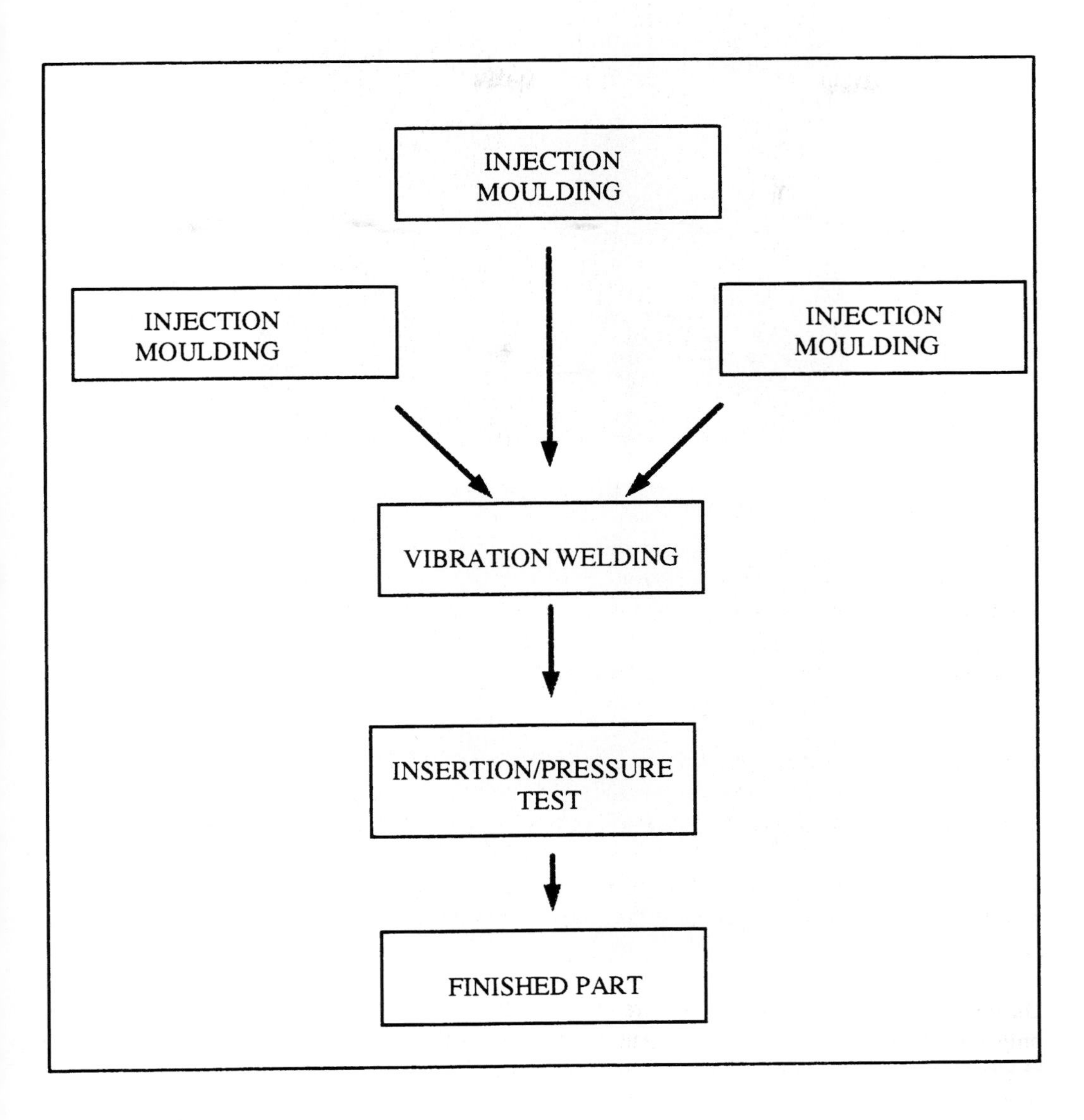

The photographs shown below are the four pieces of the vibration welded Rover K Series Engine air intake manifold.

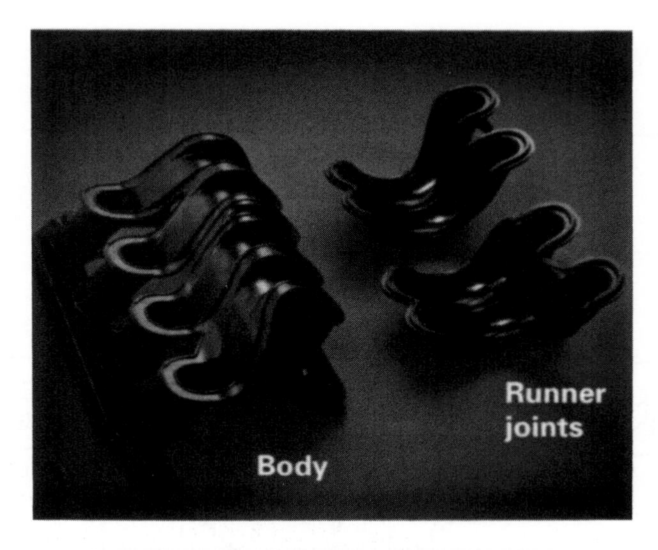

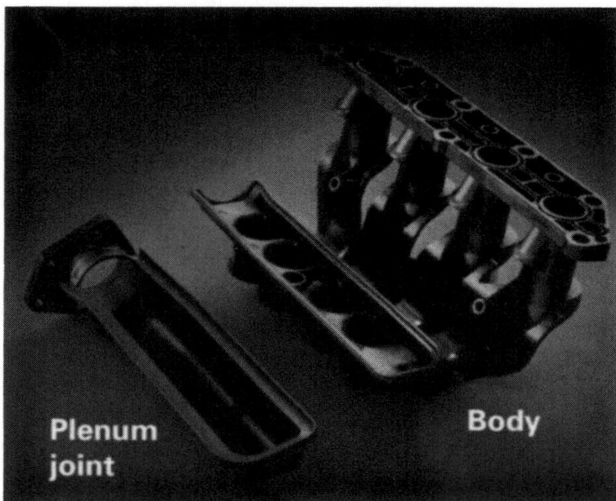

In the case of vibration welding manifold production there is an injection moulding machine for each component of the assembly.

The parts are produced by conventional injection moulding and passed immediately to the vibration welder, because at this time they are still flexible and not fully crystallised. This helps welding because it is easier to take out any warpage.

Depending on the complexity of the design there may be a number of welding operations.

General weld parameters are:

Weld depth	1.5 minimum
Amplitude	up to ± 1mm
Frequency	240 Hz
Pressure	2 N/mm² over weld flange area

The final stage as before is the inserting/finishing process.

CHOICE OF PRODUCTION METHOD

The development of vibration welded manifolds was slower than that of lost core manifolds initially due to the question of burst strength of the welded joint, particularly if backfire was to take place. Developments in the joint design and burst pressure testing have resulted in manifolds that can withstand a burst pressure of 8 Bar and a break point away from the weld bead, which is sufficient to withstand a backfire.

The main advantage of a welded manifold is cost. A two piece welded manifold is about 20% cheaper than the same part made by the lost core process. This is the main reason for the growth of vibration welded manifolds, however as the manifold complexity increases and the number of pieces that need to be welded together increase to 4,5 or more the cost difference decreases. In the extreme there are manifolds that are too complicated to weld.

Welded manifolds offer a cost saving but on the downside they involve some design compromise.

- the tracts cannot share a joint wall, there needs to be a weld flange between.

- the weld joint has to be flat in one plane

- the tracts may need to be tapered / twisted to be able to eject the moulding

- there is a maximum wrap around angle that can be welded. Downward pressure needs to be applied to the weld flange and a rule of thumb is that the minimum angle to the vertical that can be welded is 30°.

CHOICE OF MATERIAL

The method of production also affects the material that the manifold is produced from. Virtually all lost core manifolds are produced in 35% glass reinforced nylon 6.6 because of it's higher thermal stability, which is required during the core melt out process. The material for lost core moulding is usually a special grade with a lower viscosity in order to minimise the core displacement that takes place during overmoulding.

A wide range of glass reinforced nylons can be successfully vibration welded, the final choice of material depends on under bonnet temperature and chemical resistance required.

The preferred material for welded manifolds is 30% glass reinforced nylon 6 as it has lower warpage and flatter parts are easier to weld because less of the welding pressure is used to take out distortion and Nylon 6 is a higher burst pressure.

However, if the under bonnet temperature is high or if the manifold has exhaust gas fed back into it then the material will usually be 30% glass reinforced nlon 6.6, but for some diesel engines with a higher percentage of exhaust gas then a nylon 4.6 may be required.

Another factor affecting material selection is whether cooling water passes through the manifold. If so then hydrolysis resistance is required and nylon 6.6 radiator grades are used.

In order to have both the benefits of the chemical and thermal resistance of nylon 6.6 and burst strength of nylon 6, DuPont have developed a special grade of nylon 6.6 to achieve the same or better burst pressure as nylon 6.

Surface Layer Technology

Simcha Kilim M.Sc.
Addmix Ltd
Unit 20, Cygnus Business Centre
Dalmeyer Road, London NW10 2XA

ABSTRACT
By combining the ability to predict the formation of layers in mould cavities with knowledge of the transporting behaviour of a plastic injection moulding machine screw, it is possible to produce sandwich moulded components by feeding in a controlled manner two materials in a sequence to the throat of the barrel of an injection-moulding machine.

CAVITY FILL AND LAYER FORMATION

The mould cavity is filled during injection by layers forming firstly on the skin and then inside the part. Many factors govern the variation in layer formation.

THE INJECTION MOULDING MACHINE SCREW AND ITS FUNCTION

The screw's functions are divided between melt , transport and mix. All are controllable.

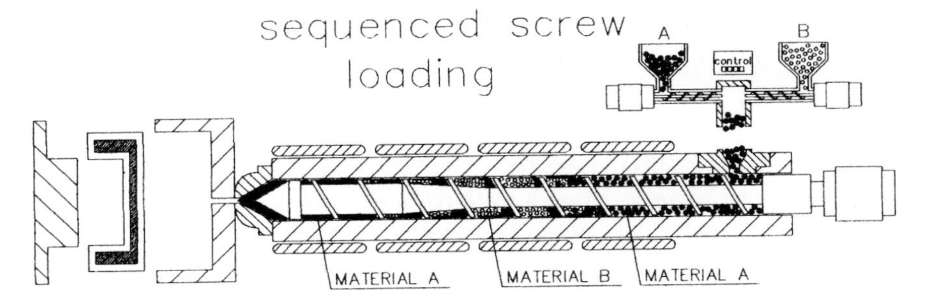

THE TECHNOLOGY

The diagram above shows a computerized dosing system which is attached directly to the feed throat of an injection moulding machine. To achieve a sandwich-moulded component, different plastic granules , A and B, are fed into the feed throat in a predetermined sequence. The material travels through the machine screw and barrel while still maintaining the feed sequence.

A small degree of mixing occurs between material A and B. In the use of colouring this means there will be a high concentration as required on the skin of the component and a low concentration at the core.

Structure Development in Injection Moulding

E. M. GIL, G. KALAY, P. S. ALLAN & M. J. BEVIS
The Wolfson Centre for Materials Processing,
Brunel University, Uxbridge, Middlesex UB8 3PH, UK

ABSTRACT
The procedures for applying macroscopic shears to solidifying melts during the injection moulding of plastics are described, and coupled with their application to a range of crystalline polymers and talc filled polypropylene, to demonstrate the structure development and physical property enhancement that is possible in injection moulding.

INTRODUCTION

The physical properties of injection moulded thermoplastics compounds are determined by composition, in terms of the polymer type and respective molecular weight characteristics, the incorporation of organic or inorganic components such as reinforcing fibres, and the microstructure within the moulding as induced by the interaction of composition, conversion process and process conditions.

The flow fields that apply in the injection moulding of plastics compounds result in a non-uniform distribution of microstructure and physical properties, as exemplified by Chivers et al [1]. The control of microstructure may be addressed in part by the application of macroscopic shears to solidifying melts, as reported in patents and papers by Allan, Bevis and co-workers in relation to shear controlled orientation injection moulding (SCORIM), see for example references[2-5].

SCORIM DESIGN CONSIDERATIONS

Macroscopic shears may be imposed on a solidifying melt by the activation of strategically placed pistons. Active pistons may be situated in the melt flow channels which lead to the mould cavity (SCORIM1) or in the mould in positions remote from the injection supply channels (SCORIM2) as illustrated in Figures 1a and 1b respectively. The arrangement shown in Figure 1b provides for local modification of structure and properties, such as erasing internal weld lines or increasing strength and/or stiffness in a particular direction, with the additional advantages of not endeavouring to modify the structure throughout the volume of a complex shaped part. This design also provides for ready application of the mould tool on any injection moulding machine of sufficient clamp force, without installation of SCORIM on the moulding machine, because SCORIM is contained within the mould tool.

The shear controlled orientation in injection moulding (SCORIM) concept may be applied to multi-cavity tooling incorporating hot runners as indicated [6] in Figure 1c, and also applied to produce specific preferred orientation distributions in a variety of cavity geometries [7], see Figure 1d.

The action of repeated shear and elongational flows during solidification provides for crystallisation under shear an/or alignment of reinforcing fibres and platelets, and may result in substantial modification and control of physical properties.

The application of SCORIM to semicrystalline polymers provides for the control of morphology which leads to substantial enhancement of mechanical properties, as represented by Figure 2 which summarises Young's modulus measurements made on SCORIM and conventionally moulded polymers. The results represented are for specific grades of the polymers [8], and deviation from the comparative modulus values given would occur for changes in molecular weight characteristics and process conditions. In general there is a substantial increase in impact strength simultaneously with tensile modulus, as represented in Figure 3 for one grade of isotactic polypropylene and a range of conventional moulding and SCORIM processing conditions. The significant difference between the morphology across the transverse cross-section of bars moulded conventionally and by SCORIM is illustrated in Figure 4.

The increases in Young's modulus shown in Figure 2 are proportionally equivalent to those realised in thermotropic liquid crystal polymers (LCP) [11] and represented in Figure 5, other than for internal weld line strength, which exhibits as a results of SCORIM an exceptional enhancement in moulded LCPs.

The proportional increase in Young's modulus resulting from SCORIM for plastics containing fibres [9] or platelets [10] is less than that for unreinforced semicrystalline polymers, because the contribution of stiffness is principally due to the inorganic component, which in conventionally moulded parts is already partially aligned through the thickness of a moulding. Application of SCORIM nevertheless provides for enhanced alignment and uniformity, with the consequent enhancement of mechanical properties and control of physical properties such as thermal expansion and toughness. An added potential benefit from SCORIM is a reduction in moulding cycle time and polymer consumption, that results from any reduction in section thickness made possible by the increase in stiffness afforded by the use of SCORIM. The latter property changes are presented below with respect to injection moulded talc filled polypropylene.

COMPOUNDING AND MOULDING OF TALC FILLED STUDY MATERIALS

The matrix polymer was homopolymer polypropylene supplied Shell Chemicals UK as grade KF6100, and the fillers were supplied by Norwegian Talc UK as micro talc IT Extra (average particle size 2.2 μm) and micro talc IT 200 (average particle size 11 μm), subsequently referred to as grades TX and IT respectively.

The compounding of polypropylene and talc (40 wt%) was carried out in the Wolfson Centre, Brunel University using co-rotating twin screw compounding extrusion. Prior to compounding the polypropylene pellets were comminuted to powder by the action of the inter-meshing twin screws operating at a set barrel temperature of 0°. The injection moulding machine used for production of 5 mm diameter bars was a DEMAG 150 NCIII fitted with SCORIM [4], and operated at a set barrel and mould temperatures of 240°C and 40°C respectively for conventional moulding and SCORIM.

CHARACTERIZATION OF MOULDINGS

Comprehensive assessment of the microstructure and physical properties of the mouldings followed [10], and included wide angle X-ray diffraction (WAXD), coefficient of thermal expansion, tensile and Charpy impact testing, micromorphology and fractography studies, to be reported in due course.

Indications of the substantial changes in micromorphology resulting from the action of SCORIM are the comparative WAXD measurements in Figure 6 for the IT filled polypropylene, as recorded for conventionally moulded and SCORIM (two live-feeds) samples respectively. The marked differences in the spectra, which represent differences in crystallinity, micromorphology and preferred orientation of filler, corresponded to significant differences in the tensile and impact properties of mouldings produced by conventional moulding and SCORIM (Table 1). The Young's modulus and yield strength of conventionally injection moulded polypropylene, tested under the same conditions were 1 GPa and 31 MPa respectively.

Table 1: Tensile test data gained at 5 mm.min^{-1} crosshead speed and 23°C temperature for PP and 40 wt% talc filled isotactic polypropylene. Standard deviations are shown in parentheses.

	Strain at Peak, %	Stress at Peak, MPa	Strain at Break, %	Stress at Break, MPa	E, Young's Mod., MPa
PPTX/C	3.04 (0.2)	33.66 (0.3)	8.84 (4.3)	21.38 (3.0)	3038 (40)
PPTX/S	5.04 (0.2)	41.42 (0.1)	40.03 (9.9)	36.32 (0.9)	3461 (77)
PPIT/C	3.33 (0.1)	32.07 (0.3)	12.68 (0.1)	23.33 (0.3)	3050 (81)
PPIT/S	6.84 (0.7)	42.23 (0.5)	49.57 (6.0)	37.05 (1.6)	3321 (158)

The TX and IT filled polypropylene mouldings produced conventionally, PPTX/C and PPIT/C, and by SCORIM PPTX/S and PPIT/S, were subjected to room temperature tensile testing at 5 mm.min^{-1}. Representative stress-strain curves are shown in Figure 7, and reflect a substantial increase in stress and strain to failure, a modest increase of Young's modulus of 15 per cent and a corresponding decrease in the coefficient of thermal expansion in the applied shear direction.

Charpy impact strength measurements for both IT and TX filled talc at 23°C and -20°C, represented more than two-fold increase in the impact strength of the SCORIM over the conventionally moulded bars at both test temperatures, and is attributed to the laminated structure induced by SCORIM as reflected in the fractographs shown in Figure 8.

CONCLUSIONS

The application of macroscopic shears to solidifying melts in mould cavities provided for substantial influence over the micromorphology of moulded semicrystalline polymers. The general application of the SCORIM technology is indicated by the provision several design options, with respect to mould design, and by the results of application to a range of semicrystalline polymers. The application of SCORIM for the structure development and physical property enhancement of talc filled polypropylene was described in more detail.

Collectively the results presented, together with previously published studies on fibre reinforced thermoplastics and thermosets, show that SCORIM provides for significant influence on the physical properties of moulded parts.

Additionally, the enhancement or optimisation of mechanical properties may provide for reduction in the section thickness of moulded parts, with consequent reductions in cycle time and mass. Optimisation of microstructure and cost may dictate that shears are applied in selected regions of a moulding.

Acknowledgement

Eva M. Gil was in receipt of a research fellowship from the European Union in support of the research reported on talc filled polypropylene.

REFERENCES

1. R. A. Chivers, D. R. Moore and P. E. Morton, Plastics, Rubber and Composites Processing and Applications, 1991, **15**, 145.

2. P. S. Allan and M. J. Bevis, UK Patent 2170-140-B.

3. P. S. Allan and M. J. Bevis, Plastics and Rubber Processing and Applications, 1987, **7**, 3.

4. G. Kalay and M. J. Bevis, Journal of Polymer Science: Part B: Polymer Physics, 1997, **35**, 241.

5. G. Kalay and M. J. Bevis, Journal of Polymer Science: Part B: Polymer Physics, 1997, **35**, 265.

6. J. R. Gibson, P. S. Allan and M. J. Bevis, Plastics and Rubber International, 1991, **16**, 12.

7. P. S. Allan and M. J. Bevis, Composites Manufacturing, 1990, **1**, 79.

8. G. Kalay, P. S. Allan and M. J. Bevis, Kunststoffe, in press.

9. P. J. Hine, R. A. Duckett, I. M. Ward, P. S. Allan and M. J. Bevis, Polymer Composites, 1996, **17**, 400.

10. E. M. Gil, Ph.D. Thesis, 1996, Brunel University, UK.

11. P. S. Chuah, Ph.D. Thesis, 1995, Brunel University, UK.

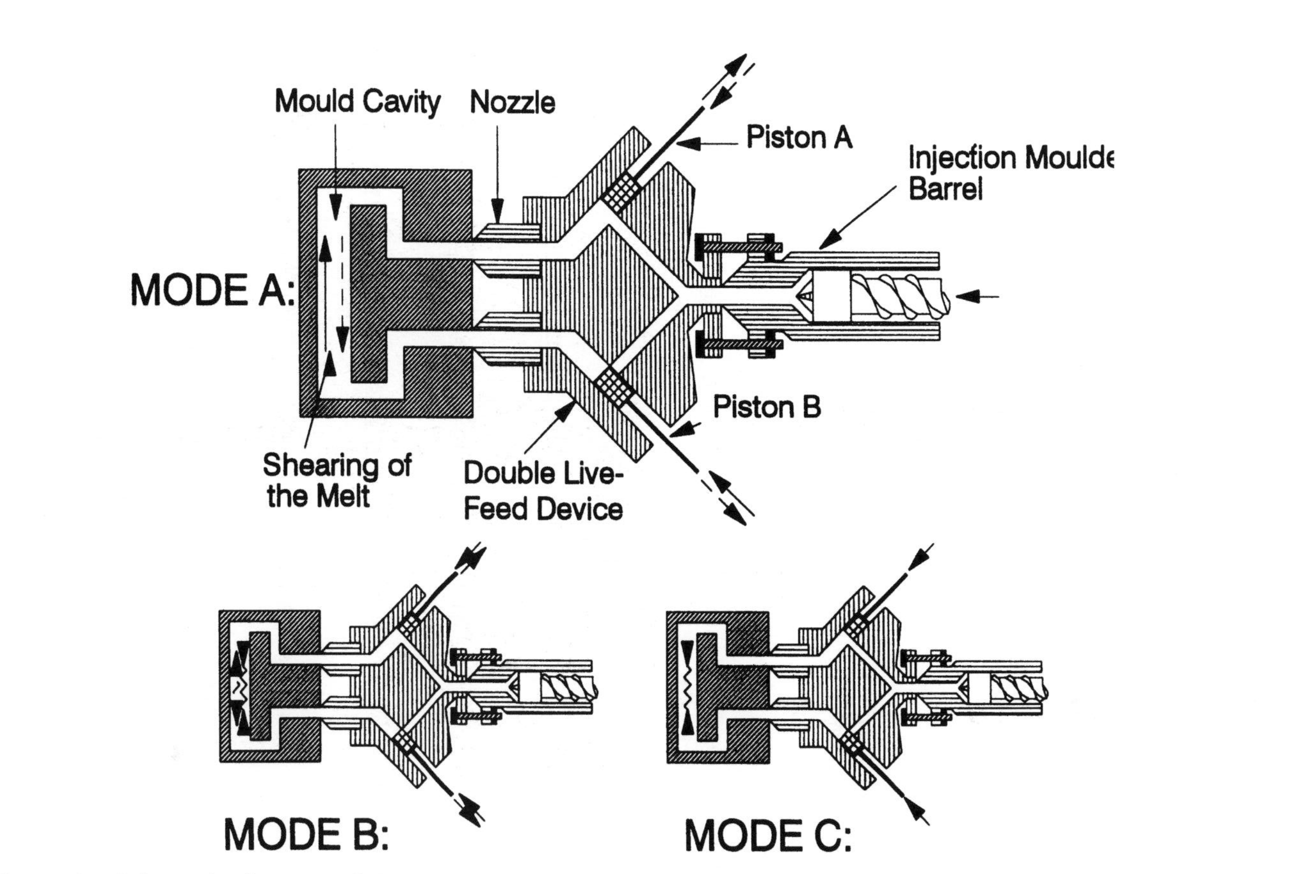

Figure 1a: Schematic diagram of the two live-feed device (SCORIM) to impart macroscopic shear during solidification of the melt in the mould cavity. To impart shear on the melt-solid interface as the interface progresses from the surface to the centre of the cavity, pistons A and B are operated 180° out of phase.

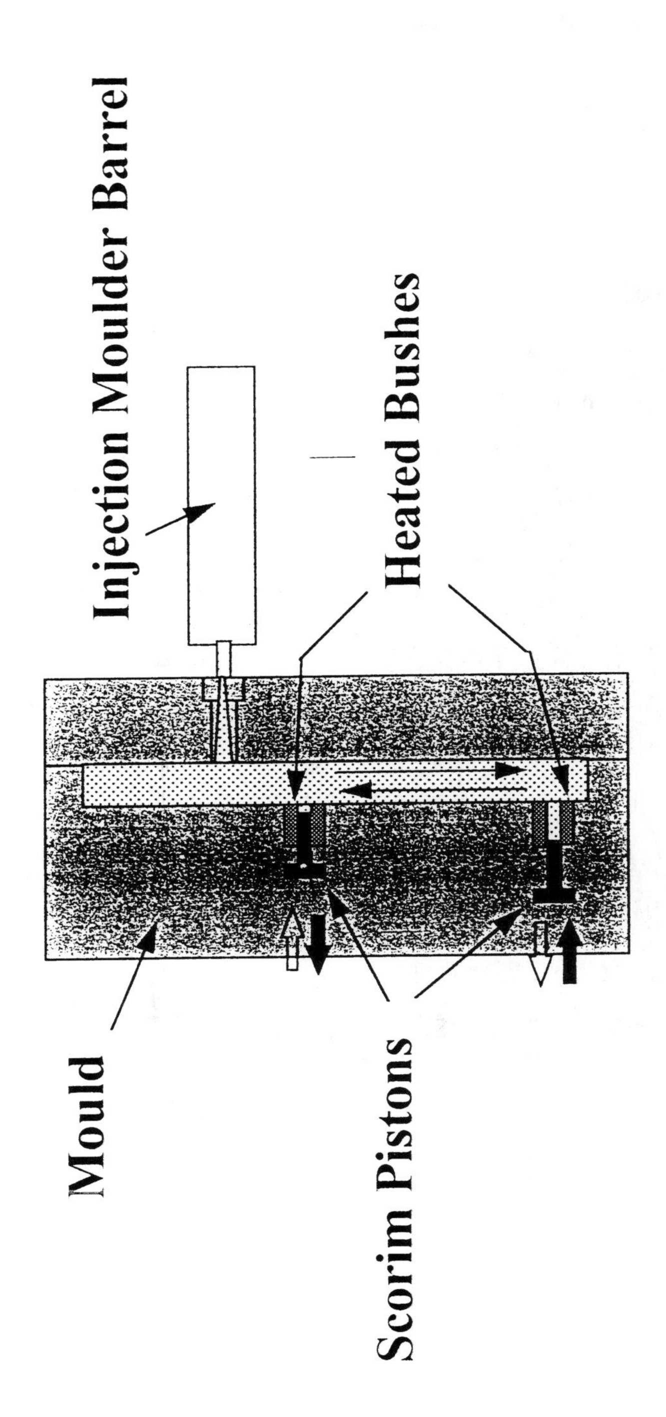

Figure 1b: Integrated SCORIM mould.

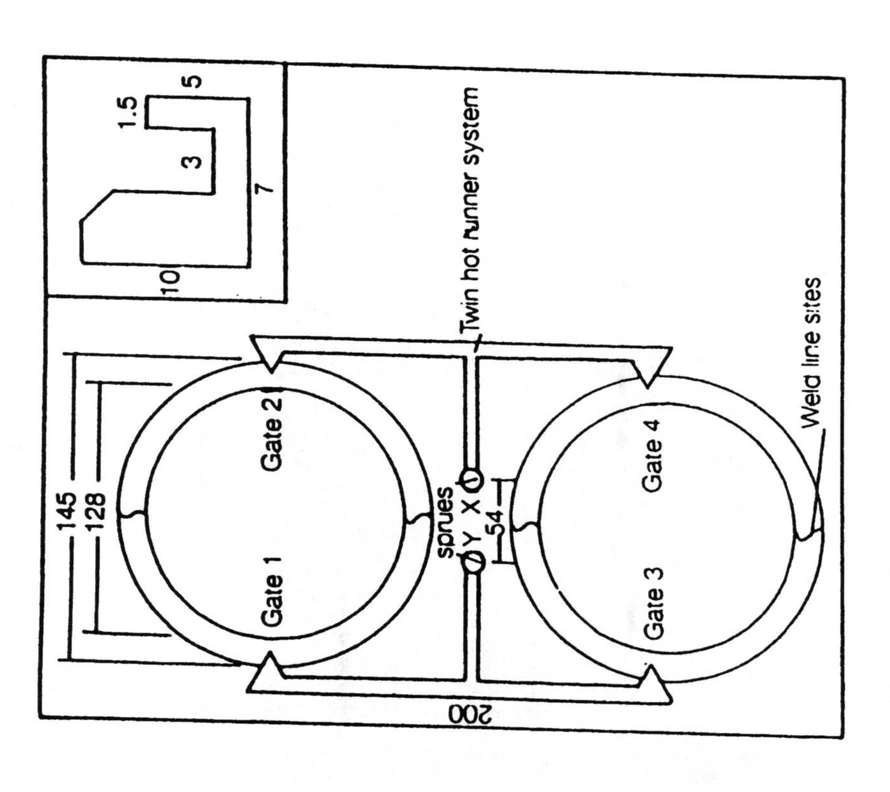

Figure 1c: Schematic diagram of multi-cavity mould successfully used with a hot runner system.

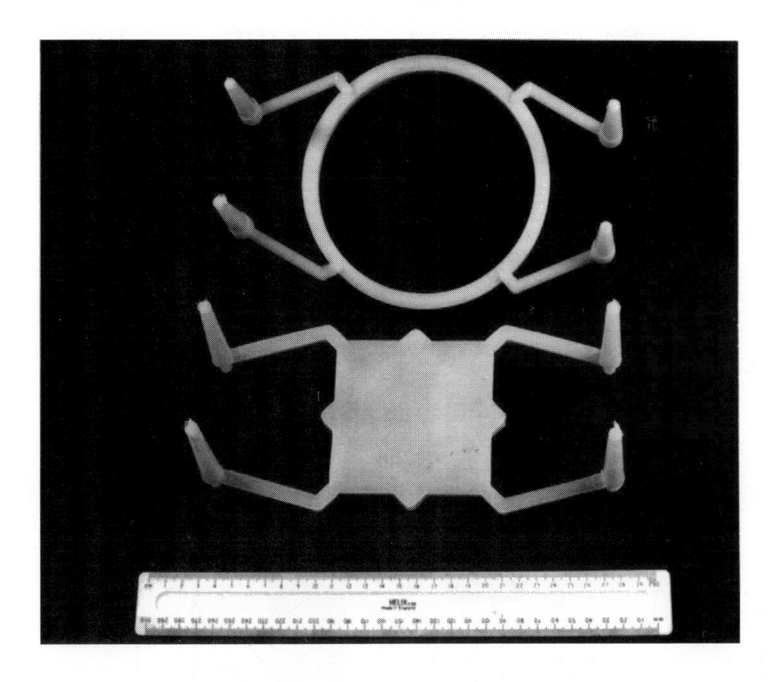

Figure 1d: Photograph of mouldings produced with four live-feeds. The circular mould exhibits a high level of circumferential alignment of reinforcing fibres, and the square moulding an in-situ formed 0°-90° laminated structure of reinforcing fibres or molecular alignment, caused by the appropriate sequencing of the activation of pistons at the four gates.

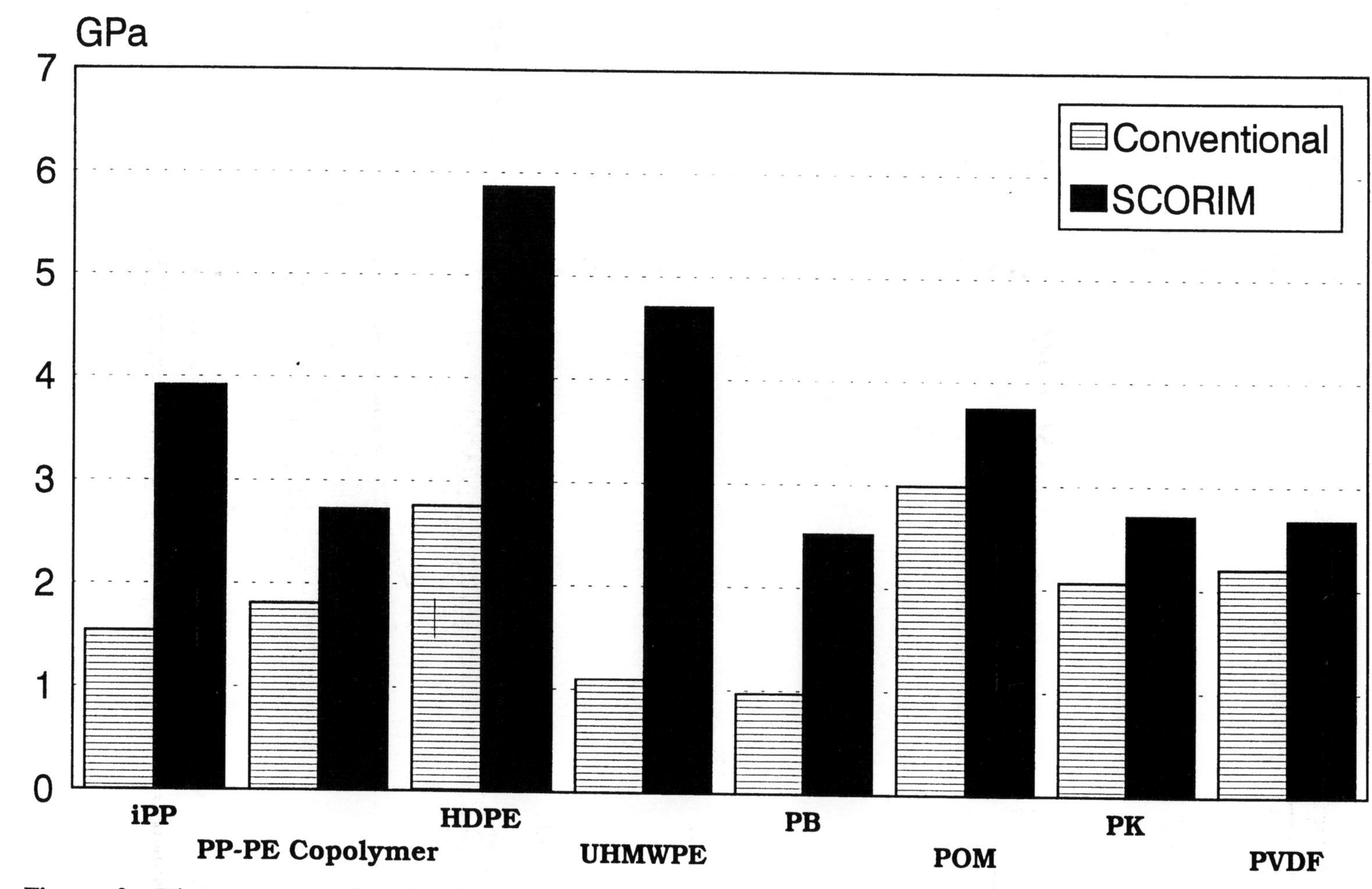

Figure 2: Diagram comparing the Young's modulus of conventionally and SCORIM moulded semicrystalline thermoplastics.

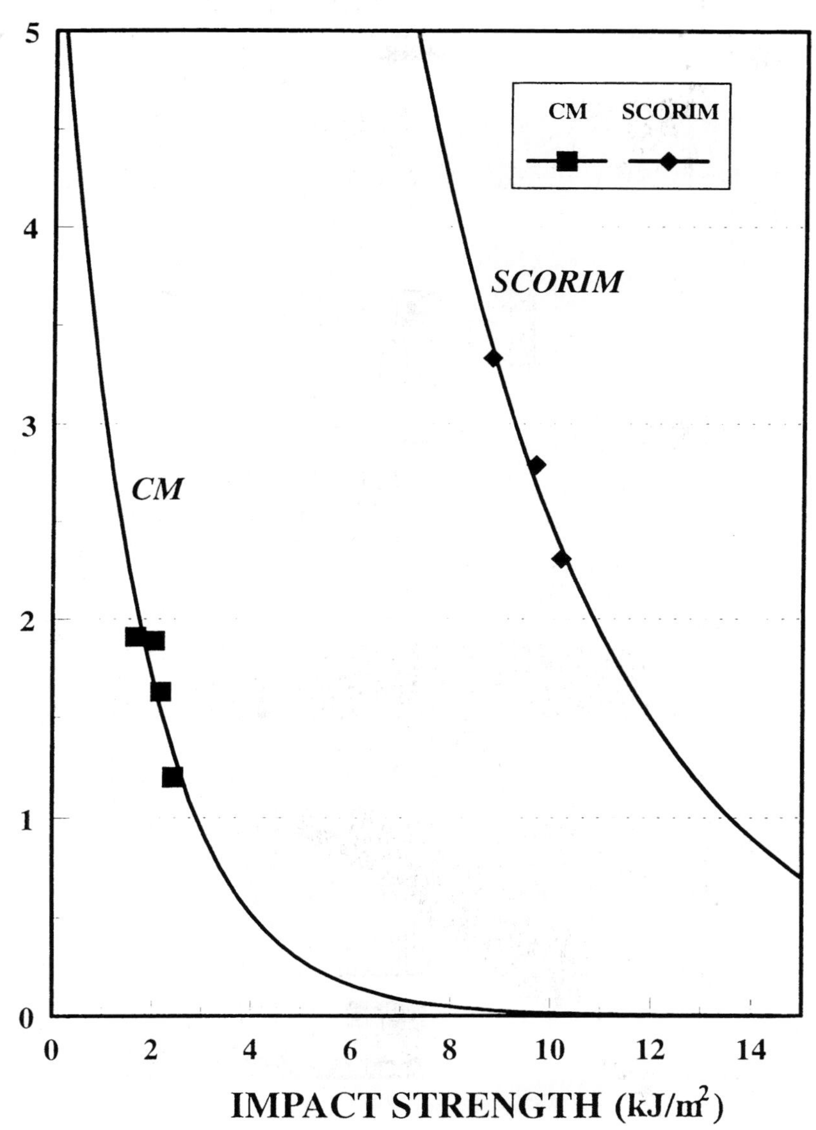

Figure 3: Young's modulus-impact strength curves for conventional and SCORIM moulded isotactic polypropylene.

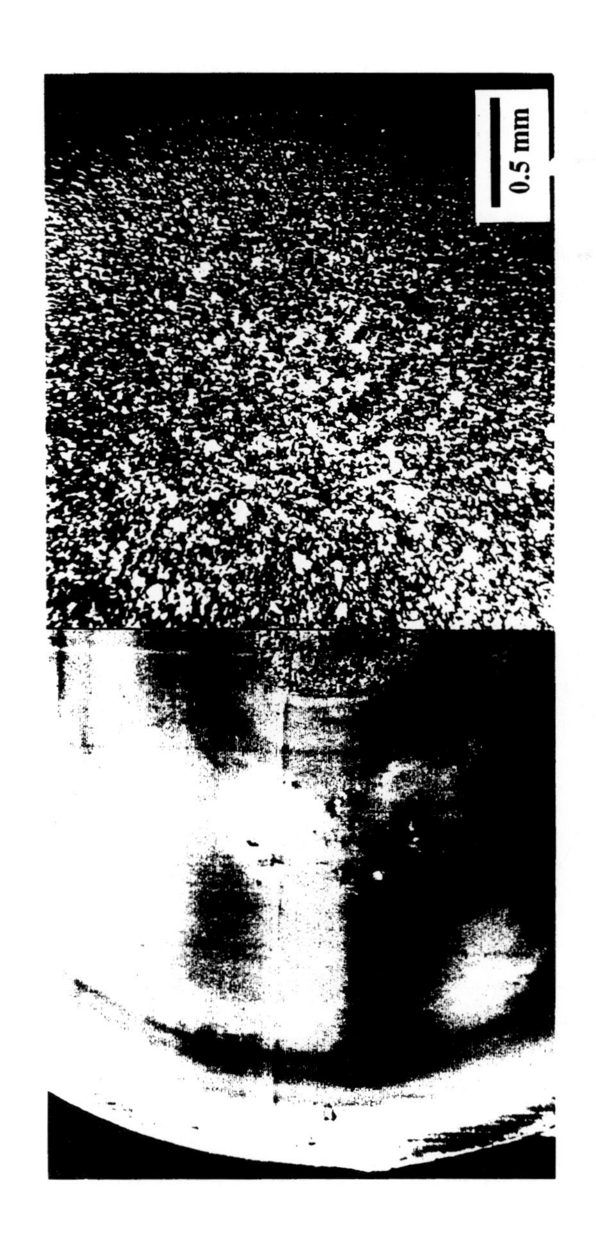

Figure 4: Transmitted light micrographs of representative transverse cross-sections of conventional and SCORIM mouldings showing the marked difference in polypropylene morphologies produced by different moulding processes. The right side of the figure relates to the conventional moulding and the left side to SCORIM, showing the much larger region of highly oriented nonspherulitic polymer that arises with SCORIM.

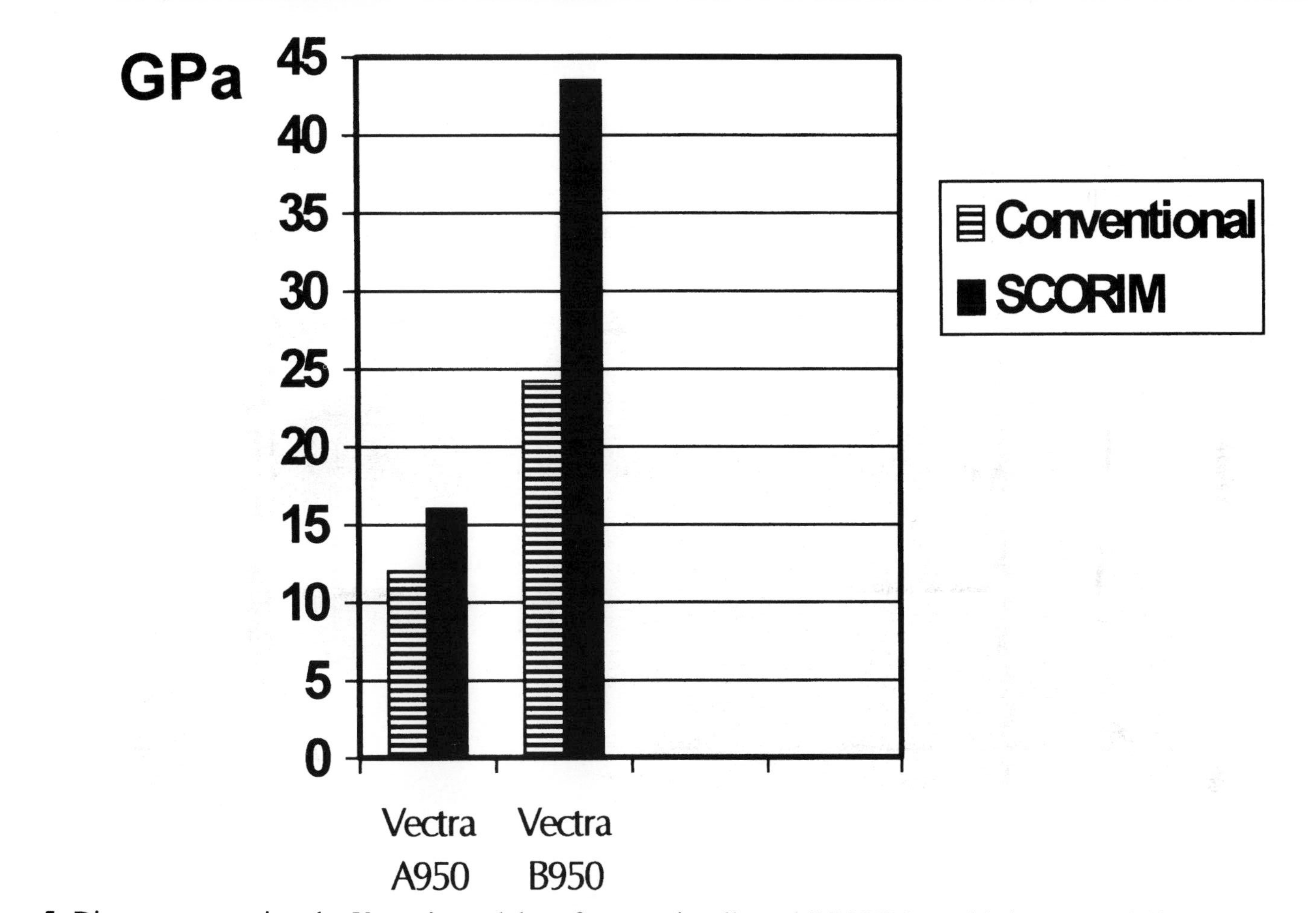

Figure 5: Diagram comparing the Young's modulus of conventionally and SCORIM moulded Vectra A950 and Vectra B950.

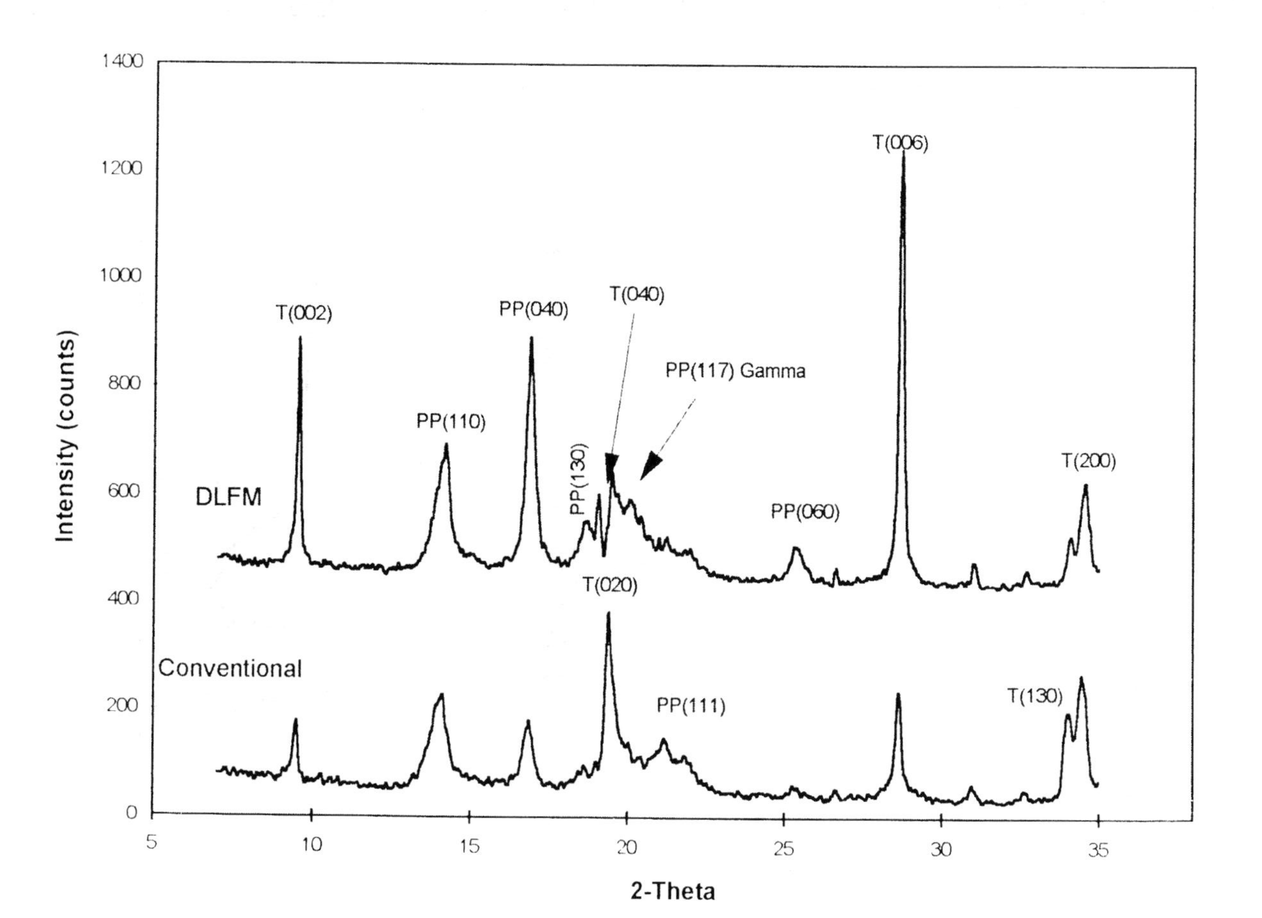

Figure 6: X-ray diffraction profiles for talc filled (PP-IT) mouldings produced by conventional and SCORIM mouldings.

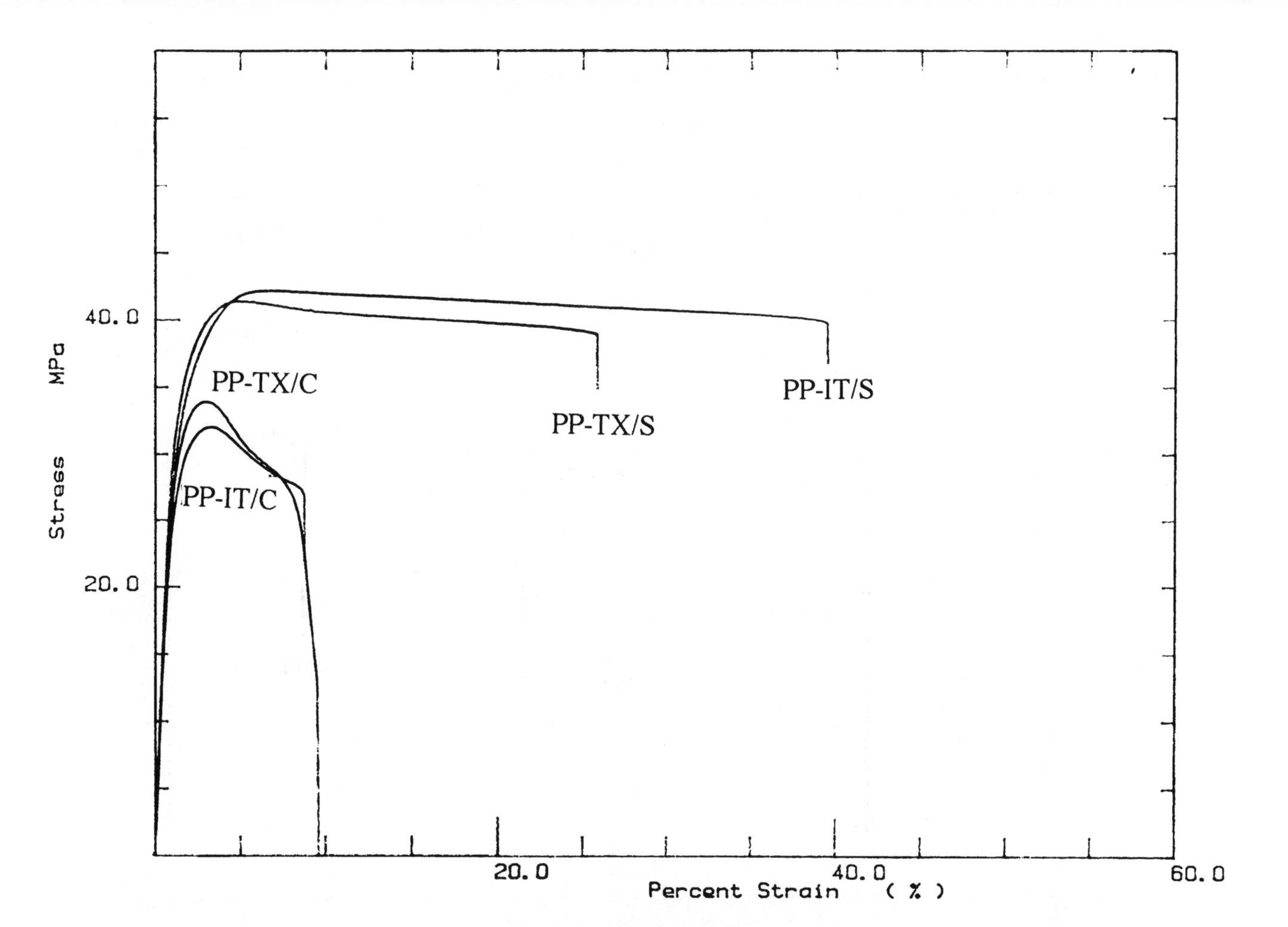

Figure 7: Stress-strain curves for talc filled polypropylene produced by conventional and SCORIM moulding.

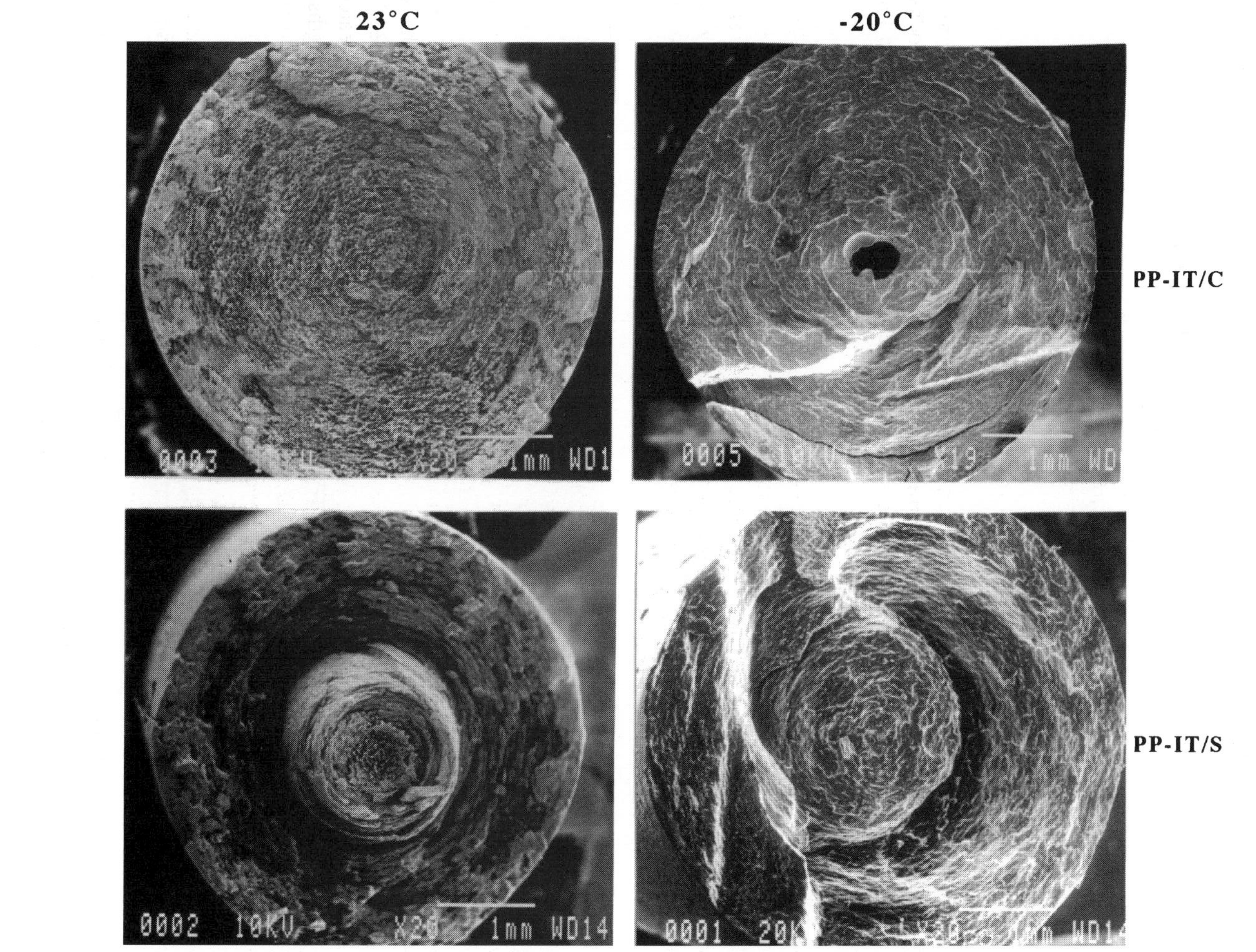

Figure 8: Fracture surfaces produced by Charpy impact testing of talc filled polypropylene (PP-IT) mouldings at 23ºC and -20ºC.

Process Control for Rotational Moulding

R J CRAWFORD

School of Mechanical and Process Engineering
The Queen's University of Belfast
Ashby Building
Stranmillis Road
Belfast BT9 5AH

ABSTRACT

Rotational moulding is generally regarded as a slow, low-technology manufacturing method for hollow polyethylene or PVC articles. Whilst this may have been true at one time, today rotational moulding is a vibrant sector of the polymer processing industry and it boasts the highest growth rate in terms of material usage.

This paper will describe the most recent developments in this sector. It will be shown that rotational moulding now has process control to rival that available in injection moulding or blow moulding. In addition, the use of pressure inside the mould will be shown to offer many advantages including reduced cycle time, improved impact strength and consistency in shrinkage values. New technologies, including internal heating and cooling, will also be reviewed.

INTRODUCTION

Engineers and designers have available to them a wide range of manufacturing methods when they are considering the use of plastics in product design [1,2]. This wide choice is advantageous in providing scope for ingenuity but it means that designers must have an awareness of the capabilities and limitations of a large selection of processing methods. For example, injection moulding is suitable for a wide range of plastics but there can be shape limitations (eg undercuts) and the number of products required must be large enough to justify the mould cost. Extrusion is suitable for a more limited number of plastics and has greater shape limitations (usually 2D rather than 3D) but is ideal for long (or continuous lengths) sections. Thermoforming has even greater restrictions on material choice because good melt strength is required as the material is formed over an open mould.

Traditionally, rotomoulding has tended to lie well down the designers list of available processing methods for plastics. This is because it is generally regarded as a relatively slow manufacturing method suitable only for hollow products in a limited range of materials. Ten years ago this would have been an accurate statement but today rotomoulding is rapidly becoming a sophisticated, precise manufacturing method for complex articles. Significant advances are also being made in regard to the range of materials which are rotomouldable. Thus, enlightened designers are able to take advantage of benefits such as large stress-free mouldings, inexpensively produced in short lead times [3].

The principle of rotational moulding is illustrated in Fig. 1. Plastic powder (or granules or liquid in some cases) is placed in a shell-like metal mould. This is then rotated relatively slowly so that all parts of the inside surface of the mould come in contact with the powder pool lying in the bottom of the mould. The rotating mould is heated and over a period of time, the plastic melts and forms a coating on the inside of the mould. When all the plastic has adhered to the mould, the mould rotation continues in a cooled environment so that the plastic solidifies to the desired shape. The mould can then be opened and the product removed. This will be a hollow moulding which is often what is required. However, in some cases a short finishing operation may take place to provide other types of shape eg a container with its lid or two non-hollow products as a result of splitting the moulding.

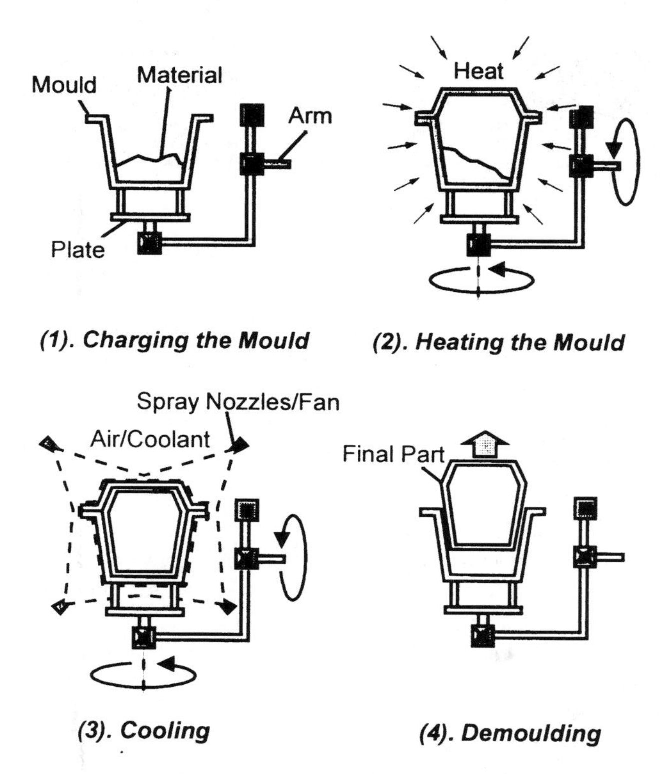

Fig. 1 Principal Stages in Rotational Moulding

It is worth noting that there are a number of unique features to rotomoulding, some of which are attractive and others which are disadvantageous. It is important to appreciate from the beginning that the process does not involve high speed, centrifugal-type casting. The moulds are generally large and have an irregular shape so that rotational speeds above about 10 rev/min would be difficult and potentially dangerous. The plastic thus coats the mould in a gentle manner without the use of pressure. This has two beneficial effects

It should be apparent, even to a casual observer, that the conditions inside the mould are likely to be quite different from those in the oven. Process control in rotational moulding normally takes place from a thermocouple sensor in one of the corners of the oven. On the basis of the reading from this, the oven burners will switch on and off to maintain a set oven temperature. The plastic inside the mould would presumably reach the oven temperature if left for a sufficiently long time. However, the mould will be in the oven for minutes rather than hours and so the plastic will never get to the oven temperature. In addition, as the oven temperature is commonly set at 300 °C or higher it would be undesirable that materials such as polyethylene (the main rotomoulding material) should ever reach this temperature or even get close to it. It is clear therefore that moulders have a complex problem in that they are dealing with dynamic thermal situation which may change in nature over the period of a working day and will certainly change over a period of weeks. How, therefore could they be expected to achieve consistent quality products?

To tackle this problem, the ROTOLOG system was developed to measure the temperature of the air inside the mould and provide this information in real time to the moulder [8,9]. Although it might appear that it is the temperature of the plastic which is required, in fact it is the temperature of the air inside the mould which provides the more fundamental information.

Fig. 2 shows a typical internal air temperature profile. Up to A, the internal air temperature rises at a steady rate which depends on the oven temperature, the wall thickness of the mould, the mould material and the amount of plastic in the mould. During this period no plastic sticks to the mould because its temperature is not high enough. At point A, the first particles of powder start to melt and because the heat input is now absorbed by the melting process, the internal air temperature rises less quickly. In the region AB the powder is melting and coating the inside surface of the mould.

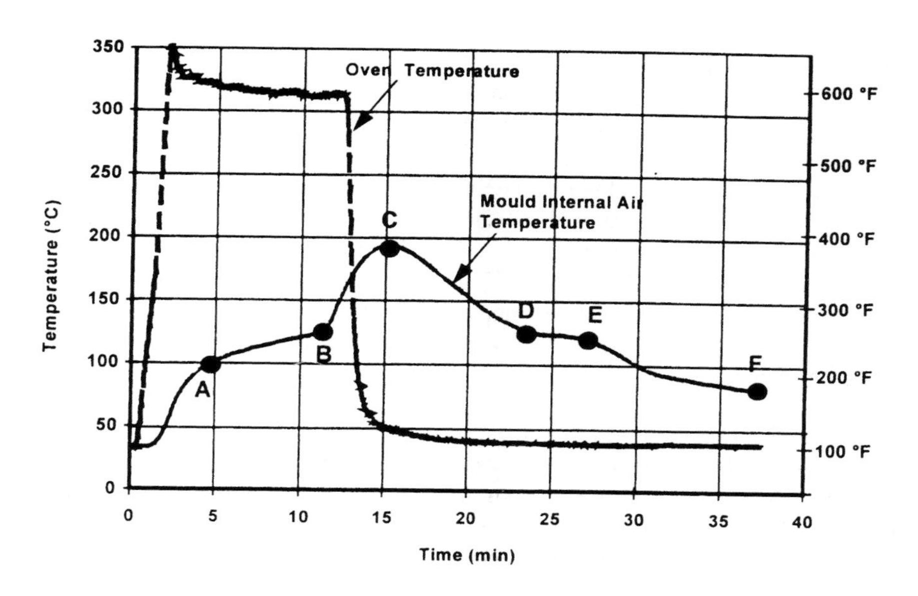

Fig 2. Temperature Profiles During Rotational Moulding

At B, all the powder has melted and so the internal air temperature starts to rise again as approximately the same rate as before. During the period BC the powdery melt coating on the wall of the mould is sintering. When the mould is removed from the oven, the internal air temperature starts to decrease again. The rate of decrease will depend on whether air or water-spray is used at this stage. Normally gentle cooling (air) is desirable to avoid warpage in the product.

At D there is a well defined "kink" in the air temperature curve which highlights the solidification phase. During this stage, heat is being released by the solidification process and so the internal air temperature decreases less quickly. Once solidification is complete, the internal air temperature continues to decrease and in this final phase it is acceptable to utilise the faster cooling provided by water-spray.

The characteristic curve shown in Fig. 2 is unique to rotational moulding and provides a very valuable insight into the whole process. In particular, the moulder is provided with a direct indication of when the powder starts to stick to the mould, when all the powder has melted, what is the maximum temperature at the inside surface and when the plastic has solidified. Thus the trial and error approach is completely removed. The curve in Fig. 2 is obtained independently of any of the machine settings, or the oven temperature, or the efficiency of the oven, or the nature/thickness of the mould.

One of the main uses of the ROTOLOG internal air temperature trace is to ensure optimum curve. In the past this important condition had to be established by trial and error. Mouldings would be produced using a series of oven times and a slice from the cross-section of the moulding would be observed to see at what point the internal air bubbles were just about to disappear [10-11]. Using the ROTOLOG device it has been found that, for polyethylenes, optimum cure occurs when the mould internal air temperature reaches 200 °C. This is independent of any machine variable, the mould material or the thickness of the plastic product.

Another very important piece of information obtained from the ROTOLOG trace is the point at which the plastic separates from the mould wall. Once this separation occurs, the cooling conditions for the plastic change quite dramatically. This is because the plastic will cool more quickly when it is in contact with the metal mould. However, if it becomes separated from the mould wall then it cools more slowly, resulting in greater crystallinity, greater shrinkage and lower impact strength. A characteristic feature of rotationally moulded products in the past has been large and variable shrinkage. This is illustrated in Fig. 3 (a) where a series of 18 mouldings were produced under nominally identically conditions and each shows quite different shrinkage.

Although the point of mould/plastic separation is not readily apparent on Fig. 2, if one measures the temperature at the inner surface of the mould, T_i, and then determines the temperature gradient (T_i - T_a) it is possible to see this point quite easily. This is illustrated in Fig. 4. In theory one would hope that this gradient should be zero but in practice this is difficult to achieve. As the plastic cools, it is evident that (T_i-T_a) tends towards zero but then goes more negative at point D when the plastic solidifies and releases heat. When the cross-section becomes completely solid, the (T_i-T_a) trace tends towards zero again. However, a second peak is observed and this co-incides with the point where the plastic separates from the mould. (Point G) Close examination of the ROTOLOG internal air trace shows that this point can in fact be seen as a change of slope on the T_a curve.

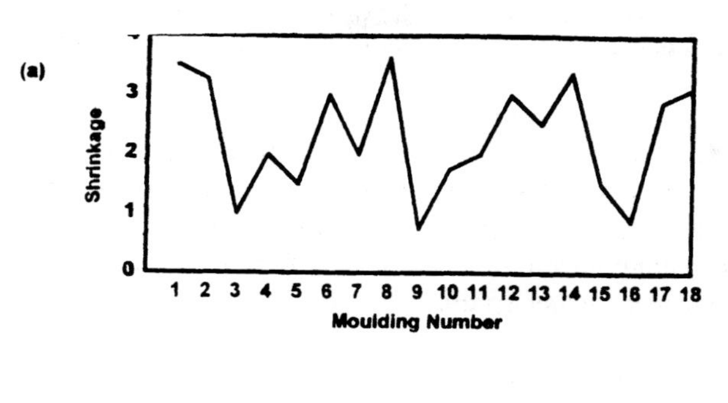

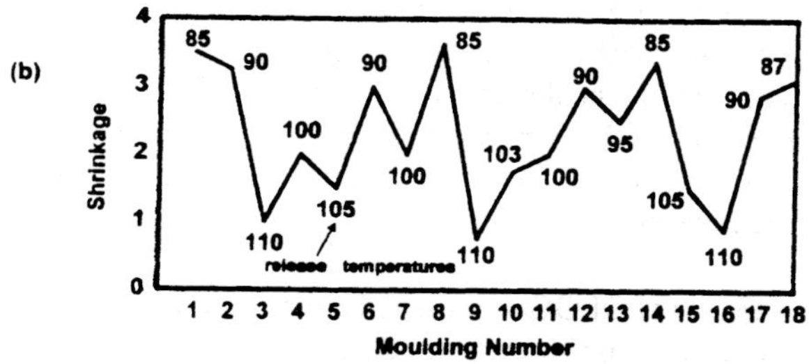

Fig 3 Shrinkage variation in mouldings produced under nominally the same conditions

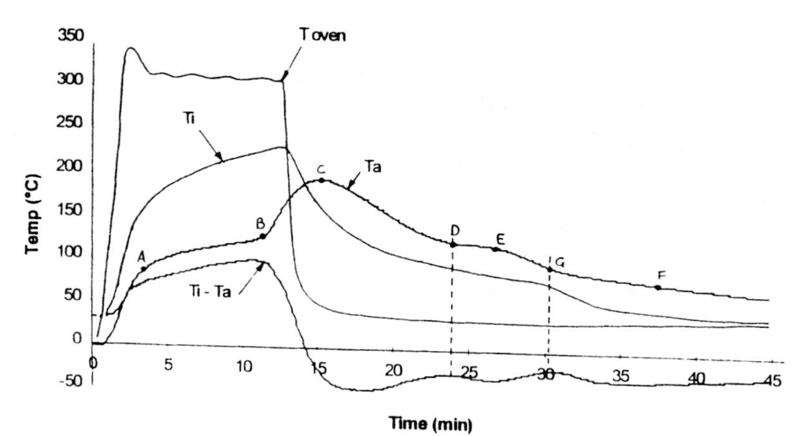

Fig. 4 Interpretation of release point using both temperature profiles

This point has enormous significance for rotational moulders. Since separation of the plastic from the mould wall occurs inside the closed, rotating mould there are no visible signs that it has occurred. Unfortunately it tends to happen quite randomly after the plastic has solidified. If one re-examines the ROTOLOG traces for the 18 mouldings in Fig. 3(a) then it is apparent that the high and low shrinkage values are simply associated with late or early release from the mould. Fig. 3(b) shows that the shrinkage data can be rationalised by superimposing the temperature at which the release occurred.

This then provides a very valuable method of getting consistent shrinkage (ie always ensure the same release temperature). In addition, high or low shrinkage can be achieved by controlling the release temperature. A convenient method of doing this is to introduce a slight positive pressure inside the mould. This keeps the plastic against the metal mould and the pressure can be relieved when release is required. The importance of pressure in the rotational moulding cycle is illustrated further in the following section.

Removal of Bubbles from Rotomoulded Products

A characteristic feature of rotomoulded products is the presence of small bubbles or pin-holes within the wall thickness and/or at the outside surface of the moulded product. These occur because pockets of gas get trapped between the powder particles as they come together [12, 13]. Under the effects of time and temperature the bubbles would eventually disappear. This occurs during the region BC (and possibly part of CD) in Fig. 2. In many cases the bubbles do not disappear completely because the time/temperature required to make this happen would cause degradation in the moulded product. Generally the bubbles are undesirable because they are not contributing to the strength of the product and when they remain at the surface they can cause problems with appearance and hygiene.

It is desirable therefore to remove bubbles if possible - as well as the effects referred to above, it is likely that the cycle time could be reduced if time/temperature is not being relied on to make the bubbles disappear. Fig. 5 shows the time/temperature effect on bubble size. At the present time, major advances are being made in the area [14]. The factors which have been shown to be important are:

- Powder particle geometry and size
- Polymer melt viscosity
- Additives
- Mould surface
- Temperature
- Time
- Atmosphere inside mould
- Surface tension
- Forces on the melt.

The parameter which has been found to have the most dramatic effect on bubbles is pressure. If a small pressure (up to 1 bar maximum) is introduced into the mould (for up to 1 minute) at a critical point in the cycle, then the bubbles can be made to disappear. This is illustrated in Fig. 6. The best time to introduce the pressure is about 2/3 of the distance between B and C in Fig. 2. If the pressure is introduced too early then some gas pressure gets trapped in the bubbles and the desired pressure differential is not achieved. If the pressure is introduced too late then its effect is minimal.

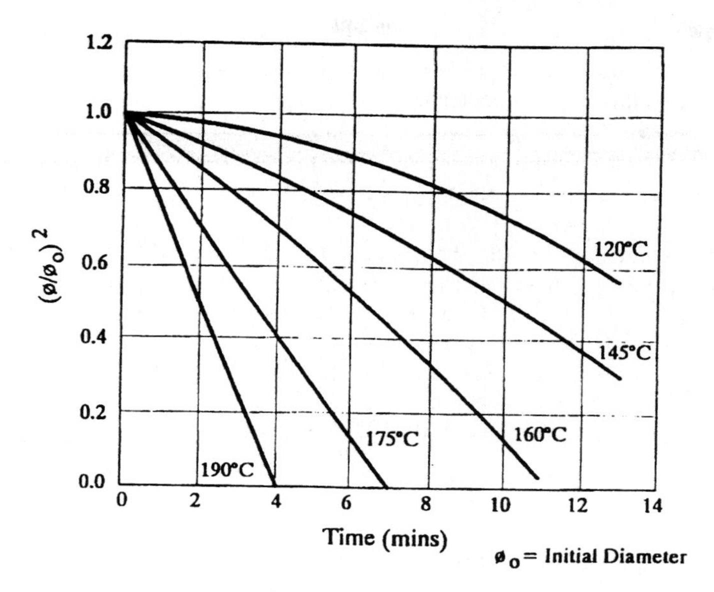

Fig. 5 **Effect of Temperature on Bubble Diameter Φ**

A major advantage of the use of pressure is that the bubbles can be made to disappear physically and so there is no need to wait for the slow time-temperature sintering process to remove the bubbles. If the bubbles are removed using pressure at 180 °C then this has two desirable effects - firstly the cycle time can be reduced by 20 → 25% because the mould can be taken out of the oven earlier. Secondly the end-product will have better impact properties because it has been subjected to the high temperatures for a shorter period of time. Typical improvements in impact strength in the region of 15% are observed.

In view of the comments in the previous section, it is apparent therefore that the use of pressure has a number of advantages in rotational moulding. The bubbles can be removed early giving shorter cycle times and greater toughness. Also, if the pressure is maintained through the cooling stages then mould/plastic separation can be controlled. This gives mouldings with consistent shrinkage and the extent of the shrinkage can be controlled over a broad range.

This leads to the concept of a Roto-Blow Moulding technology. The mould rotation is used to distribute the plastic on the inside surface of the mould and the pressure is then introduced to consolidate the moulding and control shrinkage.

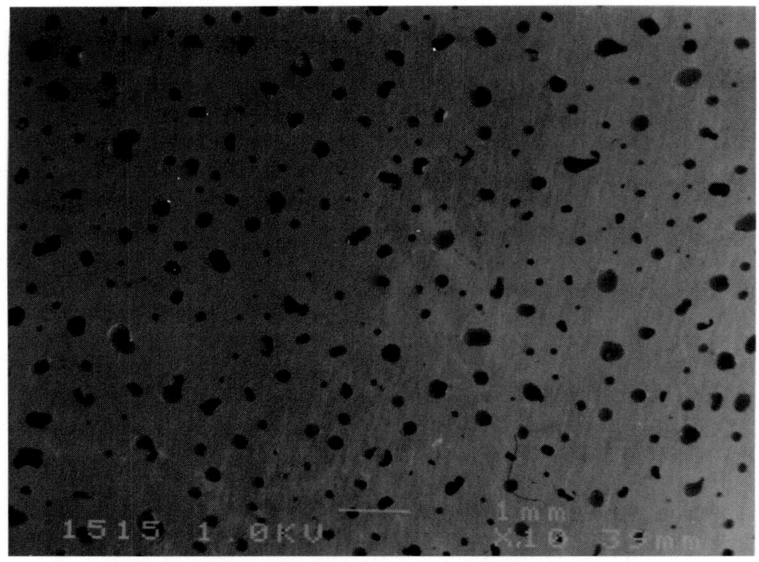

Normal Surface Porosity - No Pressure Applied

Effect of Pressure Applied for 60 seconds

Fig. 6 Effect of Pressure on Pin-holes in Rotomould Products

Conclusions

As a result of this research on the rotational moulding of plastics the following conclusions may be drawn.

(1) The measurement of air temperature inside the mould provides an accurate means of feedback control for rotational moulding.

(2) The use of pressure inside a rotational mould assists with consolidation of the melt, reduction in cycle times, increasing toughness and precise control over shrinkage.

(3) The point at which the pressure is introduced is critical. It should not occur before all the powder has adhered to the mould and the correct application of pressure is best judged from the internal air temperature trace.

(4) Inconsistencies in shrinkage in rotationally moulded products has been shown to be associated with inconsistent separation of the plastic from the mould wall.

Acknowledgements

The research work reported here has received extensive support from EPSRC, CACCIA SpA, Borealis, Exxon (Canada), Lin Pac Rotational Moulders and Tyrrell Tanks Ltd. The author also acknowledges the excellent contributions from the staff in The Rotomoulding Research Centre at Queen's University, Belfast.

REFERENCES

[1] Miller, E. "Plastics Product Design Handbook", Dekker, New York. (1983).

[2] Crawford, R.J. "Plastics and Rubber: Engineering Design and Applications", MEP Ltd, London (1985).

[3] Crawford, R.J. (ed) "Rotational Moulding of Plastics", Research Studies Press, Somerset (1992).

[4] Ramaazzotti, D. "Rotational Moulding" in Plastic Product Design Handbook ed. By E. Miller, Marcel Dekker, New York (1983).

[5] Rao, M.A. and Throne, J.L. "Theory of Rotational Moulding", SPE. J., May (1972) p. 759.

[6] Throne, J.L. "Plastics Process Engineering", Marcel Dekker, New York (1979).

[7] MacAdams, J. "How to Predict Physical Properties of Rotomoulded Parts", SPE (Society of Plastics Eng.) Reg. Tech. Conf., Chicago, USA, Oct. (1975) p.64.

[8] Nugent, P.J. "A Study of Heat Transfer and Process Control in Rotational Moulding", Ph.D. Thesis, Queen's University, (1990).

[9] Crawford, R.J. and Nugent, P.J. "A New Process Control System for Rotomoulding", Plastics, Rubber and Composites: Processing and Applications, 17, 1 (1992) p.23.

[10] Crawford, R.J., Nugent, P.J. and Xin, W. "Prediction of Optimum Process Conditions for Rotomoulded Products", Int. Polym. Process, VI (1991) p.56.

[11] Anon, "Curing", ARM Video, 2000 Spring Road, Suite 511, Oak Brook, Illinois, 60521, USA (1997).

[12] Crawford, R.J. and Scott, J.A. "The Formation and Removal of Bubbles from a Rotational Moulding Grade of PE", Plastics and Rubber, Processing and Applications., 7, 2 (1987) p.85.

[13] Xu, L. and Crawford, R.J. "Analysis of Formation and Removal of Bubbles in Rotationally Moulded Thermoplastics", J. Matl. Sci., 28 (1993) p.2067-2074.

[14] Crawford, R.J. "Formation and Removal of Bubbles during Rotomoulding", BPF Conference, Walsall, June (1993).

[15] Spence, A.G. and Crawford, R.J. "The Effects of Process Variables on Formation and Removal of Bubbles in Rotomoulded PE", Polym. Eng. Sci., 36, 7 (1996), pp 993-1010.

Illuminating the Manufacturing Process - On-line UV Spectroscopic Analysis

Henryk Herman* and Phil Hope

**Actinic Technology, 83 Crockford Park Road, Addlestone, KT15 2LN, UK*
BP Chemicals Ltd, PO Box 21, Bo'ness Road, Grangemouth, FK3 9XH, UK

The application of spectroscopic techniques to polymer production processes permits the real-time measurement of those quality variables that form the polymer manufacturing specification, such as melt index, density and additive concentration. We will describe developments using ultra-violet methods that provide on-line data on residence time distribution, and on the mixing and concentration of additives in polyethylene compounding. We will also discuss the benefits of frequent quality measurements to the QC process, and the links with multivariate spc of the process variables that leads to better product.

INTRODUCTION

The use of at- and on-line methods to characterise chemical processes is becoming more prevalent, as analytical techniques move from being laboratory-based and retrospective, to process diagnostics. In particular, the use of methods based on near infrared spectroscopy to characterise feedstocks, products and intermediates is being applied in refineries, bulk and speciality chemicals production, and in the pharmaceutical industry. In the polymer manufacturing industry, the measurement of numerous properties, including density and melt index, co-polymer structure and composition, and additive concentration, has been considered, and in some cases commercial applications have resulted [1].

The compounding and finishing, or pelletisation, stage of the polyethylene manufacturing process fulfils two primary functions: the conversion of powder from the polymerisation reactor into pellets which are more convenient for handling by converters, and the incorporation of additives, principally antioxidants and light stabilisers, to protect the polymer during conversion and in end-use service. Pelletisation is normally performed by an extrusion process, co-rotating twin screw extruders being particularly appropriate and common. The first application described concerns the development of a method for evaluating in real-time the residence time distribution in extruders, without disrupting the process, providing a diagnostic for process development and troubleshooting. The second application has much greater applicability, and concerns a real-time technique for measuring the additive levels in polyethylene melt. This is a particularly exciting development as it offers the facility to effect closed loop control of additive incorporation, and continuous quality assurance, resulting in substantial cost savings.

MEASUREMENT OF RESIDENCE TIME DISTRIBUTION IN COMPOUNDING EXTRUDERS USING UV FLUORESCENCE

The determination of residence time distribution (RTD) yields information on polymer degradation and mixing during the extrusion process, and is particularly valuable during the commissioning of new extruders and the optimisation of existing extruder performance when introducing new materials. Conventional methods, involving the addition of tracers such as carbon black, followed by off-line collection and analysis, are unsuitable for real-time use in polymer manufacturing processes for a number of reasons: the time-scale for turning around a measurement and analysis is too long, the product contamination resulting from addition of conventional tracers is unacceptable, and in most cases access to the extrudate as it leaves the die is impracticable. This is particularly true in polyethylene compounding processes, where the throughput is usually very high and the product emerges from the die through an underwater or water-ring pelletiser.

The essence of the novel technique involves the introduction of a UV fluorescent additive into the extruder feed port, and its detection at the extruder die head by means of a fluorescence probe which can be fitted into a standard pressure transducer port. As the UV additive is a commonly-available optical brightener the product suffers no visual contamination, allowing RTD to be measured without disrupting production.

Experimental Set-Up and Results

The UV fluorescence probe was constructed by modifying a commercial infra-red temperature probe (Dynisco Inc.) designed to withstand typical extrusion temperatures and pressures (350 °C, 200 bar). Considerable care was taken to maximise the light collecting efficiency, necessitating the use of silica optical fibre with a polyimide coating and appropriate adhesives. The spectra were obtained using a Perkin Elmer 3000 fluorescence spectrometer. Optimum excitation and emission wavelengths were found to be 290 nm and 441 nm respectively. The probe, fibre optic cable and spectrometer arrangement is shown in Figure 1.

Extrusion compounded samples of a high density polyethylene (HDPE), BP Chemicals Rigidex HD 6070EA, were prepared containing various levels of additive, and used to produce a calibration curve. The fluorescence intensity was found to be linear up to around 0.01% w/w additive, and temperature dependent (Figure 2). RTD measurements were performed on a laboratory scale Werner and Pfleiderer ZSK30 co-rotating twin screw extruder, by adding a few pellets containing 5% w/w additive to the extruder feed and monitoring the fluorescence at the die head as a function of time. The example in Figure 3 demonstrates good agreement between the RTD as measured on-line by UV fluorescence and off-line using a conventional carbon black addition technique. A series of measurements made at different extruder feed rates (Figure 4) demonstrates the ability of the technique to quantify the reduction in residence time and narrowing of RTD with increasing feed rate. The results were reproducible, and demonstrate the feasibility of the method for real-time process diagnostic use.

MEASUREMENT OF ADDITIVE CONCENTRATIONS DURING EXTRUSION COMPOUNDING USING UV ABSORBANCE

Extending the above technique further into the UV offers a method for real-time monitoring of the additive levels during polyethylene manufacture. As additive levels form a part of the production manufacturing specification, this technique offers considerable potential impact on production costs.

Polyethylene additives are incorporated during the extrusion compounding process, primarily to inhibit oxidative degradation during processing, storage and service. These are commonly hindered phenols or organic phosphites, added at levels usually of the order of a few hundred to a few thousand ppm. The high cost of such additives makes good control of concentration in the final product an economic priority, and the combination of real time monitoring and closed loop control to the additive feeders presents a considerable cost-saving opportunity. Currently, additives are monitored during production by off-line techniques such as liquid chromatography and x-ray fluorescence, which can be time consuming; therefore automated on-line monitoring will also result in reduced off-line testing costs. Perhaps more importantly, real-time monitoring offers improved product consistency.

The commonly occurring additives have strong absorbance spectra in the infra-red and UV regions, as well as exhibiting UV-excited fluorescence. On-line spectroscopic analysis of polymer pellets is possible, but effects such as the size and shape variation of the pellets and the presence of crystallinity makes this a demanding task as far as low additive levels are concerned. Measurement in the melt phase is easier, and if applied to a compounding extruder offers a potentially more rapid control loop. UV spectroscopic methods such as fluorescence or diffuse reflectance spectroscopy are possible, but suffer the disadvantage that light penetration depth into the polymer can be variable. For this reason it was decided to apply UV absorbance, where a fixed path length can be defined. Infrared absorbance is a possible method, but the strong absorbance spectrum of polyethylene requires a very small path length (less than 1 mm. Another benefit when using UV is that the analyser is completely decoupled from the measuring system through the use of fibre optics.

Experimental Set-Up

Trials were performed on a pilot-scale compounding extruder, a Werner & Pfleiderer ZSK57 co-rotating twin screw extruder, configured with a screw design to give scale-down operating conditions close to those used on plant production machines (see schematic in Figure 5). The melt was presented to the optical probes by incorporating them into the flow channels of a Göttfert RTR on-line rheometer, modified for the purpose. The rheometer was located on a melt plate, fitted after the 8-0 transition piece and connected via a short adapter. The polyethylene was pelletised using a Berringer water-ring pelletiser. This arrangement ensured that extruder operation was unaffected, and that melt was returned to the extruder melt stream.

Sapphire-window optical probes were supplied by Sensotron Inc., and placed collinearily across the rheometer flow channel, as shown schematically in Figure 6. The selected probe position was very close to the melt inlet from the extruder, and also allowed samples to the taken for off-line analysis from a bleed port located immediately downstream. The probes were linked to a Zeiss MCS UV spectrometer through 5 m lengths of UV-grade silica fibre optic cable. The spectrometer was controlled from a PC running Zeiss' WinAsp.

Experimental Trials

Trials were carried out using BP Chemicals Rigidex HD6007 HDPE powder as feedstock, with three different antioxidants (coded A, B and C) representing the most commonly used in current production formulations. Because the spectra for the three additives overlap to a great extent in the UV region (Figure 7), and the spectra under process conditions were not well known, it was necessary to employ a chemometrics approach to extract and manipulate the data. This approach entails building a "model" in which the on-line measured spectra are correlated with independent off-line measurements of additive content, using a series of "calibration" samples. Once the model is complete it can be checked using a separate series of blind "validation" samples, and if the model performance is found to be within the desired specification the system can be used for on-line measurements of additives.

There are a variety of algorithms for resolving data and performing correlations with measured values. The two main ones are principal components regression (PCR) and partial least squares (PLS). As a trial of this sort generates fairly limited quantities of data we chose to use PCR, which aims to express the variation in the spectral data in as few terms as possible; it can be considered a maximal data compression scheme and is particularly robust with respect to errors in the property data.

In these trials a calibration set of 24 independent samples was used, followed by a validation set of 8 samples. These were all prepared by weighing out additive/powder concentrates, then diluting to the desired final concentration by adjusting the feed rate to the extruder. The extruder was operated at a throughput of 150 kg/hr, with temperatures and screw speed designed to give specific energy inputs typical of plant production. The RTR gear pumps were operated at fixed speed, giving an estimated time for polymer from the extruder to reach the probes of around 23 seconds.

Following initial experiments in which the necessary spectral range and optimum integration time were determined, the "calibration" runs were performed. In each case the base polymer was run to allow the operator to accumulate a background spectrum, after which the weigh feeder containing the appropriate concentrate was started, and the signal monitored on the UV analyser, accumulating a spectrum every 15 seconds. When the spectra had stabilised samples were taken from both the bleed port on the RTR and from the pellet stream. The additive feed was then switched off, and the return of the UV signal to its original baseline was monitored. All the samples were analysed off-line using standard liquid chromatography (LC) with UV detection, and x-ray fluorescence (XRF), which was sensitive to additive C.

The spectra were all of high quality, with the repeatability (repeat measurements of the same sample, obtained by stopping the on-line rheometer) being better than +/-0.001 au, which corresponds to less than 1 ppm of additive. The minimum detectable levels are shown in Table 1.

Additive	MDL (ppm.mm)
A	6
B	8
C	12

Table 1. **Minimum Detectable Level of Additive.**

Data Analysis and Results

Regression of the principal components found using PCR against off-line measurements of additive concentrations yielded the parameters shown in Table 2. The coefficient of determination indicates the strength of the relationship between measured (off-line) and predicted (on-line) values for the calibration standards, the standard error of estimate indicates the quality of fit of the regression and the standard error of prediction gives an estimate of the magnitude of the error when independent samples are predicted by the model. It is worth noting that the errors incorporate those of the (off-line) laboratory determinations.

Off-line Test Method	Additive	Coefficient of Determination	Standard Error of Estimate (ppm)	Standard Error of Prediction (ppm)
Gravimetric (initial preparation of concentrate)	A	0.982	28	32
	B	0.957	52	59
	C	0.995	52	56
LC analysis of bleed port samples	A	0.989	13	14
	B	0.957	40	46
	C	0.990	67	72
LC analysis of pellet samples	A	0.963	29	32
	B	0.892	68	81
	C	0.994	53	68
XRF analysis on pellet	C	0.993	62	69

Table 2. Performance of PCR predictive model against off-line additive determinations.

The correlation between laboratory results and the model is excellent. This is further illustrated by plots of actual against predicted values shown in Figure 8.

It is interesting to note that the technique can provide additional information on residence time of the compounding process, the mixture quality and the presence of contaminants. Figure 9 illustrates the concentration of additive A as a function of time following its introduction to the compounder, by monitoring the UV intensity at a fixed wavelength (280 nm). The heavy trace shows the time to achieve stable operation; although the additive came through within 45 seconds of starting the additive weigh feeder, it was almost 600 seconds before a plateau was reached. The short time fluctuations in additive level are also believed to be real manifestations of incomplete mixing. Finally, monitoring intensity at a wavelength where UV in not absorbed by the additives can be used to detect unexpected contaminants; in this case the peaks correspond to traces of carbon black.

APPLICATION TO MULTIVARIATE SPC

The use of on-line techniques such as the above UV spectroscopic method can result in cost savings via reduced off-line testing (which can be manpower intensive). The greatly increased frequency at which *quality* data are available with no operator involvement or sampling inefficiency now allows for the use of all those process variables, such as temperature distribution, pressure, motor power and extrusion speed, that were previously reduced to an hourly or daily average, and incorporation of these into a multivariate model of the process that tracks all the variables that lead to "good" product, thus providing a comprehensive SPC package.

ACKNOWLEDGEMENTS

The authors gratefully acknowledge the assistance given by Axel Göttfert and Karl Hartmann, of Göttfert GmbH, and Vinod Mehta of Zeiss UK, in the development of the additive monitoring device, and Scott Jackson of BP Chemicals in carrying out the compounding trials.

REFERENCES

[1] J.W. Hall, D.E. Grzybowski and S.L. Monfre, "Analysis of polymer pellets obtained from two extruders using near infrared spectroscopy", J. Near Infrarred Spectrosc. **1**, 55-62 (1993)

[2] A. Senoucci, P.S. Hope, L.A. Hilliard, A.M.L. Irvine and I. Caucheteux, "A Novel Technique for in-line measurement of residence time distribution in polyethylene extrusion", Ninth Annual Meeting of the PPS, Manchester, 1993.

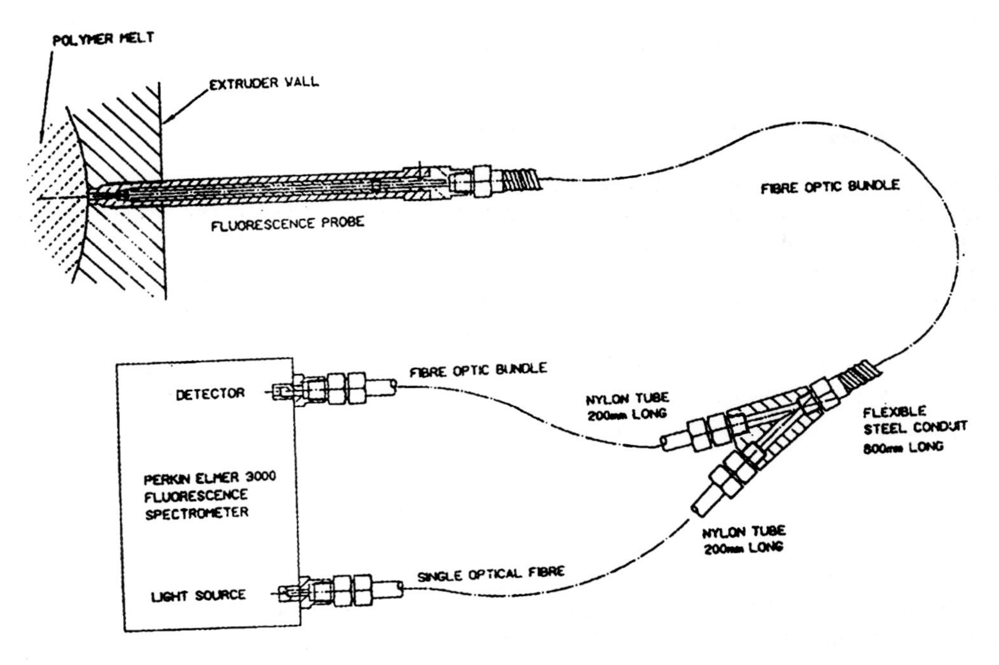

Figure 1. Optical Fibre Probe

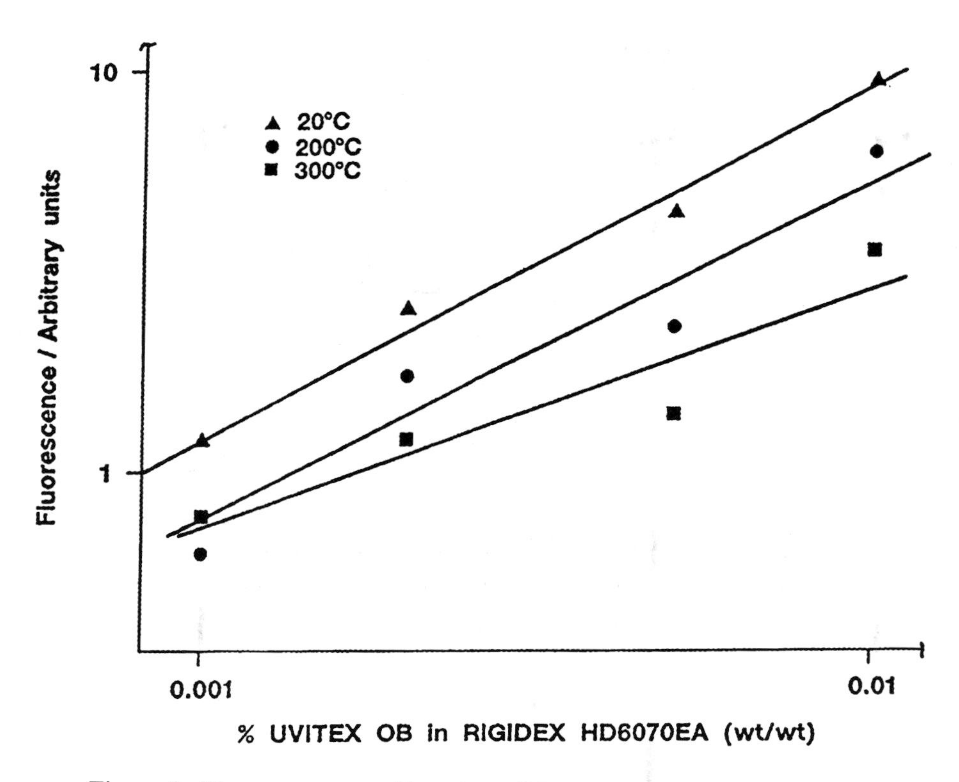

Figure 2. Fluorescence as a Function of Temperature and Concentration

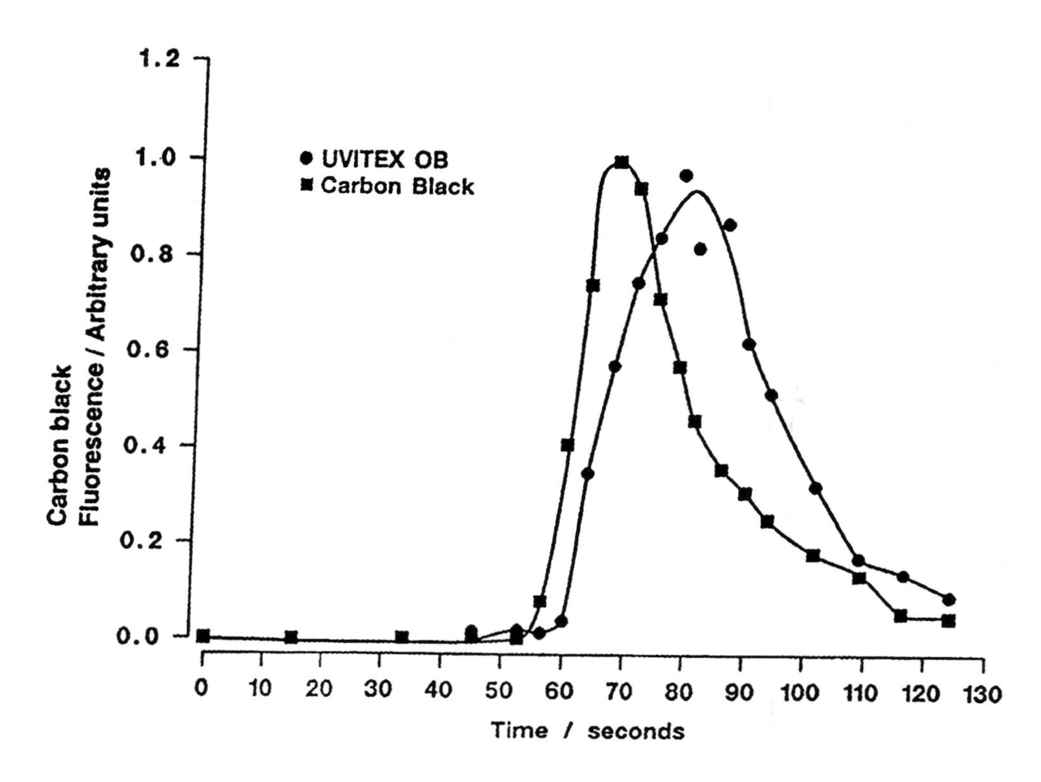

Figure 3. RTDs Measured Using Uvitex OB and Carbon Black Tracers

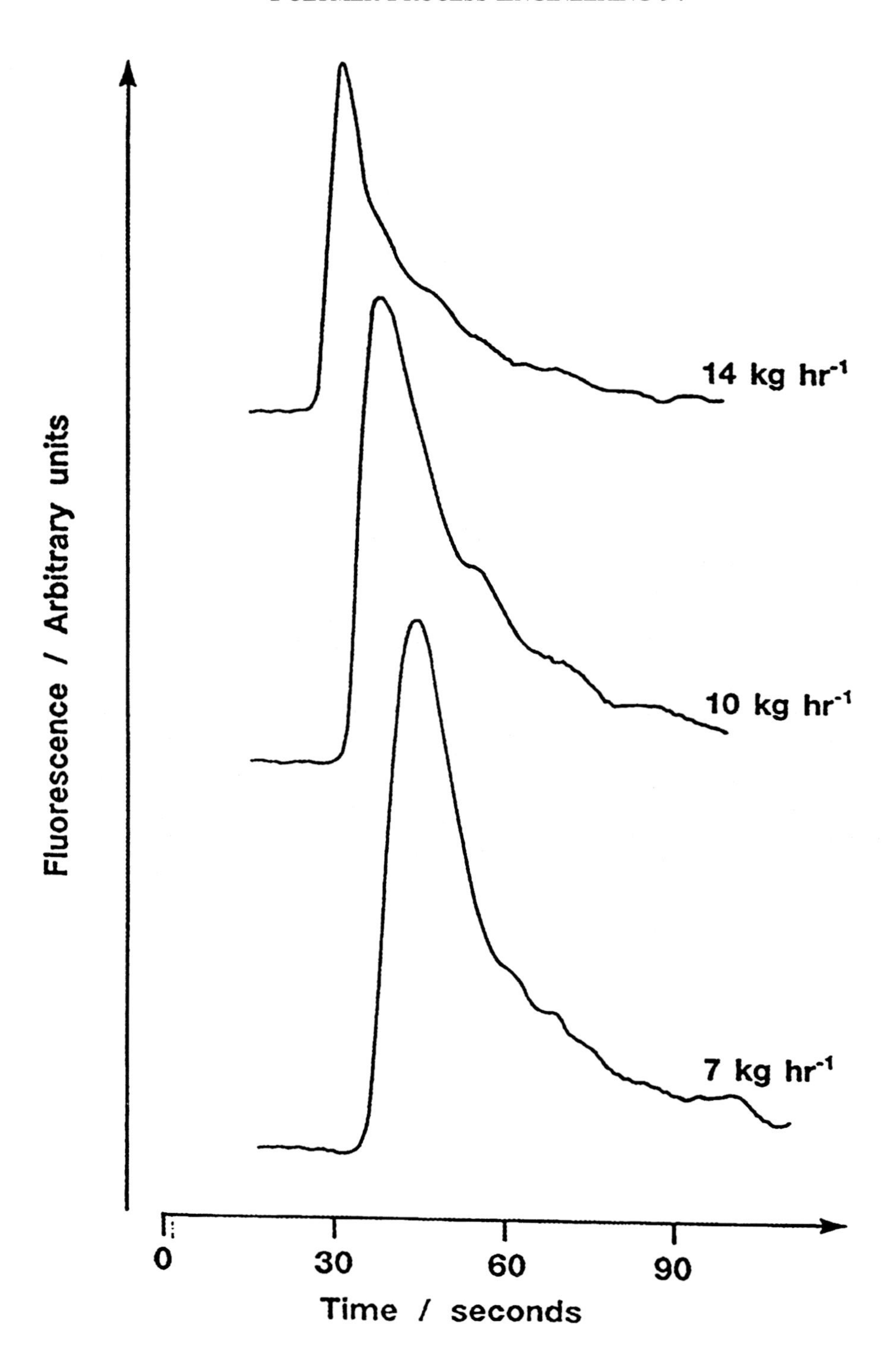

Figure 4. RTDs as a function of Feedrate

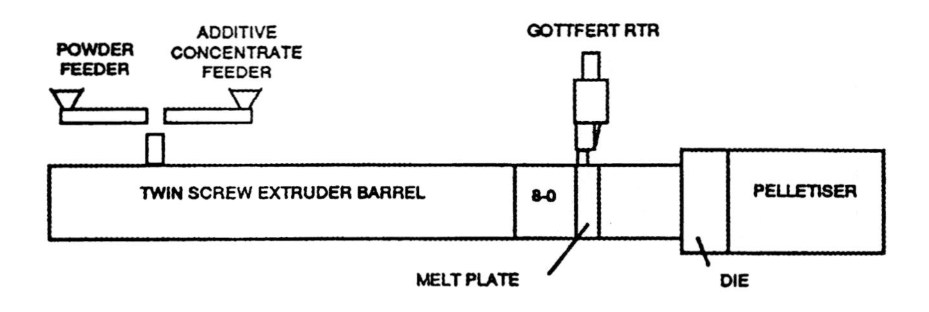

Figure 5.

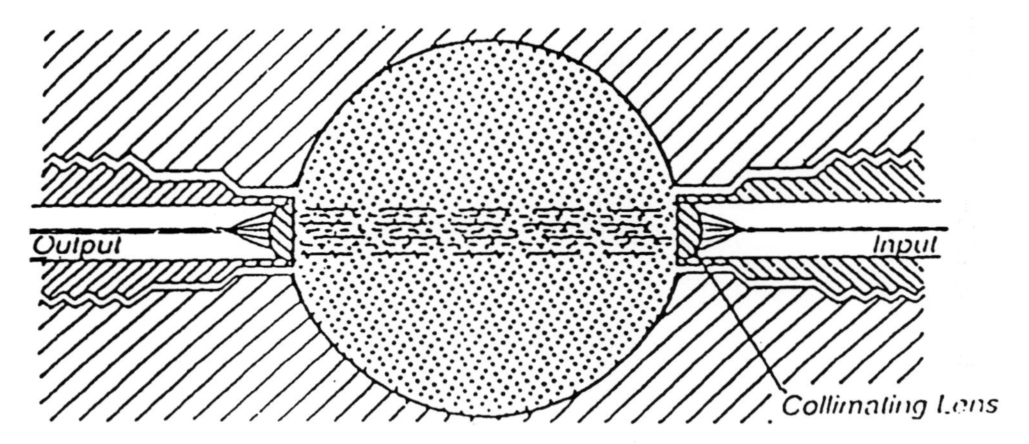

Figure 6. Placement of Sensotron probes across rheometer channel.

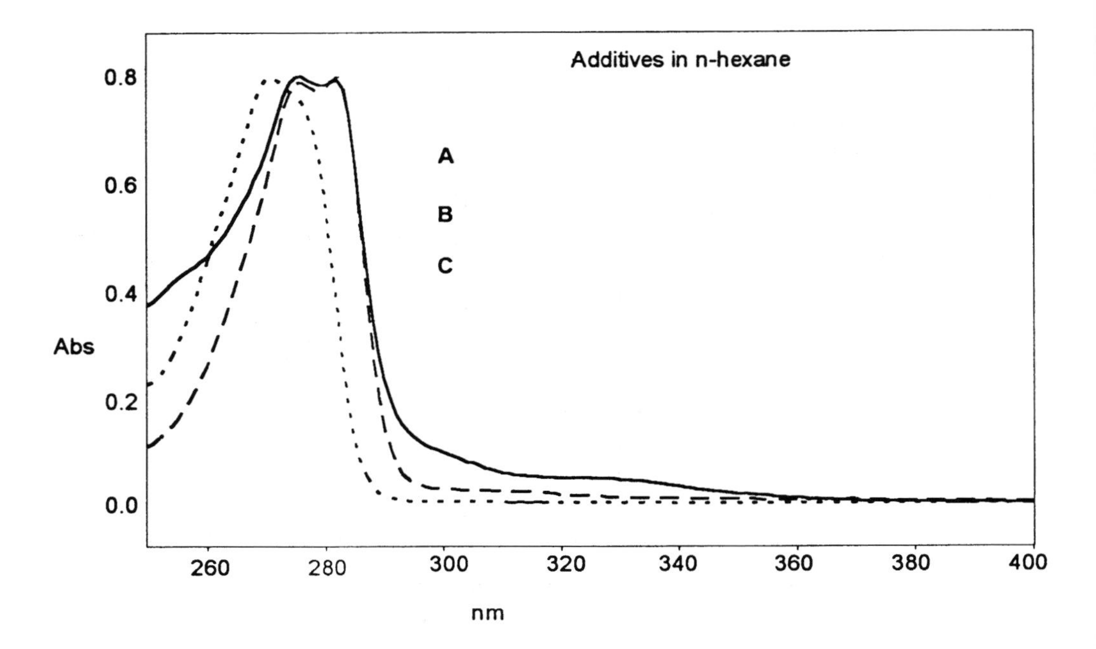

Figure 7. Absorbance spectra of antioxidant additives in n-hexane between 250 and 400 nm, normalised to peak absorbance.

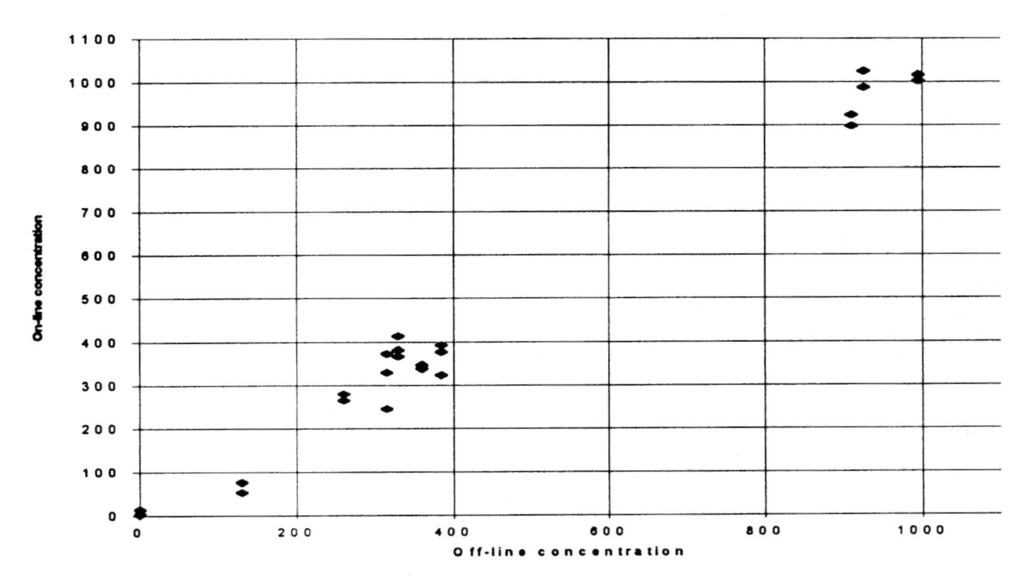

Figure 8. On-Line Predictions for Additive B against LC Determination

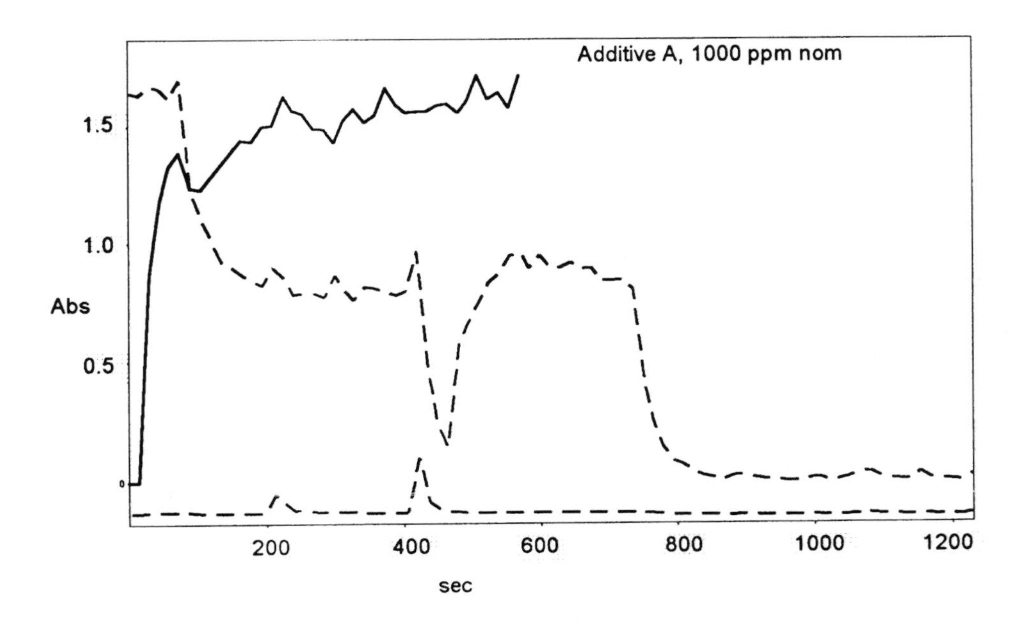

Figure 9. Additive A as a function of time a) from start of weigh feeder, and b) varying nominal concentration from 1000 ppm to 500 ppm to 0 to 500 ppm to 0.

Dual Thermocouple Loops for Extruder Temperature Control

DEREK P. FERRIS

Davis-Standard Corporation
Pawcatuck, Connecticut, USA

ABSTRACT

A method of extrusion temperature control, in which the temperature in the control zone of an extruder barrel is monitored at two critical points, and the signals from both sensors are used as controlling inputs to an extrusion process controller. A patented[1] control algorithm utilizes the dual temperature feedback signals, along with setpoint and other real-time process data, to regulate the appropriate control outputs to the zone's heater and cooler. The control system anticipates the dynamically changing conditions within the barrel zone, and responds automatically with changes in control output. It enables temperature stability to be regained after a system upset or step-change within less than one-half the time of conventional PID controllers.

THE NEED FOR PRECISE EXTRUSION TEMPERATURE CONTROL

The requirements of today's extrusion controllers are becoming ever more demanding. More complex and expensive resins to process, higher output rates sought, more involved end-products to produce - all these factors necessitate greater precision in temperature control. An extruder barrel is heated and cooled in order to raise the temperature of the raw polymer so that it will melt, and then to maintain the temperature of the extrudate under the dynamic conditions of processing. A change in extruder barrel temperature will affect melt temperature, and a change in melt temperature will affect melt pressure, which will affect the output rate as well as the characteristics or "rheology" of the extrudate.

Precise control of extruder melt temperature presents a very difficult challenge, due to the physical construction of the barrel, the design of the screw, and the processing characteristics of raw polymers. The extruder barrel presents a thick steel barrier between the inside point whose temperature is to be controlled and the heating/cooling source on the outside. This results not only in a considerable temperature differential but also a significant thermal lag. The operation of the process introduces further complications in the form of induced heating of the melt through shearing and mixing, and dynamic changes caused by changes in screw rotational speed.

CONVENTIONAL CONTROL METHODS USING ONE SENSOR

The most familiar control approach is the single-sensor PID (Proportional-Integral-Derivative) temperature control method, used in both discrete single-zone controllers as well as multi-zone definite-purpose controllers [Fig. 1].

Tuning adjustments are usually provided for gains, cycle times, proportional bands, integral and derivative time constants. With conventional PID controllers, tuning adjustments and corrections are made over time, after studying the conditions and evaluating the errors. This is true whether it is a process technician doing the evaluation and adjustment or a set of electronic circuits. So-called "auto-tune" controllers measure the gains and time constants of the system, re-calculate and then automatically adjust the PID values in a attempt to achieve the best steady-state response. Auto-tuning is implemented either with the system at rest (i.e. cold barrel, screw stopped) or during steady-state in-process condition. After the auto-tuning process is completed, the new settings may enable the controller to provide good stability under some conditions but be incapable of quickly responding to transients or changing conditions. In either case, the controller can only respond after an upset has occurred and a control error has been detected and quantified. The sophistication of some PID controllers has lately advanced to include "fuzzy logic" algorithms. Whereas these controllers can make some intelligent decisions based upon real-world conditions and anticipated changes, and can offer faster response and accuracy, they also require additional set-up and tuning parameters.

The single temperature sensor is sometimes positioned half-way between the heating/cooling source and the barrel liner, which seems to provide a stable control but is not, in fact, controlling temperature at the point where precise control is needed. At other times, the single sensor is positioned in a deep well so that the tip touches the barrel liner. This is the correct point of control, but the controller must then contend with all of the system's inherent thermal inertia and sluggish response.

OTHER CONTROL METHODS USING TWO SENSORS

Some of these control methods use the outer sensor only as an auxiliary sensor, to provide high/low alarm inputs, such as when the zone overheats, when the zone is below operating temperature, or to detect when there has been a control malfunction, such as when a change in temperature setpoint has produced no change in process temperature within a prescribed time period. Often, the auxiliary sensor temperature is not used as an input signal into the primary controller.

Other methods take the average of the signals from the inner and outer sensors, and use this signal as the control input. A deficiency of this approach is that, after the system has stabilized, there usually exists a persisting offset or "droop" between the actual and the set temperatures.

REQUIREMENTS FOR TEMPERATURE CONTROL IN A HIGH-GAIN SYSTEM

The designer of an extrusion temperature control system (in fact any closed-loop system) has to walk a fine line between designing a system with a high enough gain to be responsive to dynamic changes, but yet not so high as to be continually unstable through over-reaction. Total system gain is not only a function of the control system, but also of the mechanical components - the heater/cooler, the screw design, etc. Davis-Standard's extruders are equipped with precision-machined, bolt-on heater/cooler jackets with high wattage calrod heating elements and high-efficiency double-pass cooling tubes, connected to a closed-loop pressurized system. In the case of air-cooled extruders, the aluminum jackets have deep fins through which ducted air is blown from high-capacity blowers for maximum heat removal. For temperature sensors, Type "J" thermocouples with grounded tips are used for fast response. Screws are designed for high output rates while, at the same time, providing thorough mixing of the extrudate. With all of the mechanical components designed for high system gain, a responsive control system is of critical importance in ensuring that process stability is achieved and maintained.

ESSENTIALS OF DAVIS-STANDARD'S DUAL-SENSOR TEMPERATURE CONTROL

Davis-Standard extruders may be equipped with conventional PID controllers or with Davis-Standard's own patented[1] Dual-Sensor Extrusion Process Control [Fig. 2], hereinafter abbreviated as D-S EPC. Four important basic features are inherent in this design.

The first basic feature of this method of control is the critical placement of the two sensors - the "shallow" or "B" sensor is inserted into the heater/cooler jacket, the "deep" or "A" sensor is inserted into a deep well, so that the sensor tip actually touches the barrel liner and is, therefore, as close to the melt stream as it is possible to get.

The second basic feature is that **both** sensors play an equally important, albeit entirely different, role in the temperature controlling process while, at the same time, providing additional information for multi-purpose alarm inputs.

The third basic feature is that, while the controlling method is the same, the control system architecture may take several different forms. It is essentially a multi-zone, microprocessor-based, custom-configured process controller, but it is offered in a variety of hardware formats and software platforms according to customer preference or job requirements, all having the same unique controlling properties, however.

The fourth basic feature is the inherent "auto-tuning" nature of the control. There is virtually no need for tuning or re-adjusting the system, even when changing throughput rates, screws, or compounds. The system "learns" the new dynamic conditions, automatically adjusts its internal settings, and optimizes its set-up routines to provide stable and accurate control.

Proportional Action

Narrow high-gain proportional bands are used for maximum responsiveness, both for heating and for water or air cooling. In a conventional single-sensor PID system, this would most likely create a condition of unacceptable instability. The D-S EPC makes use of its characteristic anticipatory feature as well as the inherent thermal lag created by the mass of the extruder barrel between heater/cooler jacket and barrel liner. The controller processes each control loop every 200 milliseconds, and may make fairly dramatic changes in jacket temperature, in order to achieve the desired adjustment in process temperature as quickly as possible. The fast control response and high-gain control actions result in a high-frequency temperature oscillation in the jacket. However, the thermal mass of the barrel acts like a low pass filter, so that a high-frequency temperature variation of significant magnitude in the heater/cooler produces a minimal variation in process temperature. For example, a temperature variation in the heater/cooler of around 2.7°C will typically produce a variation in process temperature of around 0.3°C. The temperature oscillations are effectively blocked, but the desired heat/cool source temperature change is of a frequency below the cut-off frequency of the filter, and thus passes through to the barrel liner to produce the desired change in melt temperature.

As in most PID controllers, the system outputs are in the form of time-proportioned on/off control signals, which regulate the power switching devices to turn the zone heater or cooler on or off. Duty cycles and on-times depend upon the element being controlled - for example, the on-time for a water-cooling solenoid can be as short as 0.1 seconds if the "A" sensed value is close to setpoint, whereas the on-time for an air-cooling blower starter is considerably longer under the same conditions.

Since the "B" sensed value provides an accurate determination of jacket temperature, the controller is able automatically to make use of, and to compensate for, the rapid heat extraction effect of evaporative cooling in a water-cooled jacket, when the circulating water just introduced into a hot jacket flashes to steam. Conversely, in cases where extrudates are processed at low temperatures where the heater/cooler jacket temperature is often below the flash-point of the pressurized coolant, the controller automatically compensates for this also, by considerably lengthening the on-times for the cooling function.

Reset Action

The D-S EPC **Control Setpoint** is a function both of deep sensor "A" and shallow sensor "B". Taken no further, this would produce an effect somewhat analogous to single sensor control systems where the sensor is placed half-way between heater/cooler and barrel liner, or to other dual-sensor control systems which simply average the sensor inputs to form a single control input. The result of such a control input into a proportional-only controller is a large process temperature offset or "droop", once a steady-state operating condition has been reached. The Reset function is there to correct for offset [Fig. 4]. Before taking action, Reset checks for the amount of offset and for process temperature stability (minimal variations within a certain time period). Four conditions must be satisfied before Reset will occur - 1). the Control Sum Error (the sum of the deviations of the "A" and "B" sensors from setpoint divided by the weighted average of the "A" and "B" sensed values) must be within a control alarm band of +/- 5.6 °C, 2). it must have remained within these limits for at least 4 minutes, 3). the "A" deviation must have been $\geq$ 0.5 °C for at least 4 minutes, and 4). a Control Reset must not have occurred during the previous 4 minutes. If all four conditions have been met, a Control Reset action will take place. A new control setpoint will be calculated which will replace the actual (or operator-entered) setpoint or the previous control setpoint. Once this new control setpoint takes effect, a fresh temperature control action takes place, in order to attempt to satisfy the new conditions.

Derivative Action

Obviously, the faster a control system can reach a stable condition after start-up or re-stabilize conditions after an upset, the more efficient the system is, in terms of usable product output and raw material and energy saved. While the D-S EPC makes use of proportional bands for basic control function and integral factors for the reset function, the major difference between it and more conventional PID controllers is in the derivative function. Derivative action affects a control loop based upon the rate of change of error with respect to time, or $D = K * de/dt$, where D = Derivative Action, K = Derivative Gain Constant, and e = error. The derivative action modifies the effect of proportional control by changing its output value, depending upon the rate of change of the control error or, in other words, forces the proportional control output to a value which it would otherwise not have reached until some later time without the derivative function. Since the actual temperature difference between the heat/cool source and the material being controlled is continually monitored, the D-S EPC control system can anticipate the rate of change of temperature of the material following a control setpoint change, without having to resort to time-based calculations. Naturally, the greater the temperature difference, the higher the rate of change. Changes in the magnitude of the temperature difference can be caused by a process setpoint change or by a process load change. Examples of load changes could be a change in screw speed or a change in compound such as the introduction of an additive. The dual-sensor system instantly detects the temperature change, calculates the required control response, and anticipates the effect of this response. The controller can adjust the control setpoints to prevent over- or under-shoot due to a high rate of change, and quickly bring, or restore, the system to a stable condition.

Adaptive Reset

The Adaptive Reset function is a second patented[2] feature of the D-S EPC [Fig. 3]. Its purpose is to anticipate system upsets caused by changes in screw speed and to make control corrections early enough, so as to provide for rapid start-ups or to minimize the effect of a process upset and assure a quick regaining of system stability. For example, a barrel zone may be operating with constant cooling due to high shear heat generation caused by screw design and melt characteristics; if screw speed is suddenly reduced, the "A" sensor temperature will drastically under-shoot the setpoint because the heat load has been greatly reduced. A conventional control would respond in this instance with full heating.

It has been proven at Davis-Standard through exhaustive laboratory testing that the most significant parameter relating heating/cooling load to screw speed was the Control Reset or load compensation value while recognizing, at the same time, that shear heat generation is largely a function of screw RPM. The amount of reset (load compensation) is directly proportional to the amount of heating or cooling required. Adaptive Reset operates by continually monitoring screw speed, creating tabulated logs or "look-up" tables of screw speed values and their corresponding control reset values, and then associating these tables with particular recipes (stored data files of zone setpoints, control parameters, machine setups, screw types, compounds, additives, end-products, etc.). In this manner, the system learns the optimum reset values for specific operating conditions. The first time a new product is run, the load-compensating reset value is memorized for each zone over the entire screw RPM range. A thermal loading model is thereby compiled, stored, and automatically down-loaded when that same product is run again. This obviates the need for the system to wait for a process upset, then wait for the process to stabilize, in order to calculate the error and take the required corrective control reset action. It has been found that Adaptive Reset reduces the amount of control upset by 50% and offset time by as much as 66%. It is of especial benefit for those processes which require constant speed changes - such as in blow molding operations, during wind-up reel or roll changes, short experimental runs on laboratory extruders, or in recovery from an unplanned event such as an emergency stop or a system shut-down.

LABORATORY TEST REPORT

During the development and testing of the Adaptive Reset feature, extensive extrusion trials were conducted on both 63.5mm and 114mm extruders with various thermoplastic resins and screw designs, including combinations that required high cooling as well as high heating. The test data and graphs presented here are the result of a series of trials on a 114mm 24:1 L/D water-cooled extruder, having five barrel zones, and equipped with a DSB® barrier screw, processing 0.9 MI film grade LLDPE. The trial sequence was repeated three times for the three types of barrel temperature controllers under study, i.e. single-sensor auto-tuned PID, dual-sensor D-S EPC, and dual-sensor D-S EPC with the Adaptive Reset function enabled. The PID controllers were discrete units, configured with the manufacturer's recommended settings, then auto-tuned. The dual-sensor trials were conducted using Davis-Standard's EPIC II™ Extrusion Process Controller. The adaptive reset values applied in the third trial were "learned" during the second trial.

During each trial the extruder was run at 20 RPM until the process and all barrel zone temperatures, which were set at 177 °C, were stable. Screw speed was then increased to 60 RPM. After 30 minutes the extruder was suddenly stopped for 12 minutes, then re-started and ramped up to 60 RPM. All barrel zone temperatures were monitored and recorded during the trials, as was melt temperature, that critical parameter which so markedly affects the process.

0.9 MI LLDPE is a fairly high viscosity material, demanding high torque and considerable cooling from the extruder, when processed using a high-output barrier type screw. The reset values associated with screw speed for barrel zones 1 and 5 are shown in Figure 5. These are the values which were learned during the second trial and applied by the Adaptive Reset control during the third trial. The feed area zone, Zone 1, shows a heating requirement at low speed, and then a cooling demand for speeds over 30 RPM. The discharge zone, Zone 5, shows an increasing cooling requirement for all speeds.

Figures 6 through 9 show the recorded values from the deep "A" sensors (type "J" thermocouples) during the three trials with the three different control schemes, all on the same time scale. Figures 6 and 7 are of Zone 1, Figures 8 and 9 are of Zone 5. Figures 6 and 8 show a screw speed change from 20 to 60 RPM, Figures 7 and 9 show a screw speed change from 60 to 0 RPM and back to 60 RPM. In all cases, the initial reaction to speed changes for all three control schemes is identical, since the deep "A" sensors are reflecting the rapid change in process load, due to the addition or loss of polymer shear heating from the screw. The differences in the curves after the initial reaction show the marked differences in the controllers.

In Figure 6, going from 20 to 60 RPM, the PID controller undershoots the setpoint as it is recovering from the initial upset and then takes a considerable amount of time (about 25 minutes overall) to reach setpoint. The dual-sensor scheme avoids the undershoot but waits at the initial upset temperature longer, then goes through a reset and reaches setpoint in about 18 minutes. Adaptive Reset shows a dramatic improvement with no lag or overshoot, coming right back to setpoint in only 7 minutes.

Figure 7, with screw speeds going from 60 to 0 RPM and back to 60 RPM, shows the same characteristic pattern. The PID controller undershoots and oscillates, the dual-sensor control hesitates and goes through an obvious reset, and Adaptive Reset quickly returns to setpoint in about one-third of the time as the other two.

Figures 8 and 9 show similar responses on Zone 5, except that the PID controller, operating on a zone with a higher cooling load, appears to be much slower in reaction, and never reaches setpoint. The other two controllers behave in a manner similar to Zone 1, with the superior performance of Adaptive Reset being obvious once again.

The effect on melt temperature is shown in Figures 10 and 11 for dual-sensor and Adaptive Reset. They show the dramatic improvement that a quick recovery to setpoint can have on the process. In both cases, Adaptive Reset brought the melt temperature back to a stable condition in about one-half the time of dual-sensor alone.

SYSTEM ALARM AND CONTROL LIMITS

Two alarm bands are established in reference to process setpoint. The outer alarm band sets limits for determining if the process measured value (zone actual temperature as measured by the "A" or deep well sensor) is close enough to setpoint so as not to require reset action. The inner alarm band sets limits for determining if cyclical deviations in the process measured value are small enough to meet the Reset function's definition of stability.

If, after repeated Reset actions, the control setpoint exceeds a limit value of 44 °C above or below actual setpoint, no further reset action is allowed; this condition would indicate either a machine or a process malfunction, and would require investigation.

A high limit is established for the heater/cooler temperature, as measured by the "B" or shallow sensor, which defines the maximum operating temperature of the heater/cooler unit and its associated wiring. If this limit is exceeded, a zone temperature shut-down is initiated.

Dead-band limits establish the upper and lower range limits of the dead band around setpoint, in which no heating or cooling action occurs.

OPERATOR INPUT

With Davis-Standard's Dual-Sensor Extrusion Process Control system, the only operator input required is entering process setpoints or recipes and monitoring system status. Depending upon the particular hardware/software format chosen for the system, the setpoints may be entered either through a membrane keypad and system status monitored on a two-line alpha-numeric display, or setpoints may be entered through, and system status monitored on, a color CRT touch-screen industrial terminal. The size (diagonal measurement) of the CRT touch-screen may range from 230mm to 500mm.

In the simpler format, up to 10 setpoint recipes may be programmed and entered through the keypad. In the more advanced format, several hundred recipes may be programmed, edited, and down-loaded through the touch-screen.

CONCLUSIONS

Extruder barrel temperature control greatly affects extrusion rate, melt temperature and pressure, product quality, and system efficiency. An accurate and responsive temperature controller is, therefore, a critical factor in successful and profitable extrusion. Processes that involve speed changes or frequent start-stop sequences require fast recovery of control and stability. Conventional single-sensor PID controllers, even of the "auto- or self-tune" variety, can only be adjusted for current conditions, whether for those prior to start-up or under a particular load situation. Should the situation change (and it will), the controller will require some time before it can re-calculate its tuning parameters. Davis-Standard's dual-sensor extrusion process control with Adaptive Reset has been shown to provide dramatic improvements in control stability, response, and precision, through a continually active system of monitoring, learning, and then applying the optimum control parameters for changing conditions.

References

[1] L. Faillace; US Patent 4,262,737 1981.

[2] L. Faillace; US Patent 5,149,193 1992.

Acknowledgments

Graphs and associated laboratory test data included herein were taken from a technical paper by William A. Kramer of Davis-Standard, entitled "Extruder Temperature Control with Adaptive Reset", which was presented at ANTEC '95.

CONVENTIONAL PID CONTROL

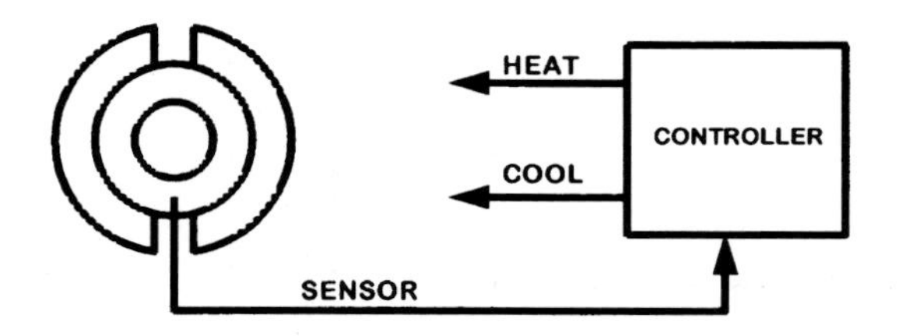

Figure 1

DUAL SENSOR CONTROL

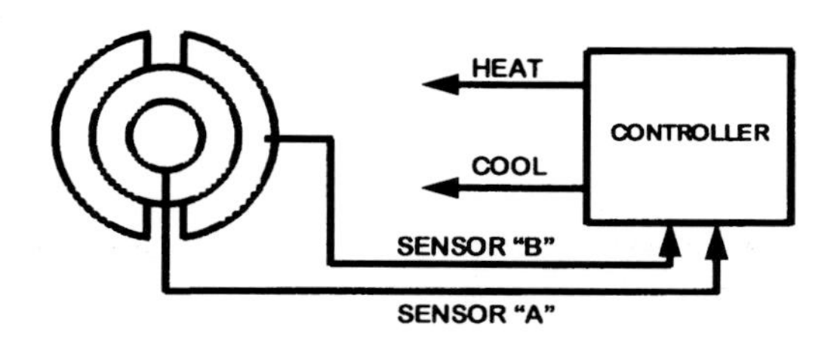

Figure 2

DUAL SENSOR CONTROL WITH ADAPTIVE RESET

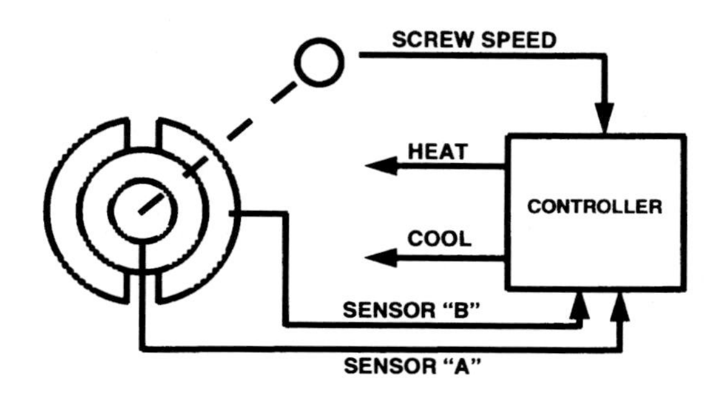

Figure 3

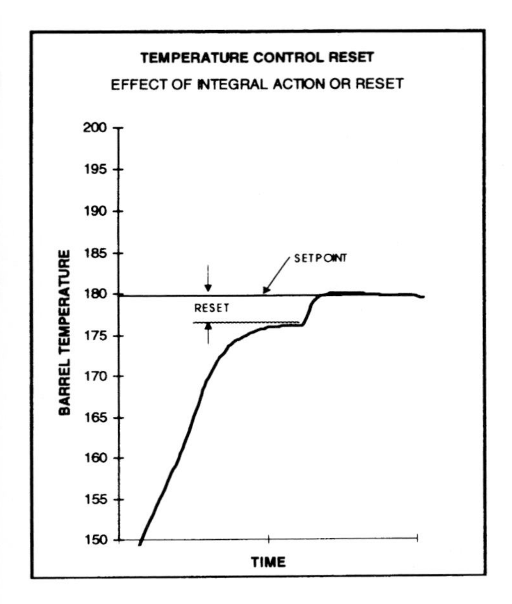

Figure 4

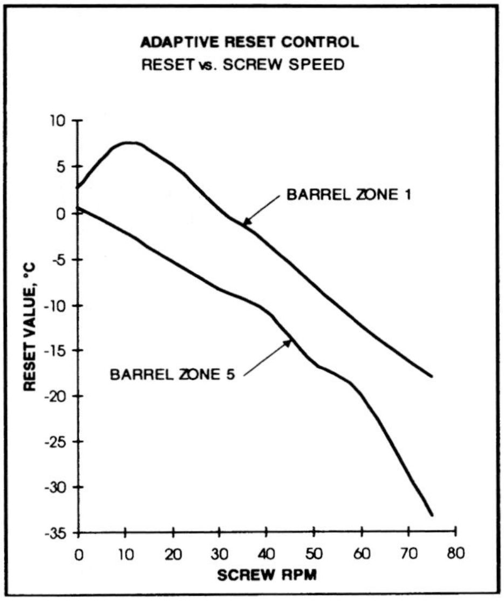

Figure 5

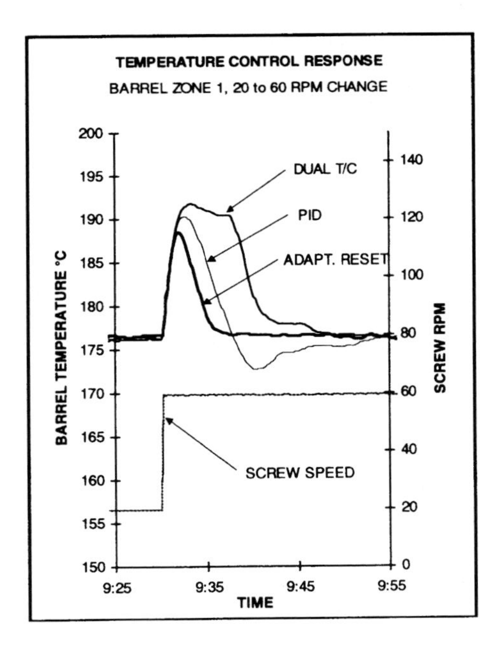

Figure 6

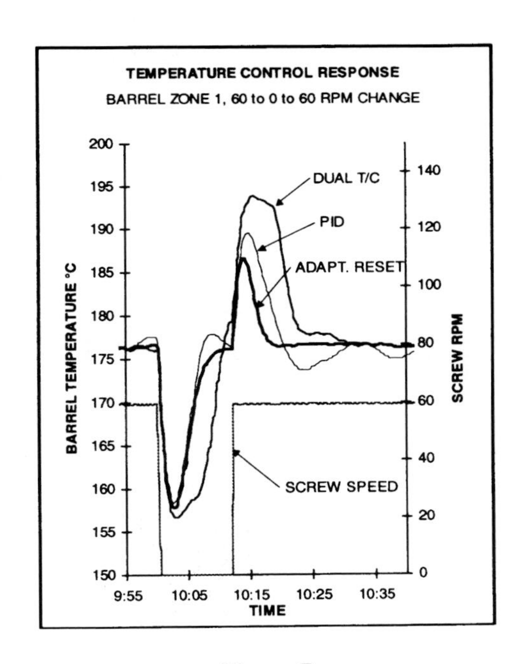

Figure 7

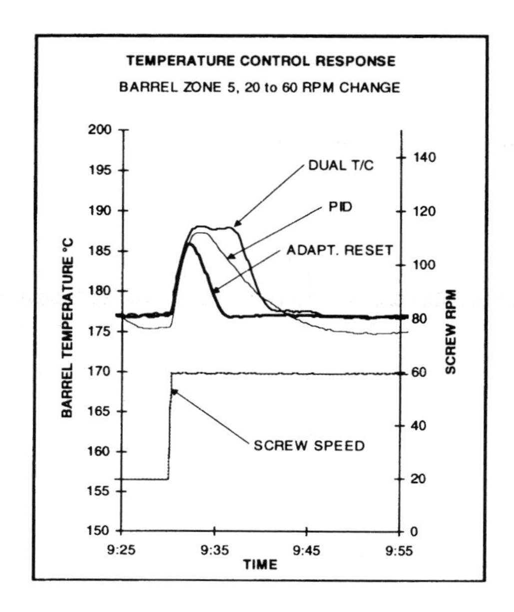

Figure 8

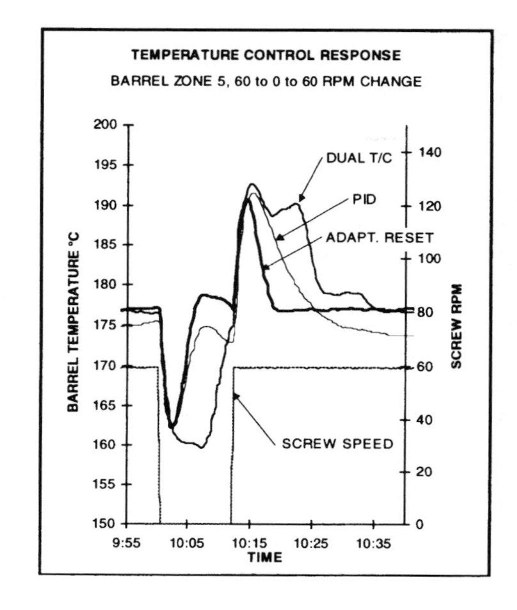

Figure 9

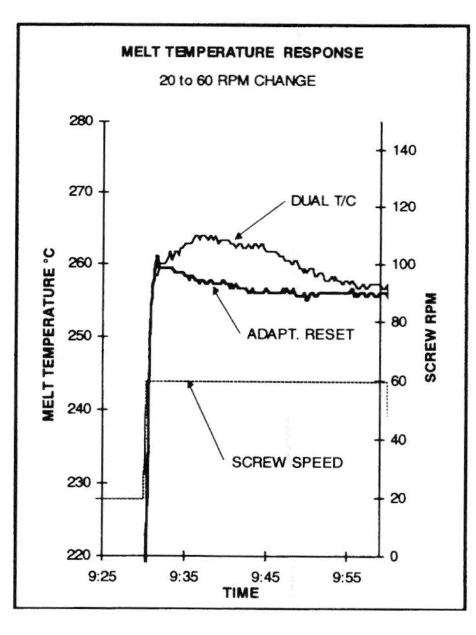

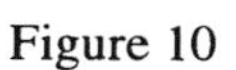
Figure 10

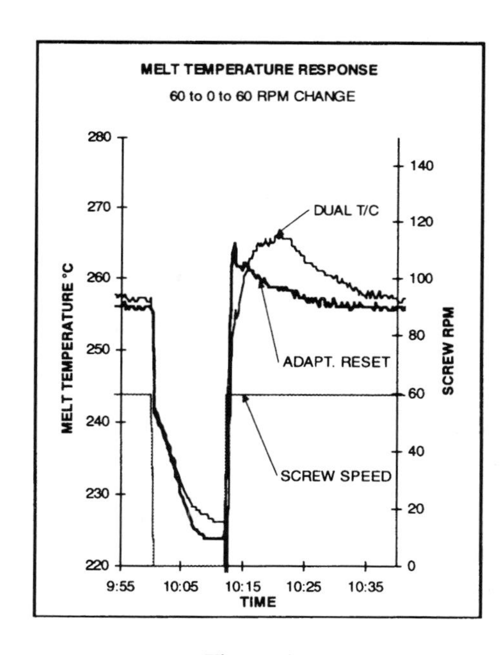

Figure 11

In-Line Measurement of Copolymer Composition and Melt Index

Marion G. Hansen and S. Vedula

Department of Chemical Engineering, 419 Dougherty Building, University of Tennessee, Knoxville, TN 37996-2200, USA

ABSTRACT

In-line monitoring of chemical processes is desired for its numerous advantages, such as lesser waste, lower developmental cycle time, and lesser costs. In this study, a methodology is presented for estimating polymer rheological properties using fiber-optic near-infrared (NIR) spectroscopy. Predictive calibration models are developed for simultaneous, in-line monitoring of polymer melt flow index (MI) and linear viscoelastic (complex viscosity) response, in conjunction with estimation of comonomer concentration for a system of poly(ethylene vinyl acetate) copolymers. The NIR spectra of flowing, molten EVA copolymers were collected in a flow cell attached to a single-screw extruder. The linear viscoelastic measurements were carried out on molten EVA polymers at 200 °C. Multivariate statistical regression analysis is presented for correlation of rheological functions and absorbance in the methylene (C—H) stretch, first overtone NIR wavelength region (1,620-1,840 nm). Results are discussed in detail for simultaneous, real-time monitoring of comonomer composition and rheological properties.

INTRODUCTION

NIR spectroscopy has numerous advantages, such as remote data collection capabilities coupled with rapid data analysis methods, availability of fiber-optics, and lack of sample-handling problems. Consequently, it has proved to be an ideal analytical technique for chemical process monitoring. In-line measurements of chemical composition, polymer relative viscosity, and polymer morphology have been carried out in earlier studies [1,2]. Feasibility studies have been conducted of composition measurements under extreme process conditions, for example molten flowing polymers during extrusion, with focus on development of fiber-optics, optical probes, and near real-time measurements [3]. This technology is now being employed in the commercial industry and is no longer restricted to measurements made using off-line laboratory instrumentation. Attempts are being made to extend the applicability of this technology to estimate industrially important processing parameters, for example MI and dynamic rheological properties in near real-time [4].

The following work will demonstrate the feasibility of extending this technique to make simultaneous measurements of polymer rheological properties and chemical composition of molten EVA samples. A procedure will be described using multivariate statistical tools for developing calibration models, which will be utilized for *real-time* estimation of rheological properties during extrusion.

THEORY

Correlation of Rheological Properties with NIR Absorption Spectra

In this work, in-line NIR measurements were carried out on molten, EVA copolymers in an extrusion process. During polymer extrusion, the rheological flow behavior of molten polymers is anisotropic. The extent of anisotropic nature of flow is governed by the level of molecular orientation and the orientation distribution of flowing polymer melt. Under nominal extrusion operating conditions involving more or less constant shear rates, these anisotropic effects will be strong functions of polymer weight-averaged molecular weight ($\overline{M}_w$), *MWD*, chain length, entanglements, and branching parameters. It is expected that these varying levels of orientation and orientation distribution in the optical measurement volume would affect the NIR absorption spectra in some manner. These rheological effects on the NIR spectra, which are attributed to anisotropic rheological flow, are *lower-order magnitude* effects.

In the NIR absorption spectra, variations in chemical composition are *dominant* effects, and in this study, these variations of comonomer concentration are termed as the *primary factors of variation*. Compared to the strong, primary effects of chemical composition, the rheological flow effects are only subtle variations in the NIR spectra, and these lower-order variations are termed as *secondary factors of variation*. It is noted that these rheological effects contain important information about physical properties that directly correlate with the molecular weight parameters. One such industrially important parameter is the polymer MI. In a very generic sense, MI is inversely related to the polymer molecular weight parameters, such as the $\overline{M}_w$ values [5]. At moderately low stress values, the MI resembles an "inverse" viscosity, and a high MI value implies a low zero-shear viscosity, η_o, or a low $\overline{M}_w$ value. Another important parameter is the linear viscoelastic material response, such as the complex viscosity response ($|\eta^*(\omega)|$; ω, the angular frequency) that is obtained from small amplitude, oscillatory shear experiments for these copolymers. In general, for EVA random copolymers, these rheological properties are not related to the comonomer ratio of ethylene and vinyl acetate (VA).

Therefore, the primary effects (due to chemical composition) and secondary effects (due to rheological flow) on the absorption spectra are independent; and these effects can be separated to provide simultaneous information about the chemical composition and rheological functions. In the following sections, experimental studies will be described and results detailed for quantitative analysis for calibration of MI and $|\eta^*(\omega)|$ response with the NIR spectra.

APPARATUS

Figure 1 shows a schematic of the system for in-line molten polymer analysis used in this study. Molten polymer is produced by a 3/4-inch Brabender single screw extruder, with an L/D ratio of 25:1. The polymer from the extruder is pumped to a gear pump. A uniform mass flow rate of the polymer is supplied from the gear pump to a variable pathlength flow cell. For the rectangular flow channel in the flow cell, the nominal shear rate was estimated at ~ 25 s^{-1}. All EVA samples were extruded at more or less constant shear rates. A temperature gradient is maintained in the extruder to produce homogeneous melting and proper conveying of the polymer along the screw. The temperature and pressure conditions are regulated by a temperature and pressure control unit. Dual transmission fiber-optic probes commercialized by Sensotron Inc.® are used to transmit light through the polymer flow stream [6]. Details of the fiber-optic probe development have been reported in an earlier work [3]. The optical pathlength in the flow cell, or the distance between the probe windows, is varied by using mechanical spacers of different dimensions. In the current setup, the pathlength can be varied between 1 mm and 9.5 mm. A pathlength of 2.5 mm was used for the current set of experiments. This pathlength was chosen to keep the maximum absorbance at less than 1.0. A 1.5 mm diameter optical beam is produced by the Sensotron® NIR transmission probes.

The fiber-optic probes are connected via 500-μ single-fiber fiber-optic cables to an Analect® Diamond-20 (DS-20) Fourier transform near-infrared spectrometer (FT-NIR) [7]. This spectrometer is equipped with an interferometer that works on the moving-wedge principle. For measurements in the NIR region of the electromagnetic spectrum, the wedges are made up of calcium fluoride (CaF_2). The spectrometer uses a quartz halogen lamp as the source and an Indium Arsenide (InAs) detector with CaF_2 windows. The wavenumber resolution used for the current experiments is 8 cm^{-1}.

EXPERIMENTAL

In the available sample set, the VA concentrations varied from 0 wt.% to 32 wt.%, and the MI values varied from 0.34 g/10 min to 370 g/10 min. The in-line NIR transmission spectra of EVA samples were collected in the melt, in the wavelength range of 1,000 nm to 2,500 nm. Ten replicate spectra were collected for each sample, and each spectrum was averaged over 64 scans. *Figure 2* shows the chemical structures of the monomers and the EVA copolymer. *Figure 3* shows a plot of the overlaid absorbance spectra of EVA samples in the NIR wavelength range of 1,620 nm to 1,840 nm. The absorbance band in this wavelength region corresponds to the first overtone vibrational mode of methylene (C—H) stretching and is a common feature for all macromolecules. (Note the presence of methylene in EVA copolymers in *Fig. 2*). This difference in the absorbance values allows for the quantification of VA. Subtle changes are observed in the shoulders of the methylene stretch doublet.

The linear viscoelastic measurements were performed on molten EVA samples at 200 °C with a dynamic stress rheometer. The samples were compression molded into disks, 25 mm in diameter, which were melted in the rheometer at the desired temperature. A 25-mm diameter cone-and-plate viscometer with a gap angle of 0.1 radians (rad) was used for these measurements. Constant amplitude, sinusoidal, shear stress at 200 Pa was applied to the molten sample, and rheological measurements were made over a wide range of angular frequencies. At a stress of 200 Pa, the EVA samples exhibited rheological responses in the linear viscoelastic region over several decades of frequency. The properties of interest were measured at twelve frequency values per decade. A continuous purge of dry nitrogen gas prevented any sample degradation, or moisture absorption, from the ambient atmosphere. Data in the frequency range, $\omega = 0.38$ rad/s to $\omega = 122$ rad/s, were considered for analysis. *Figure 4* shows an overlaid plot of the complex viscosity, $|\eta^*(\omega)|$, response, with varying

ω, for the EVA samples. These measurements were used as the reference values for the data analysis.

DATA ANALYSIS

In attempts to correlate the rheological properties with the absorbance spectra, it was observed that the range of MI values in the available set was very large. Samples were available with a wide range of MI values, from very high viscosity (MI ~ 0.34 g/10 min) to very low viscosity (MI ~ 370 g/10 min). This implies that there is significant variation in the $\overline{M}_w$ parameters among the samples. Because such large variations lead to only subtle effects of lower order magnitude on the NIR spectra, it is assumed that a *first-order response* from the $\overline{M}_w$ parameters-absorbance relationship can be obtained by expressing the $\overline{M}_w$ parameters in terms of their natural logarithms. Such data preprocessing also amounts to "linearizing" the variables used for regression. Therefore, in conjunction with the above assumption, any rheological parameter, such as MI, must be processed before regression using the same correlation of natural logarithms. The theoretical justification for using natural logarithms for such rheological properties comes from the definition of material functions, for example zero-shear viscosity in dynamic rheological experiments. For polymer melts with molecular weights greater than the critical molecular weight M_c, the zero-shear viscosity is given by the proportionality:

$$\eta_0 \propto \overline{M}_w^{\,a} \qquad (1)$$

where η_0 is the zero-shear viscosity, and a is a constant exponent. Berry and Fox [8] observed that for linear homopolymers of narrow molecular weight distribution, a = 3.4. The above expression implies that the relationship of $\overline{M}_w$ with most rheological material functions is non-linear. Taking the natural logarithm of the expression on both sides, the following relationship is obtained:

$$ln(\eta_0) \propto a\, ln(\overline{M}_w) \qquad (2)$$

In the previous section, it was mentioned that the MI variation with $\overline{M}_w$ parameters and η_0 was of an inverse proportionality. This relationship could be expressed as follows:

$$ln(\mathrm{MI}) \propto -k\, ln(\eta_0) \qquad (3)$$

Therefore, for developing the MI calibration model, ln(MI) values were used instead of MI. In a similar manner, $|\eta^*(\omega)|$ values are replaced by $ln(|\eta^*(\omega)|)$ during calibration.

Multivariate Analysis

The absorbance peaks in the NIR spectra are broad and cover a large wavelength range. Unlike the narrow, sharp peaks in the mid-IR region that correspond to fundamental vibrational bands, the NIR region consists of first and higher overtones of these fundamental bands. Therefore, the NIR regime comprises of broader bands and considerable band overlap, unlike the mid-IR region. This makes quantitative analysis more difficult and simple multiple linear regression (MLR) techniques cannot be used. Hence, mathematical modeling for NIR spectra requires multivariate techniques such as partial least squares (PLS) [2,3,9]. These methods involve data reduction of these highly correlated spectra into a few factors that explain much of the data variance.

In PLS, no *a priori* assumption is made of the relationship between the property being calibrated, **Y** (e.g. VA concentration, melt index, etc.) and the absorbance spectra, **X**. While building a calibration model for the data sets using PLS, the first few *factors* (also called *principal components* or *eigenvectors*) that explain much of the variance in the covariance matrix, $\mathbf{X}^T\mathbf{Y}$, are evaluated, where $\mathbf{X}^T$ is the transpose of **X**. Details of this procedure are given in ref. (9). The latter principal components can be rejected because they explain small variances, usually just random noise in the data. A qualitative interpretation often results in relating the initial principal components directly to the original variables. Thus, PLS is a very powerful tool for both quantitative and qualitative analyses, and helps in the formation of robust calibration models.

A calibration model that uses an optimal number of factors is desired. Using fewer factors in the calibration model would mean excluding important information from the model, and using more factors would imply including noise and thus, higher residuals in the predictions. Therefore, the choice of an optimal calibration model with minimum residuals is critical. Several statistical terms are introduced which help determine the optimal number of factors [2]. These terms are described below:

1) PRESS (Predicted Residual Sum of Squares) value

$$\text{PRESS}_j = \sum_{i=1}^{N} \left(\hat{y}_{i,j} - y_i\right)^2 \tag{4}$$

where $\hat{y}_i$ and y_i are the predicted and actual property values respectively, of the i-th sample. N is the number of samples used for calibration, and j is an index for the number of principal components incorporated into the model. The PRESS value is a measure of the residual error, and the optimal number of factors is decided by a minimum PRESS value criterion.

2) SEC (Standard Error of Calibration)

$$\text{SEC} = \sqrt{\frac{\sum_{i=1}^{NC}\left(\hat{y}_{i,c} - y_{i,c}\right)^2}{NC - 1}} \tag{5}$$

where $\hat{y}_{i,c}$ and $y_{i,c}$ are the predicted and the actual property values of the i-th sample in the calibration set, respectively. NC is the number of samples used in the calibration set.

After the calibration model is developed, it is used to predict the desired property of interest, **Y** of an independent test set (also called the prediction set). A residual error term, the standard deviation between the laboratory value and the predicted value (calibration model), is defined for these predictions, as follows:

3) SEP (Standard Error of Prediction)

$$\text{SEP} = \sqrt{\frac{\sum_{i=1}^{NP}\left(\hat{y}_{i,p} - y_{i,p}\right)^2}{NP}} \tag{6}$$

where $\hat{y}_{i,p}$ and $y_{i,p}$ are the predicted and the actual property values of the i-th sample in the prediction set, respectively. NP is the number of samples in an independent prediction set.

RESULTS AND DISCUSSION

With suitable data preprocessing in the form of natural logarithms, multivariate techniques were used to correlate rheological properties with the absorption spectra. Calibration models were developed on first-derivative absorbance spectra in order to remove baseline offsets. Separate calibration models were built to predict VA concentration, MI, and $|\eta^*(\omega)|$ response.

Because the VA content is a *strong, dominant feature* in the absorbance spectra, it is expected that a one-component principal component would be sufficient to estimate VA concentration [3]. On the other hand, because the rheological flow effects are secondary effects, additional principal components are expected for optimal predictions. Based on the PRESS criterion, a three-factor model was developed for predicting *ln*(MI). Table 1 shows the variance explained in the **y**-blocks for the VA and MI calibration models. From this table, it is observed that three factors are required to explain about 99.6% variance in the MI data, while only one factor is sufficient to account for about 99.7% variance in VA content. For MI-calibration, the inclusion of a second principal component explained an additional variance of 59.1%. Hence, the second principal component is most significant during MI-calibration and contains maximum information about MI variance.

Table 1: Total **y**-block variance explained for different factors

Number of factors retained	VA calibration model Total y-block variance explained (%)	MI calibration model Total y-block variance explained (%)
1	99.7	36.5
2	99.9	95.6
3	99.9	99.6

An external prediction set was formed using spectra of samples, which were not included in the calibration set. The calibration models developed above were used to predict the samples in the prediction set. The SEP values, as evaluated from Eq. 6, for the two models were 0.46 for MI predictions (three-factor model) and 0.62 wt.% for VA predictions (one-factor model). The in-line predictions for *ln*(MI) and VA concentration for the final calibration and prediction sets are shown in *Figures 5* and *6*, respectively. Excellent agreement is observed between the actual, laboratory primary analysis and the in-line, NIR predicted MI and VA values.

It was observed that a calibration model for estimating $|\eta^*(\omega)|$ response required four principal components for optimal predictions. *Figure 7* shows the in-line NIR predictions for the complex viscosity response for EVA samples in the calibration set. The SEP value for this calibration model was 0.80.

Real-time Predictions During In-line Extrusion

Finally, the stability of the predictions using the calibration models was tested for simultaneous, real-time predictions of rheological properties and VA concentration during extrusion. EVA samples were continuously fed through the extruder, and test scans were taken. Data preprocessing of these spectra was carried out in a similar manner to the sample spectra used for calibration. The two calibration models built earlier were then used to estimate the desired properties of VA content and MI. The stability of predictions of the calibration models were tested on six EVA samples belonging to two different classes were used for real-time predictions. Class A consisted of EVA samples with widely varying VA concentrations, but similar MI values. These constituted the first three samples in Table 2. Four EVA samples were included in Class B. These samples consisted of ~ 28 wt.% VA and had very different MI values. The samples in the two classes were continuously fed through the extruder in the order in which they are presented in the table.

Table 2: EVA samples for real-time monitoring

Sample Number	Class Identity	VA Concentration (wt.%)	*ln*(MI) values
1	A	9.11	1.86
2	A	12.12	2.05
3	A & B	28.49	1.80
4	B	27.84	5.91
5	B	28.59	4.96
6	B	28.37	3.16

The real-time predictions for VA content and MI for Class A and Class B samples are illustrated in *Figure 8*. It is observed that most predictions lie within the limits set by the SEP values. This implies that the predictions for both VA concentration and *ln*(MI) lie between the actual laboratory value ± SEP. (In the figure, these limits are defined as the upper and lower detection limits, UDL and LDL, respectively.) It is demonstrated that the calibration models can successfully estimate samples belonging to both classes. This reiterates that the PLS calibration models can *separate* the two independent factors of variation in the EVA samples, namely MI and VA concentration. Another observation is made in the smooth transition region between any two successive samples. In this region, a "blend" of the two samples is expected in terms of both VA content and melt viscosity conditions. This "blend" would progressively change in both VA composition and MI from the previous sample to the new sample. In *Fig. 8*, it is observed that the calibration models capture the expected behavior in the transition region, with regard to intermediate properties of VA content and melt viscosity in the blend. The transition region is also an indicator of the average residence time of the sample in the extruder, gear pump and flow cell.

For real-time predictions for $|\eta^*(\omega)|$ response, eight EVA samples from two different sources were used for in-line estimations. Samples from a second source were utilized for testing the validation of the calibration models. In *Figure 9*, these samples are referred to as Group I and Group II sets, where the estimated in-line NIR predictions are displayed for the complex viscosity responses for EVA samples from both groups.

CONCLUSIONS

During extrusion of molten polymer melts, anisotropic effects associated with the rheological flow translate into variations in the NIR absorption spectra. These effects allow correlation of rheological properties, such as polymer MI and $|\eta^*(\omega)|$ response with the NIR spectra. In this study, a methodology is demonstrated for simultaneous in-line monitoring of both VA content and MI values in molten EVA copolymers. It is established that the variation of MI in EVA copolymers is independently observed in the NIR absorption spectra. The robustness of MI calibration models was tested by real-time predictions on EVA samples belonging to two categories: (1) EVA samples with similar MI values but widely varying VA content, and (2) EVA samples with widely varying MI values but similar VA composition. The ability of the models to estimate samples in these two distinct groups further validated the robustness of calibration.. The predictions for both VA concentration and MI lie within detection limits decided by the SEP limits (0.62 wt.% and 0.46, respectively). The complex viscosity calibration models were validated by real-time measurements on EVA samples from a second source.

In the above experiments, one disadvantage is the use of isotropic light as the source. Unlike polarized light spectrometry, which allows a more detailed understanding of the mechanisms involved in molecular orientation during flow (information in the form of flow birefringence and dichroism), the rheological flow effects observed in isotropic spectroscopy experiments are small, and at best, can be observed only in higher order factors. Therefore, powerful mathematical tools, such as multivariate techniques, have to be used for data analysis and model development. Additional variables, such as temperature, should be incorporated into the models for further accuracy in the predictions.

REFERENCES

1. M. P. B. van Uum, H. Lammers and J. P. de Kleijn, *Macromol. Chem. Phys.*, **196**, 2023 (1995).
2. C. E. Miller, Ph.D. dissertation, University of Washington, Seattle (1989).
3. A. Khettry, Ph.D. dissertation, University of Tennessee, Knoxville (1995).
4. D. Marrow, paper 459 presented at *Pittsburgh Conference on Analytical Chemistry and Applied Spectroscopy*, Chicago (March 1996).
5. J. D. Ferry in *Viscoelastic Properties Of Polymers*, John Wiley & Sons, Inc., New York (1980).
6. Sensotron, Inc., 5881 Engineer Drive, Huntington Beach, CA 92649.
7. Axiom Analytical, Inc., 18103 Sky Park South, Unit C, Irvine, CA 92714.
8. G. C. Berry and T. G. Fox, *Adv. Polym. Sci.*, **5**, 261 (1968).
9. H. Martens and T. Naes in *Multivariate Calibration*, John Wiley and Sons Ltd., New York (1989).

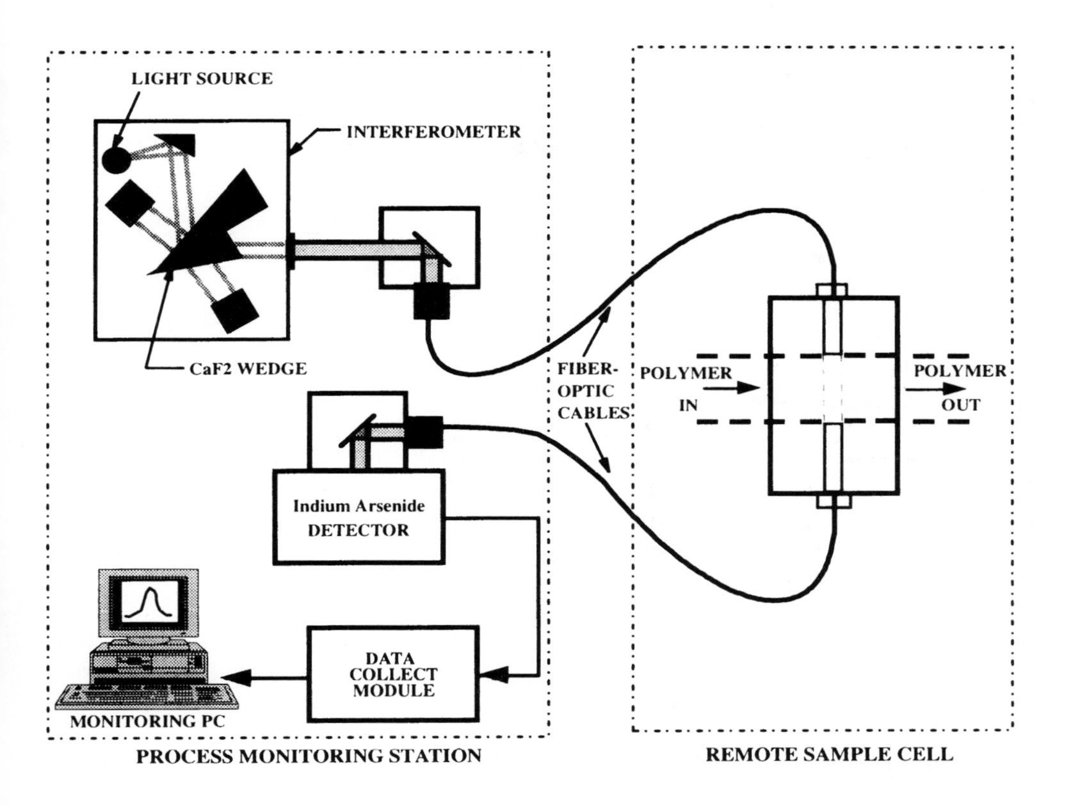

Figure 1 Schematic of the experimental setup for in-line Fourier transform Near-Infrared spectrometry of molten, flowing polymers

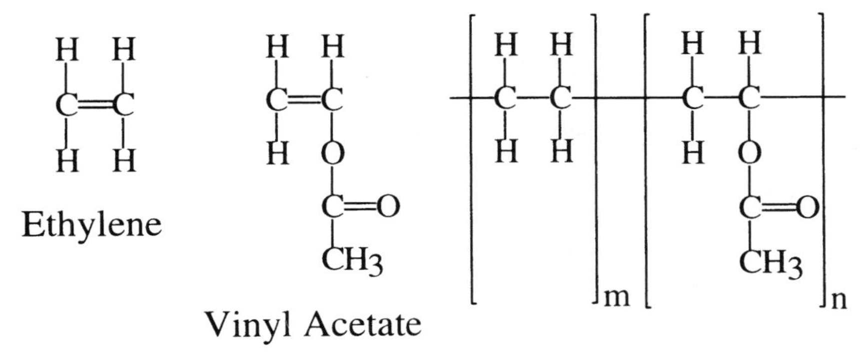

Figure 2 Chemical structure of ethylene and vinyl acetate monomers and poly(ethylene vinyl acetate) (EVA) copolymer.

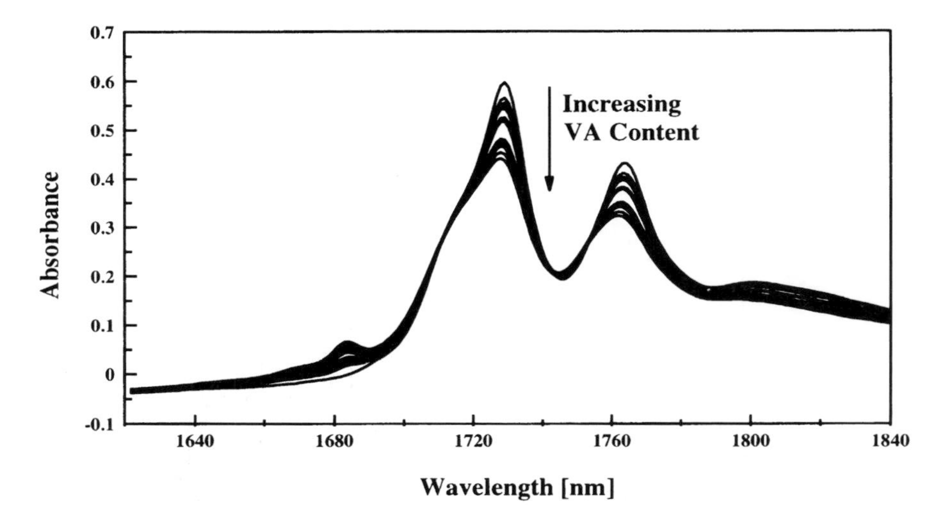

Figure 3 Overlaid absorbance spectra of EVA samples in the NIR wavelength region of 1,620-1,840 nm. The absorbance band is attributed to methylene (C—H) stretch, first overtone vibrational mode.

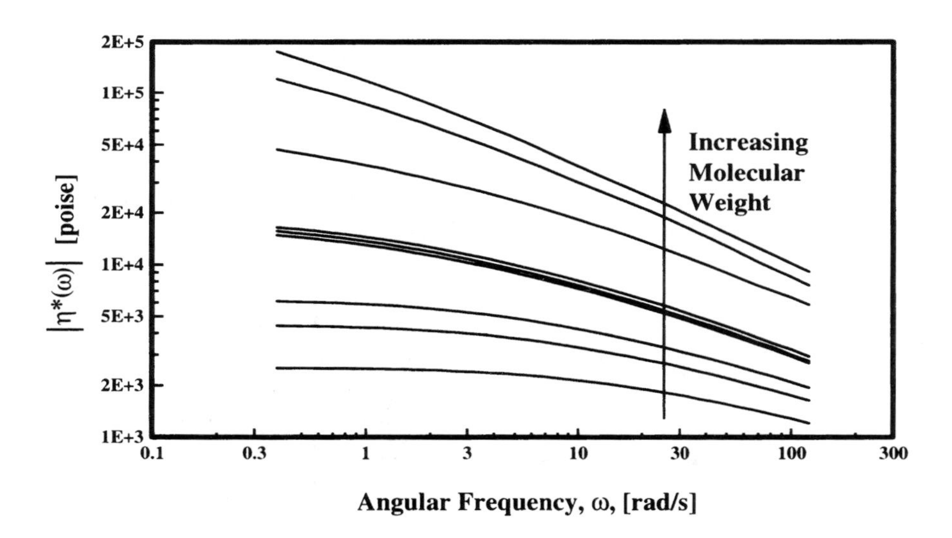

Figure 4 Overlaid plot of the complex viscosity response for EVA samples over an an frequency range of 0.38 rad/s to 122 rad/s. The measurements were carried out un shear stress of 200 Pa and at 200 °C.

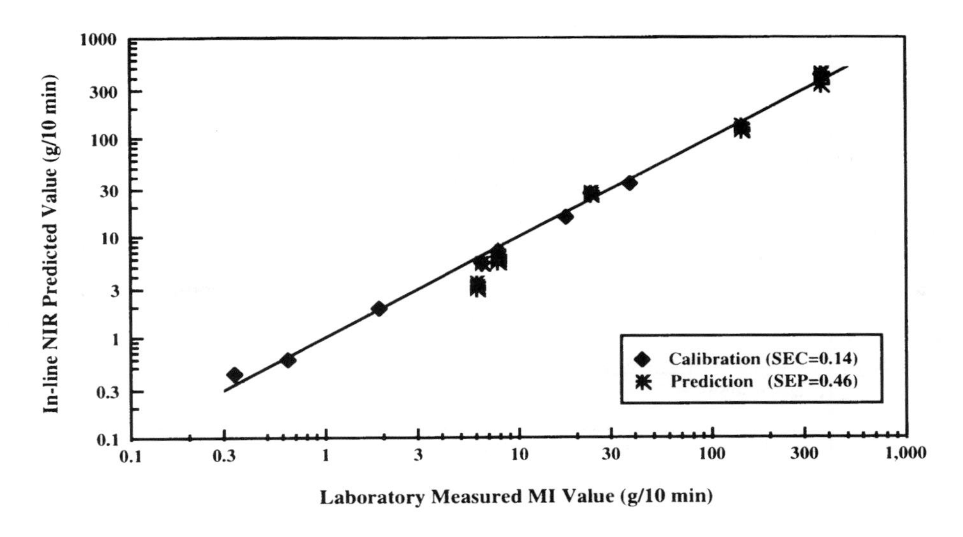

Figure 5 Plot illustrating the NIR predictions for melt index, MI. The MI calibration model was built using *ln*(MI) values. A three-factor model was used for optimal predictions for *ln*(MI) using the PRESS criterion.

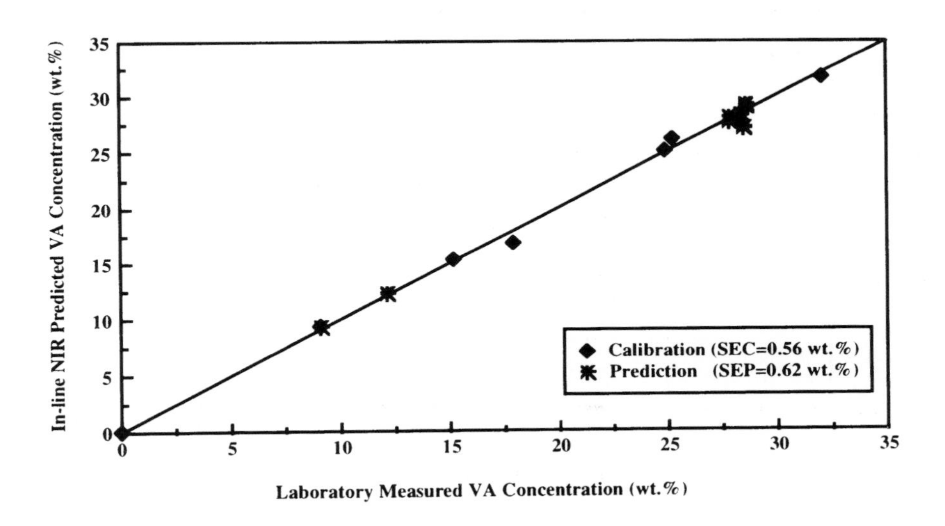

Figure 6 Plot illustrating the NIR predictions for vinyl acetate, VA concentration (wt.%). A one-factor model was developed for predicting the VA content using the PRESS criterion.

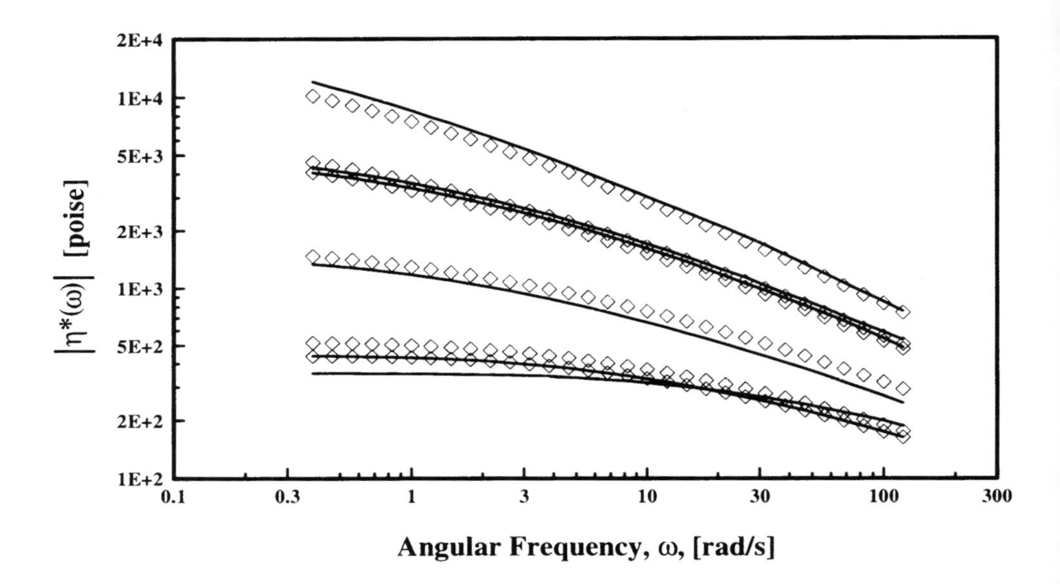

Figure 7 **Prediction Set:** In-line NIR predictions for rheological response - $|\eta^*(\omega)|$ versus ω - for EVA samples. The standard error associated with the prediction set is SEP = 0.80. The solid curve represents the experimental curve, and the symbol ◇ represents the PLS-predicted curve.

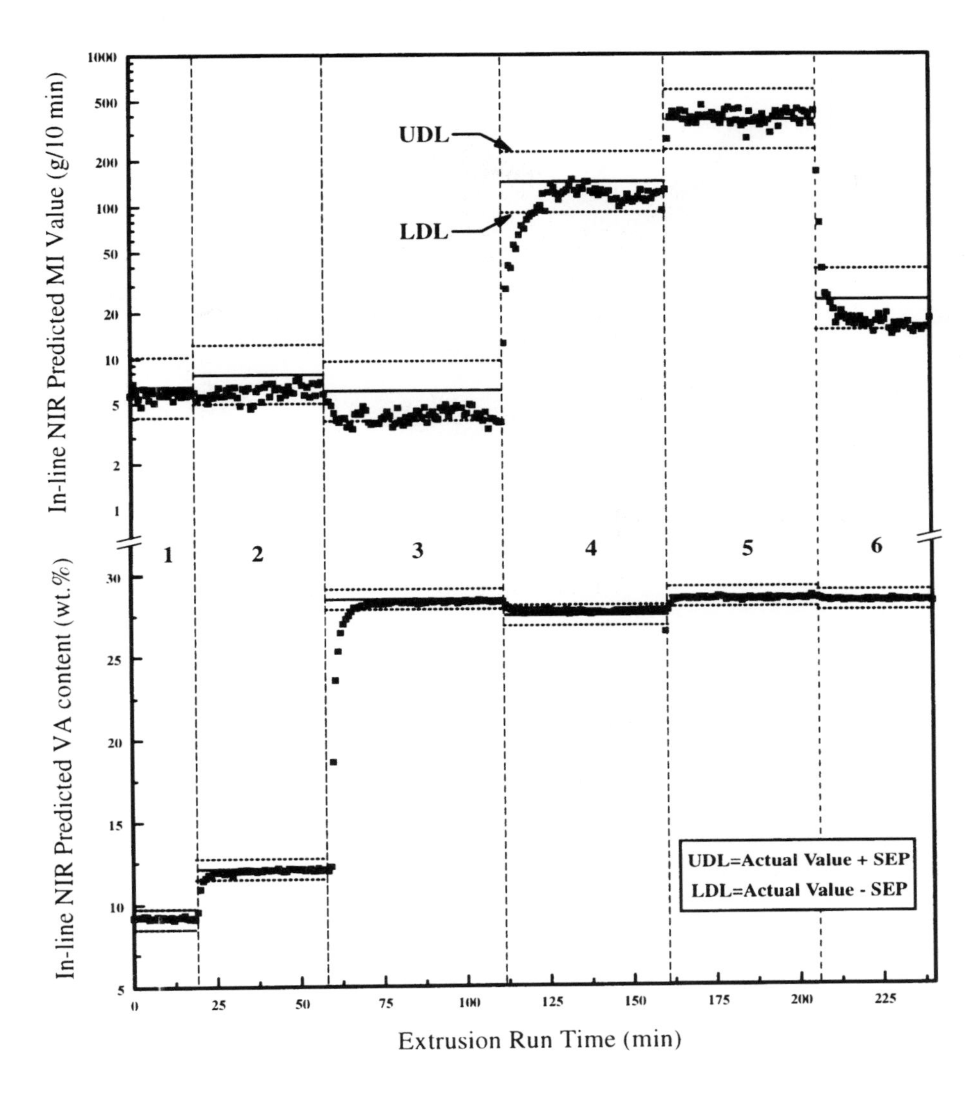

Figure 8 In-line, real-time predictions for MI and VA content in Class A (samples numbered 1, 2, and 3) and Class B (samples numbered 3, 4, 5, and 6) EVA samples. The predictions for both primary (VA content) and secondary (MI) variables lie within SEP limits associated with the calibration models. [SEP for VA-calibration model = 0.62 wt.%; SEP for MI-calibration model = 0.46 on a natural log scale]
[UDL: upper detection limit; LDL: lower detection limit]

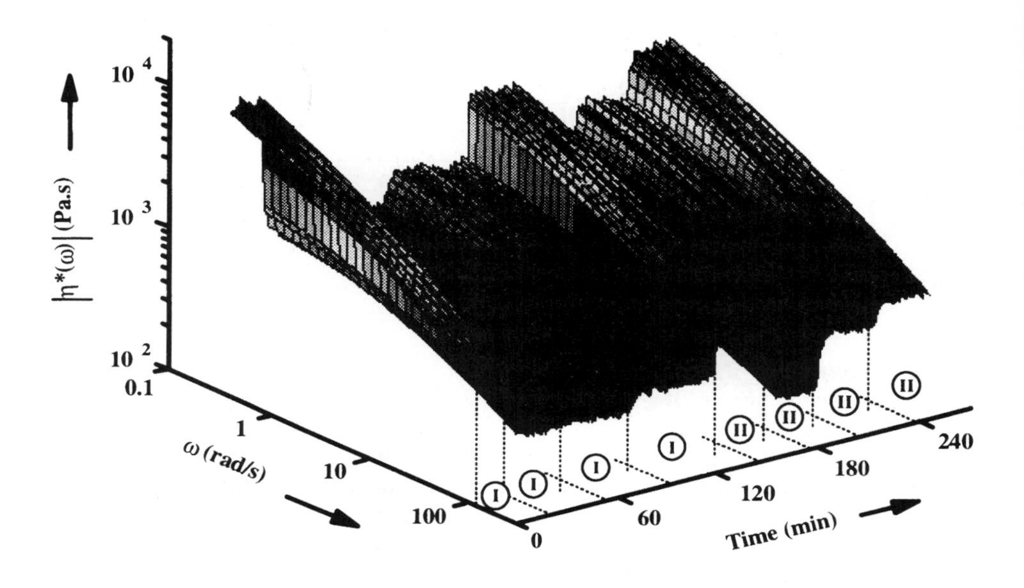

Figure 9 Three-dimensional plot showing the real-time predictions for the rheological response of Group I and Group II samples (these samples were not included during calibration): The complex viscosity curves are estimated from the PLS calibration models. The rheological data used for correlation with NIR spectra had only physical factor of variation.

Quality Regulation Techniques for Injection Molding

DAVID KAZMER
Assistant Professor
University of Massachusetts
Eng. Lab. Bldg.
Amherst, MA 01003

ROLAND THOMAS
V.P. Technology
Moldflow Inc.
259 Colchester Road
Kilsyth, Victoria 3137

GAL SHERBELIS
Manufacturing Manager
Moldflow Inc.
259 Colchester Road
Kilsyth, Victoria 3137

ABSTRACT

Global manufacturers of thermoplastic molded parts increasingly require 100% quality inspection levels. To identify the technical issues associated with this goal, the injection molding process is described utilizing a control systems approach. Afterwards, four different methods of quality regulation are compared for injection molding: open loop quality control, statistical process control, trained parameter control, and intelligent process control. For each strategy, the level of quality observability and controllability are determined.

The discussion indicates that only intelligent process control has the underlying design architecture to deliver 100% quality assurance across a diverse set of application characteristics (quality requirements, material properties, mold geometries, and machine dynamics). This article has not claimed that intelligent process control has been perfected. It does, however, identify the design intent necessary to achieve the levels of quality control and process robustness needed by industry.

INTRODUCTION

Injection molding of thermoplastics has emerged as a premier vehicle for delivering high quality, value added commercial products. Perhaps due to this success, continued global competitiveness has increased standards for product quality while requiring reduced product development time and unit cost. Despite advanced product design methods and new process technologies, it is becoming apparent that the injection molding process is neither flexible nor sufficiently robust to meet these industry requirements. The lack of robustness is sometimes evidenced by long product development cycles, excessive tooling costs, low process yields, and inferior product quality. As thermoplastic materials continue their thrust into advanced technical applications with multiple stringent requirements, the risks of proving out the injection molding process are becoming excessive. In fact, several industry managers have independently testified that "we are already starting to see the migration of customers to other manufacturing processes for time-critical applications."

Fundamentally, the difficulties associated with injection molding arise from the lack of simple and consistent relationships between the machine inputs, part geometry, material properties, and molded part quality. In product and tool design, numerical simulations have been developed to aid the design engineer. In tuning and regulation of injection molding, however, no method has had similar success in aiding the process engineer.

Polymers exhibit extremely complex material properties – non-Newtonian, non-isothermal, thermoviscoelastic rheology together with highly temperature and pressure dependent thermal properties. During processing, the material undergoes temperature and pressure increases, significant shear deformation, followed by rapid decay of temperature and pressure in the mold cavity, which leads to solidification, and locking of residual stress, orientation, and other part properties that determine the molded part quality.

While process complexity makes it difficult to attain the desired part properties during start-up, process variability causes difficulty in maintaining part quality during production. The difficulty controlling this process' phenomenological behavior is further complicated when coupled with diverse application requirements encountered in industry, ie. every commercial application generates a unique set of process dynamics and quality issues. The objective of this research is to fill this gap by developing a novel method for tuning and automatic regulation of injection molding processes.

Modern injection molding machines utilize sophisticated systems for control of machine parameters and plastic process variables. The relationships between these variables and selected aspects of molded part quality have been widely studied. While these approaches have produced some interesting correlations regarding molded part quality, they can not generally or reliably guarantee satisfactory parts without time consuming and uneconomical application-specific research. A commercial system has been developed to address industry needs [1]. The product is composed of: 1) an initial input utilizing flow simulation for product and process optimization, 2) an expert system developed for defect elimination on the production floor, 3) a design of experiments approach to verify process stability and model the quality dynamics which provides information to 4) a continuous and automatic production quality monitor and controller.

SYSTEM DESCRIPTION

This article focuses on quality regulation for injection molding; tuning and process set-up for injection molding has been described elsewhere [1]. The problem will be introduced by discussing the control system schematic shown in Figure 1. Afterwards, the remainder of the article will develop and compare alternative quality regulation strategies.

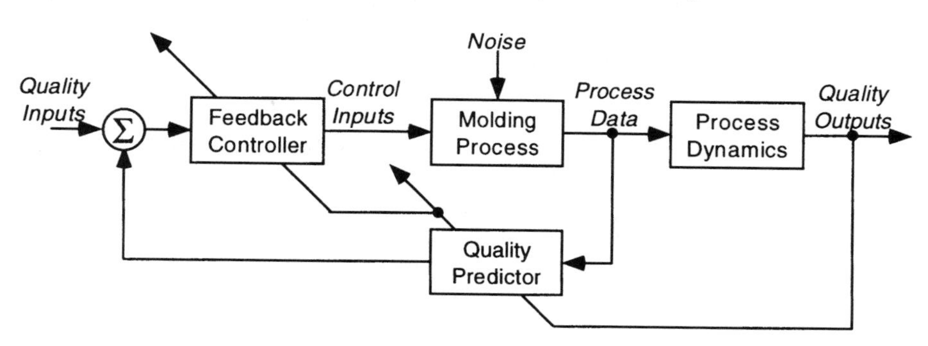

Figure 1: General Quality Regulation System Schematic

The central element of Figure 1 is clearly the molding process, which has been broken into two sub-systems for convenience. The primary inputs to the molding process include machine parameters (such as temperature, velocity, and pressure profiles), material properties (rheological, thermal, and compressibility behaviors), and mold geometry (part topology, configuration, and parameters). An undesirable but second set of inputs to the molding process is noise, both intrinsic (natural variation about the process means) and extrinsic (longer term fluctuations of uncontrolled parameters). Noise can be observed but is generally expensive to convert to a controlled input. For example, relative humidity in the plant air might have an observable effect of the molding process and molded part quality. Moreover, relative humidity can be controlled albeit at a fairly significant cost.

All the molding process inputs are converted to process conditions that act on the polymer melt. Many of these process conditions (such as melt temperature, melt pressure, cavity pressure, etc.) are observable and related to the molded product quality. However, the exact relationships between the process conditions and the output quality are not always available. Rather, it is the internal process dynamics within the mold cavity which determine the molded part quality. These process dynamics are not designed or controlled. Instead, this transfer function is determined by the laws of physics.

Definition I: A system is controllable if every output mode is connected to a control input [2].

This definition is not met for the quality outputs of the injection molding process. This means that molded part quality is not directly controlled but rather influenced through the process conditions. As such, regulation control for injection molding must act on the observed process data for feedback to the control inputs.

This architecture is shown explicitly in Figure 1. A quality predictor converts the observed process data to estimates of molded part quality. The internal transfer function may be arbitrary and diverse in nature: empirical or analytical in derivation, statistical or knowledge-based, extensive or brief. The arrow leading from the observed quality outputs through the transfer function indicates a potential learning mode for the quality predictor. This means that the coefficients and output of the quality predictor can be improved with feedback of the observed quality.

The predicted output quality is then compared to the desired product quality. The feedback controller converts any discrepancies in molded part quality to a set of control inputs that should produce the desired output quality. The arrow through the feedback control function indicates additional potential for learning from previous control actions and the subsequent changes in molded part quality.

To this point, the schematic and discussion has been entirely general with the goal of describing all existing and possible regulation strategies. The remainder of the article will develop and compare alternative quality regulation strategies.

REGULATION STRATEGIES

Before delving deeper into quality regulation strategies, the difference between quality regulation and process regulation should be clarified. The following definitions explain that quality regulation addresses the setting of process inputs while process regulation addresses the maintaining of process inputs.

Definition II: Quality regulation is the control of process inputs to produce molded product that meets customer specifications.

Definition III: Process regulation is the control of internal process variables to ensure that process conditions coincide with the desired process inputs.

Open Loop Quality Control

The large majority of industrial molders practice open loop quality control, shown in Figure 2. In this regulation paradigm, the mold tooling and control inputs are qualified to meet customer specifications. There is no active cycle-to-cycle quality regulation in production – the machine parameters are set and the resulting molded parts assumed to have sufficient quality. Intermittent and acceptance sampling are used to verify molded part quality with any subsequent issues resolved by the process engineer.

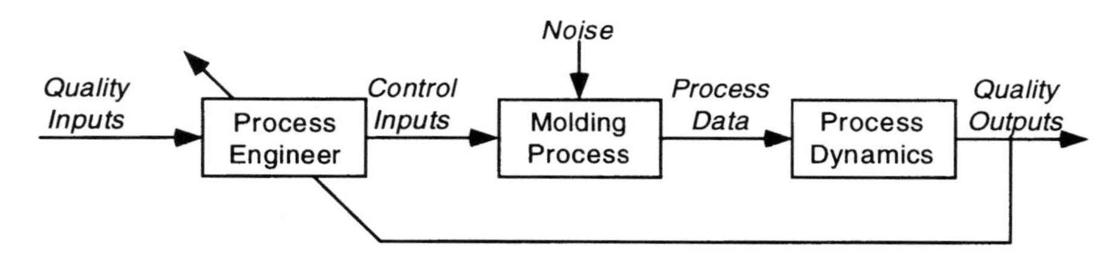

Figure 2: Open Loop Quality Regulation

There are some grave concerns associated with open loop quality regulation. In particular, many molders utilize press operators to provide quality assurance and feedback regarding the molded products. Two significant issues are apparent with this technique. First, operators are not always sufficiently trained, let alone have adequate time, to accurately identify quality defects. Second, visual inspection only permits identification of a restricted set of defects (mostly aesthetic). Unfortunately, it is non-visual defects (such as warpage, dimensions, and stress) which often result in costly later failure during assembly or end-use [3].

From a fundamental perspective, Deming [4] has shown that a sample size *n* taken from a lot of size *N* provides no information about the number of remaining non-comforming items in the lot. This means that sampling and diagnosing quality for a small set of parts provides no guaranteed correlation with the quality of the non-sampled parts. As such, the molder should practice either 0 or 100% quality inspection after considering the specification requirements, process capability, and cost of accepting defective samples [5]. 100% quality assurance is only commercially feasible with on-line, closed-loop, quality regulation systems.

Statistical Process Control

Statistical process control (SPC) techniques are becoming standard on injection molding machinery [6, 7]. As such, an increasing number of molders are utilizing SPC for quality regulation. Figure 3 shows a control schematic for SPC. In this regulation paradigm, process data are assumed to be representative of molded part properties. The quality inputs to the system are not quality attributes but rather sets of upper and lower process specification limits.

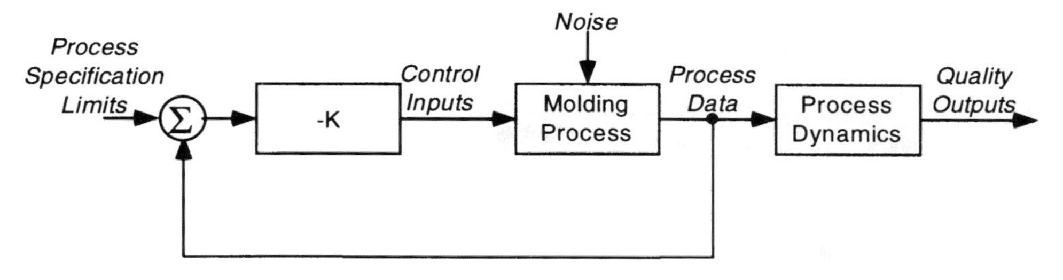

Figure 3: Statistical Process Control Quality Regulation

A capable SPC quality regulation system will guarantee that the process data are maintained within the stated upper and lower bounds. Herein lies a difficulty: definition one states that the molded part properties are not controllable. As such, there is no guarantee that specification limits on the control inputs will deliver the desired part properties. The primary challenge with SPC quality regulation, then, is the identification of significant process controls and the specification of their limits.

Definition IV: A system is observable if its modes can be deduced by monitoring sensed outputs.

Ignoring issues of quality controllability, there is a second major issue regarding the use of SPC for injection molding: observability. SPC only governs the controlled machine variables. As such, there are many uncontrolled variables that influence molded part quality. For instance, relative air humidity, screw/barrel/hydraulics wear, and material properties are significant process variables that can not be controlled with SPC.

Moreover, the interaction between these uncontrolled variables and the control specification limits are not precisely defined. Thus, it becomes the job of the trained process engineer to track part quality and continually refine the process specification limits in an outer feedback loop not included in Figure 3. Few molding houses maintain the level of required expertise to successfully leverage SPC for quality control. As such, SPC is more indicative of the capability of process regulation than quality regulation.

Trained Parameter Control

More recently, independent researchers have related specific molded part attributes to individual process measurements. For instance, quality regulation strategies have been proposed to utilize process data for cavity pressure [8,9], infrared melt temperature [10], hydraulic pressure ratios [11], tie-bar deflection, and other signals [12, 13]. The typical control schematic for an advanced system of this type is shown in Figure 4.

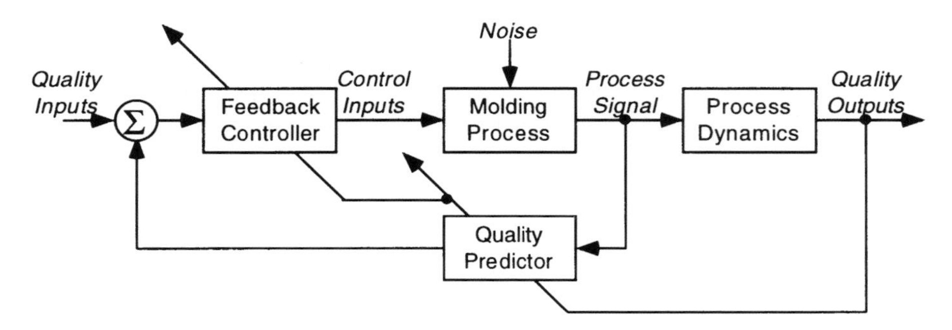

Figure 4: Trained Parameter Control Quality Regulation

In this quality regulation paradigm, previous knowledge has gained some insight between a specific process signal and molded part quality attributes. The assumption, similar to SPC, is that the process signal can be related to molded part properties. Generally, this assumption is better founded in significant empirical and analytical research. Figure 4 represents an advanced system in that learning has been included for quality prediction and feedback control. As such, the system may be able to adapt to different application requirements. For instance, a cavity pressure based system may relate peak cavity pressure to existence of flash. The exact level of cavity pressure which causes flash as well as the remedy for correction may be learnable on-line.

Different systems have utilized artificial neural networks [14, 15], fuzzy logic [16, 17], expert systems [18], and regression [19] for quality control. While these approaches are fundamentally different, all rely upon developing a relevant quality model for the specific application. The creation of the internal forecasting model can be performed either on the basis of a test plan or historical process observation. Once the model has been developed, the quality characteristics of the molded product are typically forecasted from the measured process data during production.

There are two intertwined shortcomings in all quality regulation systems that utilize trained control of a single process signal. The primary issue concerns observability, i.e. it is not feasible to deduce all quality attributes a single process signal for every existing molding application. The second issue is also related to observability. Since only one process signal is controlled, all other process inputs become extrinsic. Thus, the originating sources of variation in the process signal and molded product quality can not always be uniquely and precisely identified. As such, the regulation system has to be re-trained in the new process domain after significant process changes have occurred. These shortcomings prohibit high levels of quality assurance across multiple quality attributes.

Intelligent Process Control

Definition V: Intelligent process control enables the manufacture of world-class quality product without any expert knowledge of plastics by the design, tooling, or molding personnel.

Intelligent process control (IPC) is a quality regulation strategy that determines the fundamental relationships between control inputs, process data, and quality outputs automatically on-line. Quality regulation in IPC has been developed with the philosophy that:

- no bad parts should be accepted, and
- a minimal number of good parts should be discarded.

These two goals are competing in that an increasing number of good parts will be rejected as the quality criterion becomes more stringent. Therefore, the effectiveness of the system to assess the molded part quality is critical – the quality costs induced by poor quality prediction can quickly exceed the possible benefits of the quality control system [20].

As shown in Figure 5, intelligent process control uses existing knowledge and results from process simulation as a starting point, but then continues to develop quality relationships and feedback rules on-line. Moreover, intelligent process control utilizes all available process data from a vector of sensors rather than relying on any one process signal. For these reasons, it is unlike statistical process control or trained parameter control.

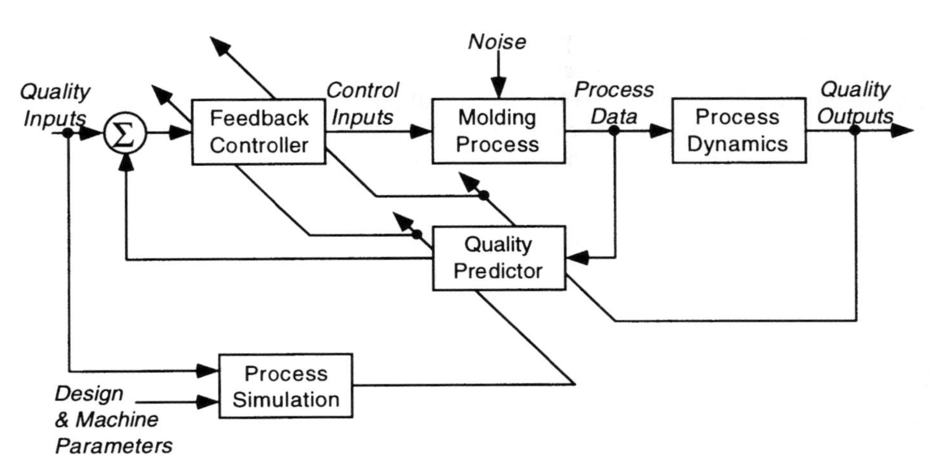

Figure 5: Intelligent Process Control Quality Regulation

To function properly, IPC requires model development and validation on-line. This is performed through an automatic design of experiments technique which analyzes the process data and, with measured quality feedback from the operator, determines a suitable process model. Model development is not trivial, requiring two transformations as shown in Figure 6. Moreover, solving the inverse problem to change the control inputs based on discrepancies in product quality is not trivial, often requiring solution of ill-formed matrices as well as resolution of conflicting quality parameters.

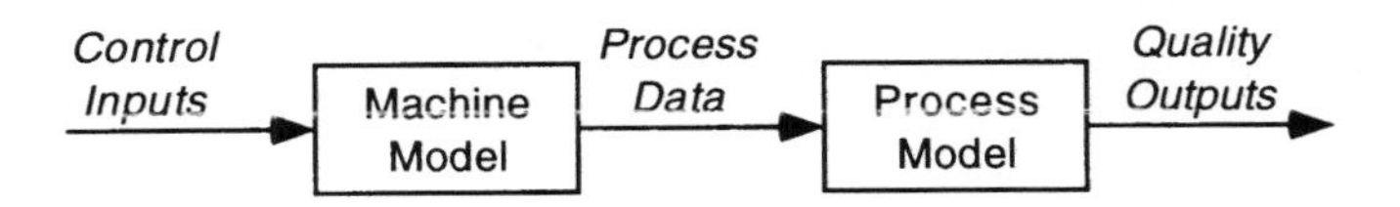

Figure 6: Needed Transformations for Intelligent Process Control

Good molding practice dictates that the process window around the nominal operating point is sufficient to tolerate natural variation in the molding process without producing defects. As such, the implementation examines the repeatability of the molding machine. From the process examination, the behavior and distribution of each of the process variables is automatically characterized. The required size of the process window is then calculated as representative of the process capability index, Cp [21]. This approach is robust in that the process variance is not assumed but rather measured for each application, i.e. representative of the actual material properties, mold geometry, and machine dynamics in the molding process. The more variation a given machine exhibits, the larger the size of the process window which musty be verified to produce good parts.

In establishing the quality model, successive sets of experiments are utilized to verify production suitability of the current operating regime and move the process set-points to an improved area when possible. An appropriate experimental design is developed given te number of process variables being investigated, resolution and order of models, and number of part quality specifications. The flexibility of this approach permits the operator to specify a small set of process variables to investigate, such as injection speed, injection stroke, and pack pressure, or a much larger set that might include nearly all molding parameters. With the design of experiments fully specified, a series of molding trials are automatically executed. The process set-points are automatically modified to investigate the process space. Subsequent experiments are performed until the quality relationships are known and no defects are in the vicinity of the nominal set-points.

Quality regulation in IPC is similar to Trained Parameter Control in that it uses process data to predict molded part quality. It is vastly superior, however, in that it utilizes a vector of quality indicators (rather than one) to monitor the molding process and determine if any quality changes have occurred. The list of quality indicators is substantial and may include averages, derivates, integrals, instantaneous samples, and ratios of ram velocity, ram position, injection pressure, cavity pressure, hydraulic pressure, and melt temperature throughout the injection molding cycle. Each quality indicator has been selected a-priori to possibly correspond with process, material, and part property shifts. This knowledge has been gained through extensive molding experiments across several years.

Recalling the shortcomings of Trained Parameter Control, moreover, it should be noted that these quality indicators are not assumed to be indicative of molded part quality. Rather, the validity of each quality indicator is truly ascertained on-line during the previously described process investigation. Quality indicators with poor correlation to molded part quality are automatically discarded. The boundaries for the remaining quality indicators are then initially derived directly from the process investigation, and used to estimate the molded part quality.

Given the vector of quality indicators and their specifications, parts are molded continuously with two levels of quality diagnostics. The first level of diagnostics determines the quality of the shot just molded and ensures that the impact of random process variation is small. The second level of diagnostics examines trends in the molding process and eliminates defects due to systematic process disturbances. The system utilizes statistical process control and auto-regression techniques to assess the performance and stability of the system. If trends (due to material or mold temperature variation, for instance) are seen in the data, the production quality monitor will attempt to automatically correct the process to prevent molding defective parts, using regression coefficients which have been developed in the process investigation and later in production. If instabilities (due to barrel temperature cycling, etc.) or other trends occur which the quality monitor can not correct, the system will alert the molder of the difficulty before many possibly defective parts are molded.

A picture of one screen in the developed quality regulation system is shown in Figure 7. The screen is divided into three primary areas. A control chart of each quality indicator consumes the majority of the screen, and can be used to indicate the current molded part quality relative to the specified attributes. Above the control chart are two windows. In the left hand window is a list of quality indicators which are currently being monitored – the user can select any quality indicator for display at any time. In the right hand window is a listing of capability indices: a univariate process capability index [21] as well as an aggregate, multi-variate index as defined in [22].

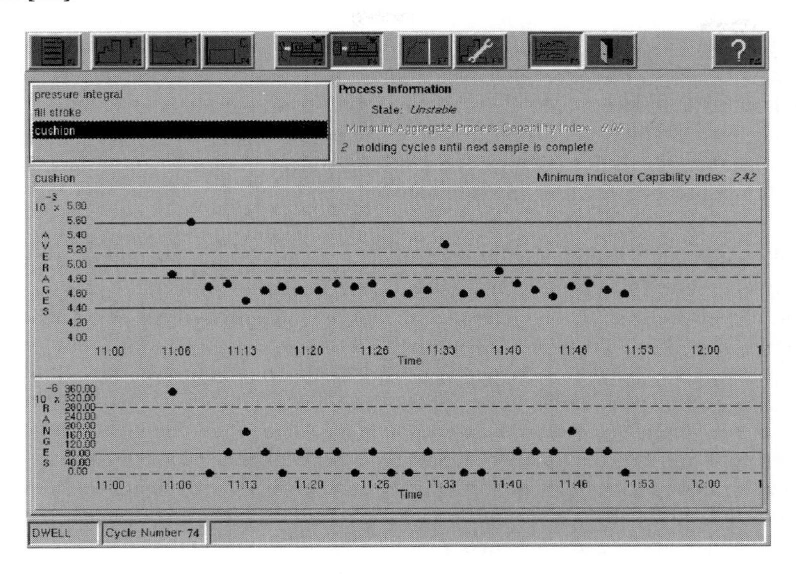

Figure 7: Production Quality Monitor

DISCUSSION

This paper has discussed four different methods of quality regulation for injection molding. Table 1 is a direct comparison of the characteristics of each strategy. There are three significant axes of comparison. With respect to quality outputs, only trained parameter control and intelligent process control estimate and control molded part quality from measured process data. As such, open loop control will never provide 100% quality assurance while statistical process control will provide quality assurance in the rare cases when the selected process variables and their specification limits happen to coincide with 1:1 mappings of quality. Trained parameter control may be able to provide assurance when molded part quality is directly affected by the controlled process signal.

Table 1: Comparison of Quality Regulation Strategies

Characteristic	Open Loop	Statistical Process Control	Trained Parameter Control	Intelligent Process Control
Quality Control	Open	Closed	Closed	Closed
Expertise Required	High	High	Low	Low
System Learning	No	No	Yes	Yes
# Inputs	Many	Many	Few	Many
# Quality Outputs	N/A	None	Few	Many
100% Quality Assurance	No	Rarely	Maybe	Yes
Sensitivity to Extrinsic Noise	High	High	Medium	Low

Among the quality control strategies, only intelligent process control is able to repeatably identify and compensate for extrinsic noise. This is due to its system design, which incorporates multiple process signals with simulation results and on-line model validation and learning. The other quality regulation strategies may not be able to cope with changes extrinsic to the system. For example, consider a cavity pressure based and trained system. A change in the melt temperature may produce a reduction in viscosity that would result in a detected reduction in the cavity pressure traces. Without observing the external variables, however, it is difficult assess the cause between melt temperature, screw rotation rate, relative air humidity, or material variation for a diverse set of resin characteristics and mold geometries. Moreover, the compensation strategy to correct quality problems is not always obvious or consistent between applications.

This example argues strongly for multiple process measurements as well as on-line learning capability. Of the quality regulation strategies shown, only intelligent process control has the underlying architecture to support these functional requirements. This article has not claimed that intelligent process control has been perfected. Rather, several different quality regulation strategies have been examined in support of illustrating the design intent of intelligent process control.

Development plans for IPC's quality regulation involve three primary lines of work. First, further research is being performed to improve the efficiency and accuracy of on-line model development. Naturally, there is a trade-off between number of process variables to include in the model determination and the time required to develop the model. Improvements in efficiency and capability are related to a second area of development: knowledge application. There is much available, yet imperfect, knowledge of the molding process that exists in numerical simulations and human expertise. The performance of IPC would be greatly enhanced if this knowledge could be better applied. Finally, IPC has been implemented with the design intent to be as modular and refined as possible, permitting arbitrary application of experimental designs, sensor technologies, and quality attributes. Further development and validation is required to propose an industry-standard quality system specification.

CONCLUSION

The continued thrust of thermoplastics into advanced technical applications has resulted in end-user requirements which can exceed standard product development and manufacturing capabilities. The lack of process robustness is sometimes evidenced by long development cycles, excessive tooling costs, low process yields, and inferior product quality. Injection molding process simulation has provided major benefits to the product development cycle by facilitating evaluation of design candidates before tooling steel. At this time, however, the need for greater robustness in the molding process remains critical.

This paper has compared quality regulation strategies. As described, intelligent process control is completely generic and caters well to each application's diverse set of process dynamics and quality specifications. The quality system develops models between quality attributes and control inputs on-line, then continuously monitors the process: detecting and discarding defective parts due to random disturbances while adjusting the process for systematic errors. It should be clear that the system has significant potential to achieve the levels of quality control and process robustness needed by industry. Subsequent analysis [23] has shown that such on-line quality control systems can generate added value through four mechanisms, by:

- improving product quality to command a price premium or remove cost from the product,
- improving the production yields of acceptable parts,
- reducing the cycle time without adversely affecting part quality, and
- reducing mold commissioning and set-up times.

However, the goals and expectations of implementing such an on-line quality control system must be considered within each molding operation prior to adoption of the technology. The described methods can become standard operating procedure for any molder. If implemented well, they will directly result in greater customer satisfaction as well as significant reduction in manufacturing costs.

ACKNOWLEDGMENTS

The authors would like to acknowledge the contributions of all members of the Moldflow Intelligent Process Control team. Their genuine dedication and foremost ability have realized a new and revolutionary technology of benefit to the plastics industry.

REFERENCES

1 D. O. Kazmer, J. C. Rowland, G. Sherbelis, Journal of Injection Molding Technology, v. 1., p. 44, 1997.

2 R. E. Kalman, ASME Journal of Basic Engineering, v. 83, p. 34, 1960.

3 D. O. Kazmer, Dynamic Feed Control for Injection Molding, Doctoral Dissertation of Stanford University ME Design Division, 1995.

4 H. S. Gitlow, The Deming Guide to Quality and Competitive Position, Prentice-Hall , Englewood Cliffs, N.J, 1987.

5 R. E. DeVor, T. Chang, J. W. Sutherland, Statistical Quality Design and Control: Contemporary Concepts and Methods, Macmillan Publishing Company, New York, 1992.

6 W. A. Shewhart, Journal of the American Statistical Association, v. 8, p. 546, 1925.

7 W. A. Shewhart, Economic Control of Manufactured Product, Van Nostrand, New York, 1931.

8 M. R. Kamal, Macromolecular Symposia, v. 101, p. 167, 1996.

9 F. Gao, W. I. Patterson, M. R. Kamal, Polymer Engineering and Science, v. 36, p. 1272, 1996.

10 C. Smith, European Plastics News, v. 20, p. 33, 1993.

11 P. D. Coates, R. G. Speight, Part B of Proc. IMECHE, v. 209, p. 357, 1995.

12 K. Wilczynski, International J. Polymer Science and Technology, v. 22, p. 79, 1995.

13 C. T. Burke, D. O. Kazmer, Proceedings from the 1994 Annual Technical Meeting of the Society of Plastics Engineers, v. 53, p. 572, 1994.

14 S. L. B. Woll, D. J. Cooper, B. V. Souder, Polymer Engineering and Science, v. 36, 1477, 1996.

15 V. Witgosky, Plastics Engineering, v. 50, p. 25 (Oct. 1, 1994).

16 L. Zadeh, Information and Control, v. 8, p. 338, 1965.

17 S. T. Kewswani, R. J. Waswski, Ceramic Engineering and Science, v. 16, p. 74, 1995.

18 S. Kaeoka, N. Haramato, T. Sakai, Advances in Polymer Technology, v. 12, p. 403, 1994.

19 D. G. Kleinbaum, Applied Regression Analysis and Other Multivariable Methods, PWS-Kent Pub. Co., Boston, MA, c1988.

20 W. J. Morse, Measuring, Planning and Controlling Quality Costs, National Association of Accountants, Montvale, New Jersey, 1987.

21 F. Alsup, R. M. Watson, Practical Statistical Process Control, Van Nostrand Reinhold, New York, 1993.

22 D. O. Kazmer, P. Barkan, K. Ishii, Proceedings of the 22nd Annual ASME Design Engineering Conference, 1996.

23 J. R. Rowland, D. O. Kazmer, Proceedings from the Society of Plastics Engineers Annual Technical Conference, 1997.

Injection Process Monitoring and Process SPC

A Dawson, A Key, M Kamala, R M Rose and P D Coates
IRC in Polymer Science & Technology, Mechanical & Manufacturing Engineering, University of Bradford, Bradford UK

ABSTRACT

The use of Viscosity Index, VI, obtained from primary injection specific pressure integrals in a low noise region of the injection moulding cycle, has been used in this work as a convenient and simple, yet meaningful, way of quantifying the injection moulding process, for use in quality assessments such as SPC.
The clear dependence of VI on set injection velocity and set melt temperature has been shown for an ABS and an HDPE. Long production run data was obtained in the form of VI values from a Polymer Insights intelligent moulding modules, allowing quantification of process variation. Process variation has been shown to be greatest at start-up of the injection moulding process. Start-up data also indicate the dynamics of the process as it moves towards stability.

1. INTRODUCTION

Process variation is an important issue in precision injection moulding. Injection machine technology has continued to improve, with increasingly tighter control of event sequencing and specific machine variables. However, this complex multivariable, non-steady state process does not have a widely accepted route to closed loop control of the process. One reason for this is a lack of process knowledge, and possibly a lack of integration of different areas of expertise. Raw materials variation, together with machine variation and machine dynamic response, contribute to overall variation in the process.

One approach towards closed loop control of injection moulding has been presented elsewhere [xx] – this involves attempts to correlate product quality with real time process variable measurements. The choice of process variable in this context can include injection pressure integrals, melt temperature integrals, cavity pressure measurements, hold phase measurements, etc. Other routes include model reference adaptive control strategies, in which the aim is to use a computer model of the process to assess machine set up and to simulate mould filling, comparing this with the actual moulding process performance.

However, what are the levels and patterns of process variation which actually occur? Any control strategy needs accurate, robust information from the process, preferably related to the actual processing behaviour of the polymer. In the work reported here, we focus upon *process monitoring* using a simple injection pressure integral method [e.g. 1, 2] to follow changes in the injection moulding process in a variety of situations, in both our laboratories and in the industrial production environment. The studies described here are complemented by extensive Taguchi experimentation [3], involving product quality measurements and computer modelling comparisons, which will be reported elsewhere.

2. PRACTICAL MONITORING OF INJECTION MOULDING

One of the significant quandries faced when attempting to analyse the injection moulding process is the sheer amount of data which can be generated when monitoring several variables cycle by cycle. Despite the increasing use of computer monitoring, the danger is to be submerged by data – which, although carefully collected, is consequently of little use if it cannot be analysed in an adequate timescale. Extensive previous and ongoing experimental studies in our laboratories and in the manufacturing environment have used full computer monitoring of a range of machine and process variables. Typical data sampling rates required are 50Hz in the injection phase (up to 1000Hz if required, to follow dynamic changes, although the dynamic response of sensors limits the usefulness of such rapid data collection) and at lower frequencies in the hold/solidification/screwback phases.

The range of variables which might typically be monitored include:

- screw rotation speed,
- screw displacement and injection velocity,
- set barrel temperatures,
- shot dose,
- switchover position,
- cushion,
- hydraulic injection pressure,
- nozzle melt pressure,
- nozzle melt temperature,
- tool temperature,
- in-cavity pressures
- in-cavity temperatures,
- hold pressure (hydraulic, nozzle or in-cavity)
- screwback pressure
- sequence times (injection time, hold time, cycle time)

Clearly, it is impractical to monitor the injection moulding process in such depth at high frequency for every cycle and hope to analyse such data in a meaningful way. Some simplification is required – but the choice of what to include, exclude or condense should clearly be based on process knowledge. In the experimental work reported here, a Polymer Insights (PI) moulding module has been employed to provide a convenient and real time *condensation* of the large amounts of data captured. Figure 1 shows a schematic diagram of the use of the PI module, a microprocessor based unit for process monitoring and data analysis developed in our laboratories.

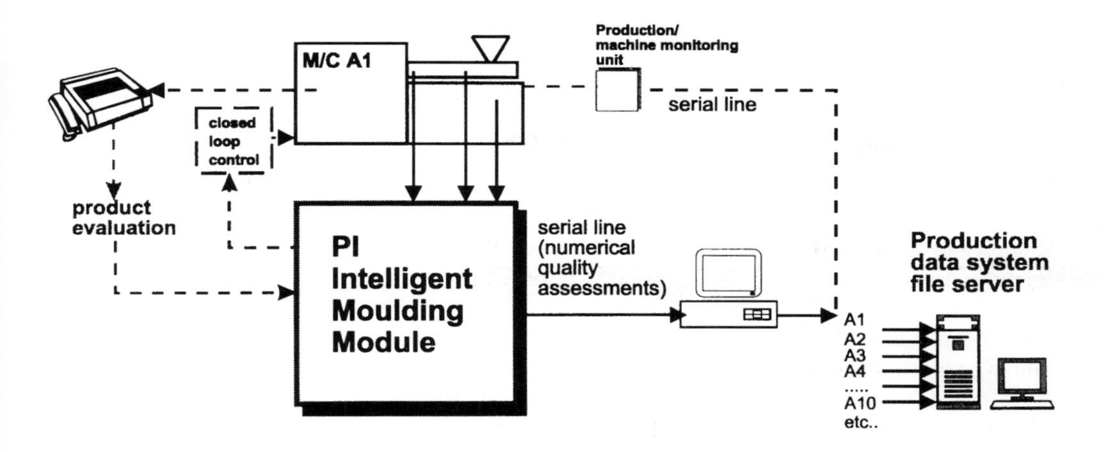

Figure 1. Schematic of Polymer Insights module links in the injection moulding process.

In its simplest form, the PI module provides the researcher or operator with a single injection pressure integral value, obtained at a low noise region of the injection moulding cycle [1] and consequently less susceptible to inherent process noise related, for example, to machine dynamics. Such injection pressure integrals can be conveniently made using nozzle melt pressure sensors or hydraulic pressure sensors, if the latter have been adequately signal conditioned. In principle, cavity pressure sensors can also provide pressure integrals, but clearly only in the times when melt is present at the cavity pressure sensor, and prior to solidification of the skin at that point. Cavity sensors also tend to be more expensive and require fitting to each tool. Even so, useful information can be obtained by this route. The basic PI module can have up to 8 analogue input signals (so could have input from hydraulic, nozzle or cavity sensors, screw position or velocity sensors, etc, for the injection and/or hold phases), with two digital outputs and serial data output to a remote PC. A constraining factor of any monitoring system which is also processing the collected data is the rate and amount of data which can be collected without missing important process data. In practise, this tends to limit the number of channels sampled at high rates.

The PI module is flexibly programmable, so that a range of integrals can be calculated, if required. For reasons of practicality, only primary phase injection pressure integrals, or Viscosity Index (VI) values, are reported from these initial studies, although combined pressure and temperature indices, and hold phase measurements have been studied and reported in part [4, 5].

Good correlations of Viscosity Index with product quality assessments have been previously reported [1, 2]. Although it is acknowledged that the VI is a simple measure, which may be reinforced by combination with melt temperature integrals, for example, or hold P measurements too, it has the advantage of being a rapidly available assessment of the status of the moulding operation, and is the first stage in providing meaningful statistical process control (SPC) for long production runs. Other simple measurements which industry have tried to use for injection moulding SPC include readily obtained machine variables such as injection time, cycle time, etc, which in general do not correlate with product quality, and hence should only be used for production monitoring purposes, where they are clearly of most value.

3. EXPERIMENTAL DETAILS

Experimental studies cover a range of polymers and several process technologies, and were conducted both in our IRC processing laboratories and in industry. Results for a semi-crystalline polymer (high density polyethylene, grades HDPE 5050 and HDPE5411, BP Chemicals) and an amorphous polymer (acrylonitrile butadiene styrene, ABS grades M3FR, Bayer and Terluran 968SM, BASF) are reported.

Processing equipment included a CM ACT30 30 tonne servo-electric machine, a Sandretto 60 tonne proportional hydraulic machine and a Stork 440 tonne servo-hydraulic machine, the former two housed in our laboratories, the latter in Birkbys Plastics Ltd. Full computer monitoring, using a PC and external bus interface (Microlink Ltd) were normally used in parallel with the PI module.

Initial experiments concern an assessment as to how changes in VI depend on key injection moulding variables. Demonstrating that VI values clearly reflect *intentional* changes to the process, here by changing set injection speed and set melt temperature over a wide range, provides quantification of the sensitivity of VI values for given moulding conditions. In turn, this provides an insight into the magnitude of VI levels and variations observed in the manufacturing process. These initial experiments were undertaken in our laboratories, using a Sandretto 60 tonne proportional hydraulics injection moulding machine. A directly gated simple tubular product moulding was used in the studies. The mould was temperature controlled at a set point of 30C (stabilising at around 50C). The process was allowed to stabilise before sampling VI for 5 cycles at each process condition, for both polymers.

Next, the use of VI to monitor large scale industry runs is examined. The PI module was used to monitor large scale production runs at Birkbys Plastics Ltd for manufacture of a two impression automotive instrument housing in ABS, on a Stork 440 tonne servo-hydraulic machine. Monitoring commenced from start-up (see below).

Finally, a region of the injection moulding process which is potentially likely to exhibit greatest variation, namely startup, was investigated, using a Cincinnati Milacron ACT30 servo electric 30 tonne injection moulding machine and a DIN tensile specimen mould. Machine parameters were set, with the tool at ambient temperature; data was then collected from the first cycle onwards, with the tool temperature controller switched off. The experiment was repeated with the tool temperature controller being switched on just as cycling started. These experiments allow monitoring of the process as it tends towards stability. Start-up data was also collected from production runs at Birkbys, for the same production run used to study long term variation.

4. RESULTS & DISCUSSION

Initial experiments showed that Viscosity Index, VI, clearly depends on injection velocity and set melt temperature for each polymer (Figures 2 and 3). In effect, Figures 2 and 3 reflect the pressure drop – flow rate behaviour of each polymer flowing through the nozzle and mould, i.e the (shear rate thinning) rheology of the material. It should be noted that this is not a viscometric flow –i.e. the flow path does not provide a simple geometry with steady velocity fields – indeed there will be both shear and extension occurring. Consequently, the Viscosity Index, although having the same units as viscosity (Pa.s) is not an actual viscosity measurement.

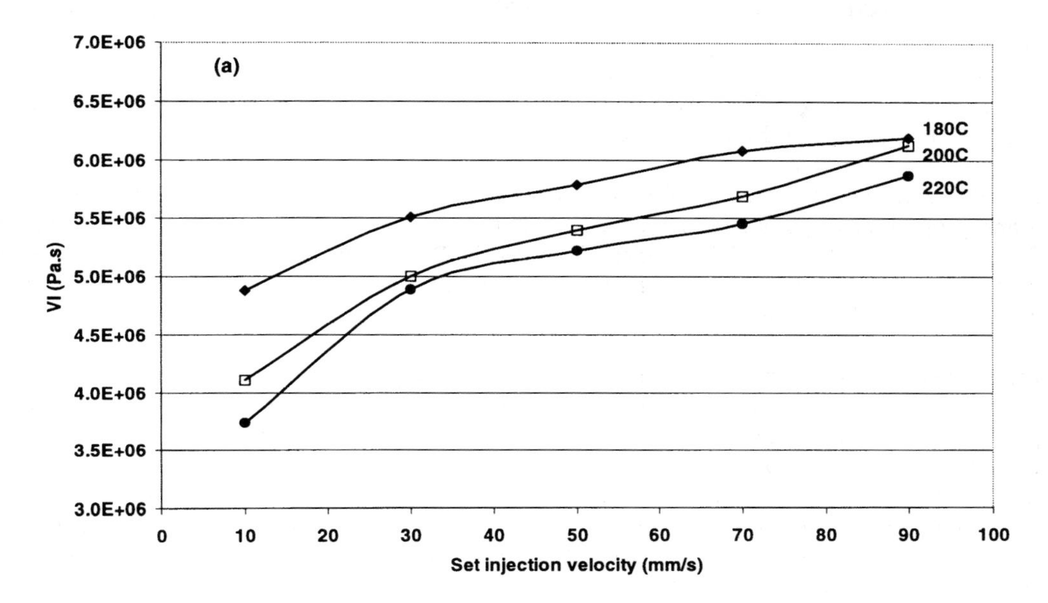

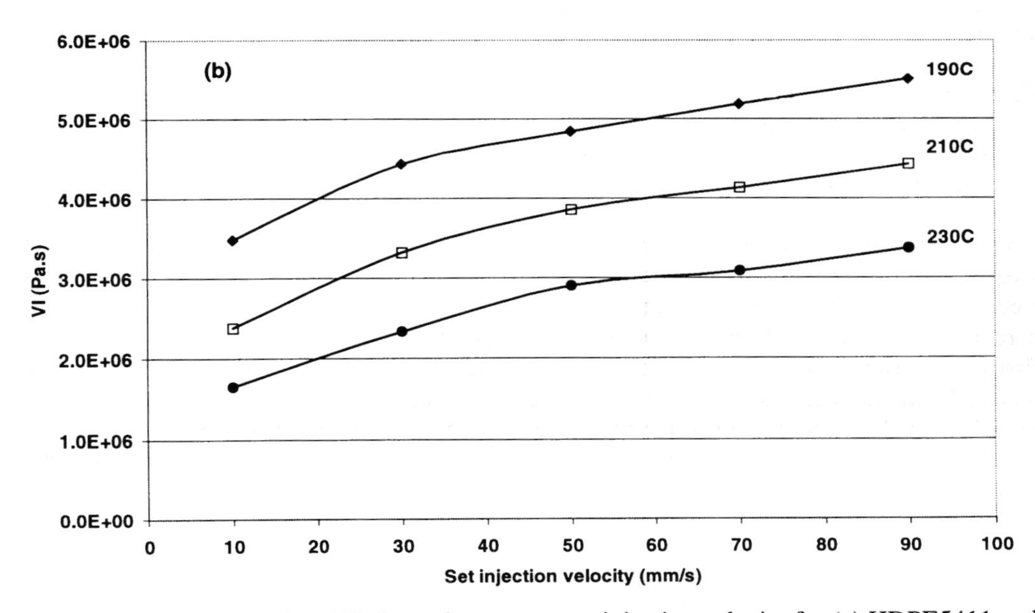

Figure 2. Viscosity Index (VI) dependence upon set injection velocity for (a) HDPE5411 and (b) ABS M3FR, for a simple tubular product.

Curve fitting of the data from Figure 2 shows a 'power law' dependence of VI on set injection velocity V_{inj}, namely:

where the coefficients C and b both depend on set melt temperature, C falling with increasing temperature, and b rising with temperature.

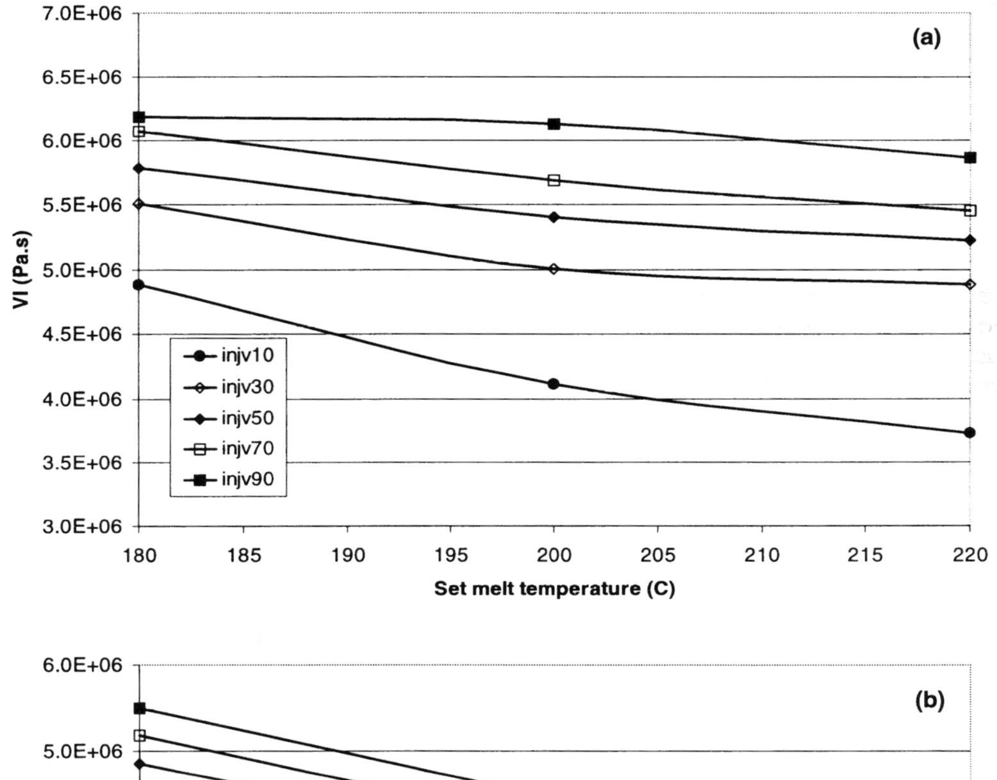

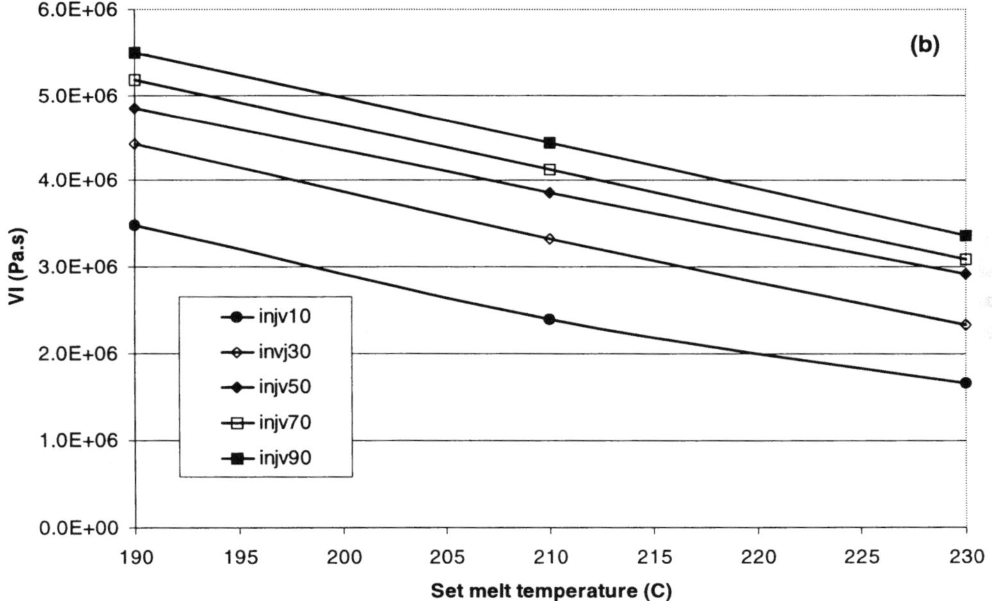

Figure 3. Viscosity Index (VI) dependence upon set melt temperature for (a) HDPE5411 and (b) ABS M3FR, for a simple tubular product.

Curve fitting of the data in Figure 3 shows a linear fall in VI with set melt temperature, T_{inj}, namely:

$$VI = G.T_{inj} + H$$

where G is the slope of the linear fit, and H is the intercept.

The results from Figures 2 and 3 show that temperature appears to have a larger effect than injection velocity on VI for ABS, whereas temperature and injection velocity have similar effects on VI for the HDPE studied. Also, overall VI values at the same set melt temperature are rather higher for HDPE. These observations are in keeping with the rheological behaviour of the two polymers.

The VI values therefore appear to reflect the rheology of the polymer melt being processed. Change in melt rheology is a key factor in injection moulding quality control, i.e. VI values provide a real time indication of the state of the polymer melt. Other important factors such as raw material variation (as-supplied, or by addition of regrind) affect melt viscosity values, and will also clearly affect VI values [ref.].

Some typical data for a large scale production run is presented in Figure 4, which shows 4,000 cycles (from actual start-up) of the dual impression automotive instrument housing tool (equivalent to 2.5 days production) – the full data run was 11,000 cycles, i.e. around one week of production. It is notable that the process appears to be running to close tolerances for the majority of cycles, as indicated by the relatively small variation in VI. Machine interruptions are clear, as are other (minor) problems from the large spikes in VI – products associated with these spikes are specially inspected.

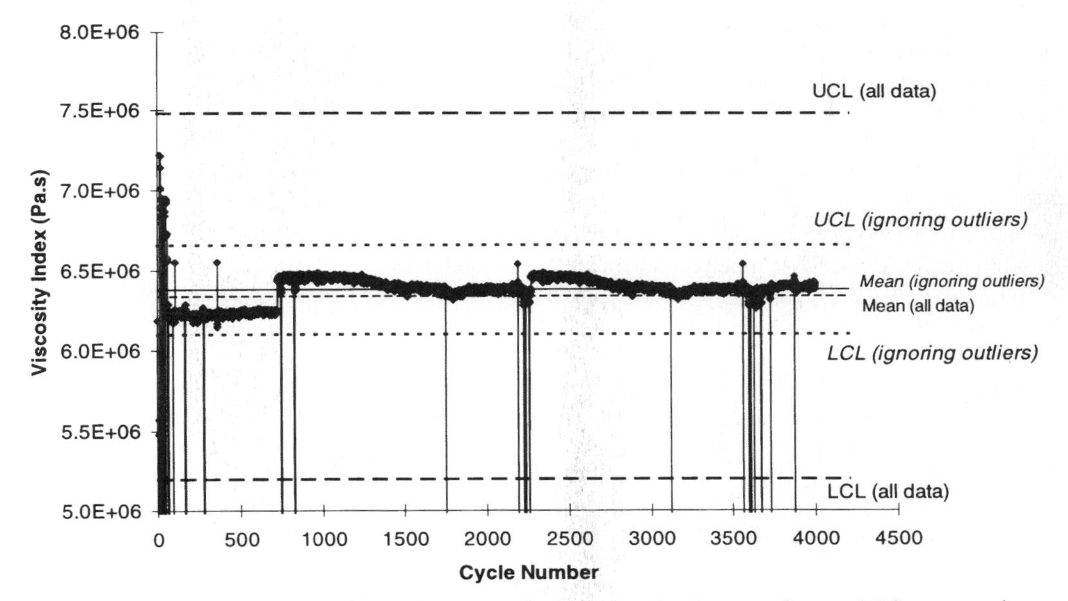

Figure 4. Viscosity index variation for part of a long production run for an ABS automotive instrument housing, at Birkbys Plastics Ltd.

If VI is a useful correlation with product quality, then a chart such as Figure 4 represents 100% automatic quality inspection; any VI value breaching set control limits can be used to trigger a robot, to set the part aside. However, setting of the control limits is an issue which needs careful attention.

The mean VI for all 4000 cycles is shown in Figure 4, together with mean VI and control limits, based on $\pm$ 3 standard deviations (σ) of the mean VI. Inclusion of the outliers leads to excessively wide control limits: the ($\pm$ 3σ/mean VI) band is 36% of mean VI. Exclusion of the outliers from the full data set produces a slightly lower mean VI, and tighter limits: the ($\pm 3\sigma$/mean VI) band is now only 8.6% of mean VI. Figure 4 illustrates the need to handle process data carefully, and in particular the need to obtain suitable, robust yet sufficiently sensitive control limits. If a shorter sample set (e.g. 100 cycles) is used to obtain control limits, this may lead to exceedingly tight limits, causing signalling of too many 'suspect' products. For example, in a steady state section (e.g. cycles 1000 to 1500) the ($\pm$ 3σ/mean VI) band is only ~1% of mean VI. It is noted that the mean VI shifts at certain points (related to specific events in the production environment). One possible approach is to calculate a local mean VI and employ a fixed control limit spread, so that the band of acceptable production moves with the mean VI. If mean VI moves significantly, this is also a cause for concern, as this may signal a change in product quality. Work on automating the calculation of control limits during production is ongoing.

Experience plays a part here: all of the products represented in Figure 4 inside the tighter limits calculated by ignoring outliers were of acceptable quality, indicating that a change of around $5x10^5$ Pa.s (=5 bar.s) in VI, i.e. a $\pm$ 3σ/mean VI band of 8.6% of mean VI could occur for this production process without detriment to part quality. This indicates that the VI measurement is, in fact, extremely sensitive, and will certainly signal quality problems when they occur.

Start-up of the moulding process is likely to be associated with greater variation, as the process moves towards stability. The data from Figure 4 is re-presented in Figure 5, with a mean VI and 3σ control limits obtained from 100 cycle sets of VI values. It is clear that process variation is greatest at start-up with a $\pm 3\sigma$/mean VI band of 157% of mean VI for cycles 1 to 100, around 54% for the next 200 cycles but settling to very small values (~1%) when the process stabilises. Special causes (related to specific events in the production environment such as machine stoppage, which affects several factors, especially temperatures) lead to regions of larger variation – which appear to be fairly consistent in their adverse effect on process variation. The time taken by a process to stabilise is clearly important from an economic viewpoint, and there is pressure in the manufacturing environment to obtain acceptable quality products in the first few cycles.

Figure 6 shows results from laboratory experiments on start-up for HDPE, conducted on the CM ACT30 machine. Also included here is the rise in tool temperature (measured by an embedded thermocouple), which indicates that the process is reaching thermal stability by cycle 150. Mean VI and control limits for the whole 150 cycles, and for the more stable section of the curve (cycles 50 to 150) are also shown. This figure reflects the observations made for Figure 5.

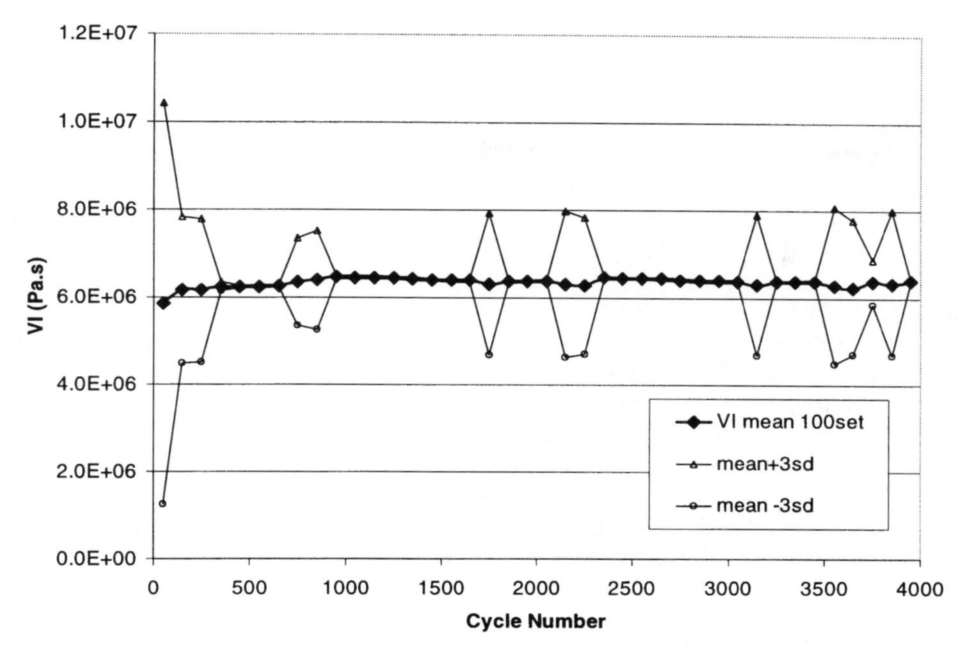

Figure 5: Mean VI, with ± 3 standard deviation (sd) limits calculated for each 100 cycle set of data, for long production run of ABS automotive instrument housings at Birkbys Plastics Ltd.

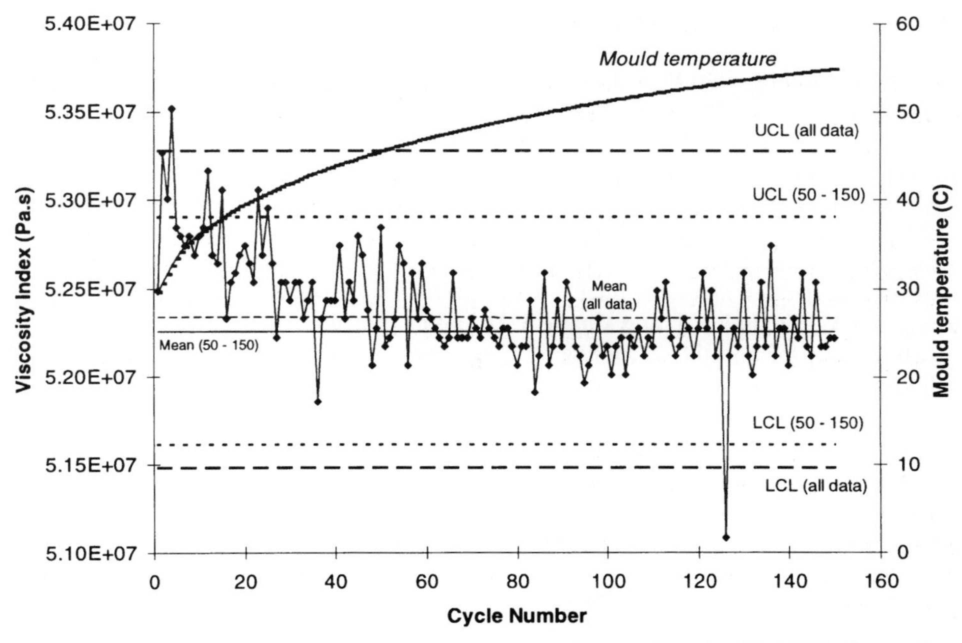

Figure 6. Startup data from CM ACT30 servo-electric machine, for HDPE5050, tensile specimen moulding.

4. CONCLUDING COMMENTS

Accurate process measurement is of major significance in developing enhanced process understanding for use in the controlled production of consistent quality parts. However, it is clear that full process monitoring is currently impractical for real time assessment of the process. The Viscosity Index, VI, obtained from primary injection, has therefore been used in this work as a convenient and simple, yet meaningful, way of quantifying the injection moulding process, for use in quality assessments such as SPC.

The clear dependence of VI on set injection velocity and set melt temperature has been shown for an ABS and an HDPE; these effects appear to reflect the rheology of the polymer melts. As such, the VI is a useful indicator of the state of the melt being processed – i.e. it has the potential for being a powerful (and sensitive) indicator of changes which could significantly affect product quality.

Long production run data can also be efficiently obtained using VI values. A Polymer Insights (PI) intelligent moulding module has been used to obtain data for an 11,000 cycle production run using ABS, which has allowed quantification of process variation. Automatic setting of control limits remains an important practical issue.

Process variation, measured by variation in VI values monitored by the PI module, has been shown to be greatest at start-up of the injection moulding process. Start-up data indicate the dynamics of the process as it moves towards stability.

References

1. **Coates P D and Speight R G**, Towards intelligent process control of injection moulding of polymers, Proc. I Mech E, part B, J. of Engineering Manufacture, Vol 209, 357-367 (1995)
2. **Coates P D, Dawson A J, Key A, Peters C and Jagger R,** Intelligent Monitoring for 100% automatic inspection of quality in injection moulding,, SPE Tech Papers, XXXXII, (1996)
3. **Khoshooee N and Coates P D,** Consistent melt production through the use of Taguchi methods in injection moulding quality control, Proc PPS 12, Sorrento, May 1996
4. **Speight R G, Yazbak E P and Coates P D,** In-line pressure and infrared temperature measurements for injection moulding process control, SPE Tech Papers, XXXXI, 647-650 (1995)
5. **Speight R G,** PhD thesis, University of Bradford, 1994

Acknowledgements

We are grateful for support of EPSRC, Birkbys Plastics Ltd, Dynisco UK Ltd, Polymer Insights and the University of Bradford for aspects of the programme presented here.

Neural Computing: Problem solving and applications

Dr John Pilkington[1], Dr Rajan Amin[2] and Mr Tim Hyde[2]
[1]Scientific Computers and [2]ERA Technology Ltd

ABSTRACT
Neural networks make excellent classifiers and regression models for applications in process control, process modelling, condition monitoring and related manufacturing contexts. Their power is in making arbitrary non-linear approximations, which include linear models as a special case. Time to develop models can be minimised, increasing the productivity of application developers.

A disciplined approach to Neural Networks is required to make the most of their potential and to avoid the commoner dangers. In the first part of this paper, we explain what Neural Networks are and how they behave as non-linear models. We follow this with a description of a development cycle which has been used successfully to deliver robust models for real engineering problems.

In the second half of the paper, we concentrate on a practical application which demonstrates how Neural Networks may provide significant benefits in process based industry. The application utilises the modelling and prediction capabilities of Neural Networks and involves an industrial drying process which is typical of the type of complex problem encountered in process engineering. The application shows how Neural Networks provide a robust solution which results in improved product quality, throughput and substantial energy savings.

WHAT ARE NEURAL NETWORKS, AND HOW DO THEY WORK?

In this section, we develop a perspective for the general understanding of Neural Networks. Here, we concentrate on the performance of neural models in recall mode. This allows us to address a common problem with understanding how Neural Networks behave as non-linear models, that is, how they represent non-linear relationships in the problem domain.

From this understanding, we can proceed to demonstrate that the training process is equivalent to a parameter estimation problem. This is a useful perspective, because it puts Neural Networks into the same frame of reference as standard statistical techniques.

The difference from standard statistical techniques is that Neural Networks use optimisation methods for parameter estimation where as standard techniques use analytical procedures. The advantage for Neural Networks is that they can behave as arbitrary non-linear correlators; the disadvantage is that unless careful control is exercised, the parameter count can increase beyond what can be supported by the data, with consequent overfitting and failure to generalise properly.

We can attempt to understand Neural Networks from both functional and structural perspectives i.e., how we should think about what they do, and understanding how they do it.

Functionally, Neural Networks are arbitrary non-linear correlators. Let us dissect this statement one word at a time!

(1) Neural Networks are correlators. Correlators extract information from existing data. To be useful, the identified correlations must exist generally in all data from the same source. If the relationships discovered by a neural network in the training data hold true in the general data corpus (and, particularly, into the future), the neural model is said to generalise. A model which discovers complex relationships which are specific to the training data is not useful. Much of the art of making good neural networks is to do with ensuring generalisation

(2) Neural Networks are non-linear correlators. We live in a non-linear universe, so many processes are not adequately modelled if there is an underlying assumption of linearity. Further, Neural Networks are arbitrarily non-linear. There are many other techniques for making non-linear models, but most of them require some prior assumptions about the kind or extent of the non-linear relationships to be incorporated. Their arbitrariness makes Neural Networks specially useful in exploratory data analysis.

So, how do they achieve this property of arbitrary non-linear correlation? Neural models map a multidimensional input space to a different multidimensional output space. Between the input and output spaces, they insert an internal model space. Points in this space are defined by the outputs of a set of basis functions. The basis functions are often referred to as hidden units. Thinking of them as basis functions helps to understand what they are doing, and also brings thinking about Neural Networks into line with other modelling techniques.

Basis functions are just functions which have a definite output value (a single number) for each point in the input space. So each point in input space generates a profile of activations in the model, representing a point in the model space. We assert that, given such a point in a suitably chosen model space, we can determine the desired output of the model. There are several theorems suggesting that, in principle, it is possible to build models like this. But unfortunately, they are existence theorems. They don't tell us how to find the specific basis functions that do the trick in a given situation. Neural Network training is about finding these functions.

When considering choice of basis function, we have to be clear about the distinction between the kind of basis function to be used, and the particular instance of each function, defined by its parameters. We also have to decide how many basis functions (hidden units) will be required.

Choice of kind of basis function is almost equivalent to choice of the kind of neural network we shall use. Restricting ourselves to models which do not have feedback, if we choose logistic (Fermi function) or hyperbolic tangent basis functions, we will end up with a classical Multilayer Perceptron (MLP). If we choose Gaussian functions which have a maximum response at some point in the input space, and a radius parameter which determines their response to neighbouring points, we get a Radial Basis Function Network (RBFN). If we choose functions which compete with each other for representing a point in input space on some distance criterion, we get a CounterPropagation model, Learning Vector Quantisation or, with some modification, a Self Organising Map. Choice of Neural Network "type" can become confusing but is essentially a secondary consideration as all these basis functions support satisfactory input/output mapping, in principle.

"type" can become confusing but is essentially a secondary consideration as all these basis functions support satisfactory input/output mapping, in principle.

The particular instances of the selected basis functions are determined by their parameters, which are given particular values by whatever parameter optimisation technique is employed during training.

Much of this general perspective can be summarised by a model for a Processing Element (PE, Figure 1). It is convenient to decompose the basis function represented by a PE into a summation function, transfer function, and output function.

The summation function always performs a vector to scalar conversion, that is, a single number is derived to represent the incoming point in problem space. The transfer and output functions are scalar. Of course, since we have (typically) an array of basis functions, each input vector is represented internally by a vector of PE output values.

All neural models have PEs with a summation function. It is not always necessary to have a transfer and an output function. We can think of "no" transfer (equivalently, output) function as being the identity function.

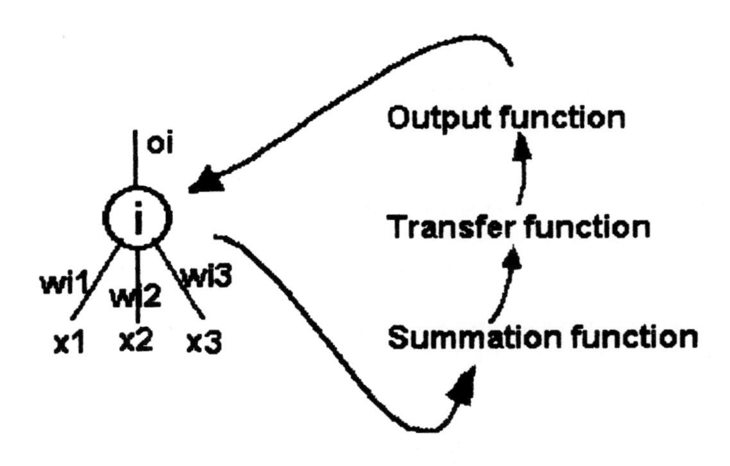

Figure 1. Model for a Processing Element

Figure 2a,b shows what the Neural Network looks like when you put all the basis functions and their parameters together.

In these figures, inputs are represented at the bottom, outputs at the top. The three circles in the middle of Figure 2b represent three PEs or "hidden units" in a single "hidden layer" in a Neural Network.

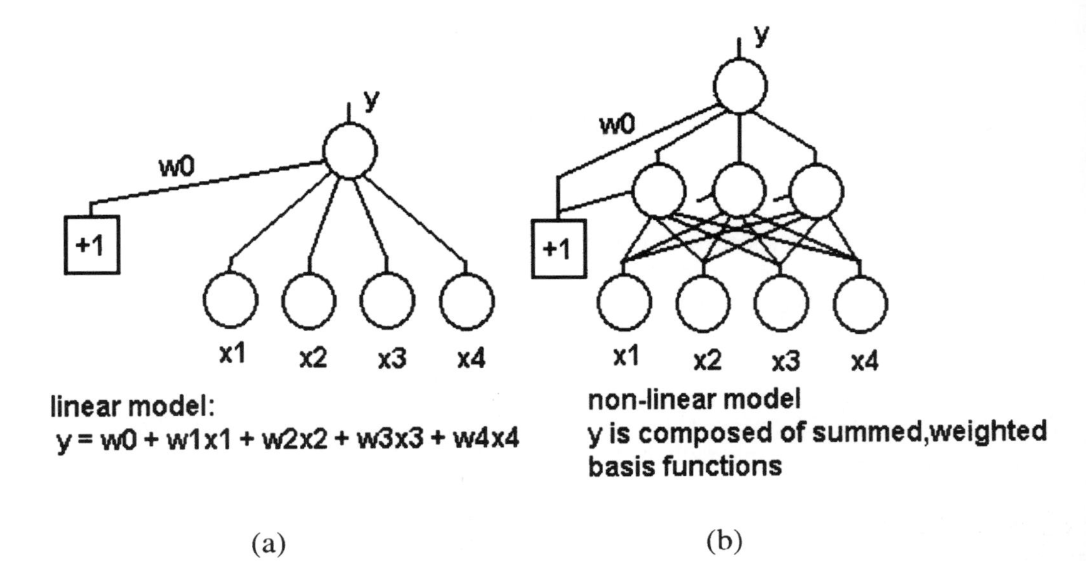

Figure 2. (a): Linear model; (b) Neural Network

Figure 2(a) shows how a linear regression model can be represented in the same notation as a Neural Network. This is to emphasise the similarity between Neural Networks and standard statistics. Notice the presence of bias in both these models. In the linear model, bias generates a constant offset. We shall examine the part played by bias in a neural model later.

Figures 1 and 2 allow us to reinforce the equivalence of "weights" and parameters; these are just the lines connecting the circles. In Neural Modelling, the values to be put on these lines are determined by the training process. Figure 2 also illustrates how the inclusion of basis functions in a model increases the parameter count. This is an undesirable feature of Neural Networks. In practical applications, we wish to construct Neural Models with minimal parameter count.

Given this general framework, how shall we understand the workings of a particular species of Neural Network, for example the Multilayer Perceptron? The resources which a Neural Network uses to make input/output mappings are the transfer functions, usually monotonic, e.g. Fermi functions or tanh in a Multilayer Perceptron, and their parameters, i.e. the weights in the model. Given a general type of transfer function, the weights determine the particular instance. The interesting thing here is that the bias weights contribute something quite different from the weights on the input values ("ordinary" weights), although they are optimised using exactly the same algorithms.

"Ordinary" weights can make a single PE look like almost a linear function, or almost a step function depending on their magnitude. Depending on the sign of the weights, a particular hyperbolic tangent function may be monotonic increasing (the way we ordinarily think of them) or monotonic decreasing.

Changing the "ordinary" weights on a uni-dimensional hyperbolic tangent function, therefore, can generate a whole family of particular instances (Figure 3).

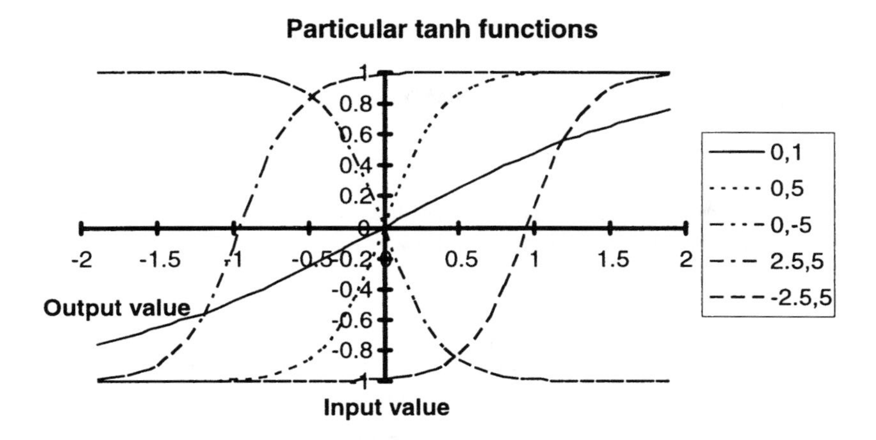

Figure 3. Particular instances of tanh functions generated using the parameters (bias, weight) shown in the legend. The curves are graphs of tanh(bias + weight*x) where x varies from -2 to +2. This is a one-dimensional example, but the principle extends to multidimensional inputs.

In a model with many basis functions, each of these instances will contribute to the output in a particular way. A set of very steep, step-like functions will lead to a classifier, while small weights will lead to a model whose basis functions never saturate; this will be suitable for constructing real function approximations.

However, it is noticeable that all the tanh instances produced by manipulating the "ordinary" weights (with bias set to zero) cut the graph at its origin (Figure 3). This is a severe constraint, which would limit the representational power of a set of tanh functions if it could not be relaxed.

The effect of manipulating the bias weights is to push the steep part of the tanh to the left (negative bias) or right. From a classifier point of view, negative bias is like positive threshold or decision boundary. From a real function approximator point of view, bias increases the representational richness of a function array by allowing arbitrary incorporation of non-linear elements at precise locations in input space - positive increasing slope, positive decreasing slope, negative increasing slope, and negative decreasing slope. It is this richness that allows MLPs to capture arbitrary non-linear relationships in the input data.

We have looked carefully at how Neural Networks "work", that is, how they represent arbitrary non-linear relationships. But we have not considered how the required parameters may be found in practice.

If we start with a bad model (random numbers for the parameters) then it is easy to see that this is a credit (or blame) assignment problem: we need to assign blame for errors to each parameter, and change the parameter values to reduce the error. The key to the "backpropagation" algorithm was the development of effective techniques for apportioning blame for errors at the output of a basis function mixture model on the input parameters of the basis functions, i.e. the weights from input to hidden PEs.

Briefly, the "blame" assigned to a weight between an input and a hidden PE is the "blame" assigned to the weight between the hidden PE and the output of the model (which is easy to find), scaled by the value of the output weight and the current slope (derivative) of the transfer function. A variety of more or less sophisticated algorithms exist to use this information to calculate changes to the weights which will improve model performance.

Given that optimisation techniques exist to find values for the parameters in a Neural Network of fixed structure, the problem still remains of how to find a good structure (number of hidden PEs) to support the particular I/O mapping hidden in the training data.

It is important to realise that this is a separate optimisation from parameter optimisation. Here, we are talking about optimising network complexity. In automated train/test cycles, parameter optimisation is an inner loop occurring within an outer loop of complexity optimisation. In older approaches to Neural Modelling, this outer loop was often badly controlled, leading to much waste of time as practitioners fiddled around wondering whether five hidden units was right, or perhaps five hundred? Or whether just one more hidden unit might provide that magic answer? It is better to adopt a principled approach to determining model complexity, preferably using an automated train/test cycle. These come in two flavours: constructive (synthetic) and destructive (analytical). Cascade Correlation is a constructive algorithm:

(1) Start with no hidden PEs (essentially, a linear model), and optimise it

(2) Build candidate basis functions to model the residuals

(3) Incorporate in model, but only establish if performance improved using test data (cross validation)

(4) Repeat from (2) until the residuals are random - no more modelling to be done.

An example of a destructive technique is Save Best with Pruning. Start with "too many" hidden units. Train until the model starts to overfit the data (test data set required). Examine each hidden PE in turn and assess its contribution to the model. Then eliminate the least important and re-train the simpler model. Repeat until all hidden PEs are making a significant contribution.

HOW CAN THESE PRINCIPLES BE APPLIED IN PRACTICE?

In briefly outlining what Neural Networks are, and how they work, we have emphasised that selection of Neural Network "type" is a secondary consideration, and that an unprincipled approach to Neural Network training can lead to a lot of fiddling about with different model structures. The temptation to waste time in this way can be avoided by employing an automated model development cycle.

In practice, this means putting a lot of the paraphernalia of traditional Neural Network development into the background, and concentrating instead on the training data itself. Data transforms and input variable selection will be particularly emphasised. It is interesting that these techniques are important to any regressional modelling endeavour, and this gives a new perspective on the general utility and position of neural modelling with respect to the other techniques currently available: because of their arbitrariness, Neural Networks are especially supports good neural models will also support other techniques should these, in the event, be preferred.

The main emphasis in a practical model development cycle should be to produce the simplest models which will just support the required input/output mapping. This is because the fewer parameters we have to estimate from a given body of data, the more robust, in general, that estimation will be.

The main stages in the model development cycle are:

(1) Partitioning the data into separate train, test, and, if there is sufficient data, validation, subsets.

(2) Search for data transforms. Neural Networks work best if the dynamic range of the data matches the dynamic range of the transfer functions, e.g. [-1, +1]. This can be achieved by a linear scaling, but linear scaling will preserve any skewness or unusual distribution in the data. On one hand, such idiosyncrasies of distribution may help in the subsequent model building process; on the other hand, it may be better to find data transforms which produce more uniform distributions. Which transform to use depends upon the data in question; typical examples are log, inverse, square, square root, etc.

(3) Input variable selection. One of the temptations with Neural Modelling is to throw all the data items at the model in the hope that this will produce good results. But one of the most effective ways to produce minimally parameterised models is to reduce the number of input variables. So, before a final model is made, it is a good idea to explore subsets of the input variables (and their candidate transforms) to see if the input dimension can be reduced. Exhaustive search of all possible combinations is impossible for all but low dimension problems. One practical technique for higher dimensions is genetic optimisation, which efficiently searches spaces of combinatorial complexity, with the specific objective of discovering good combinations of variables to support the desired input/output mapping. For example, the dimensionality of some data may have been reduced using Principle Components Analysis, and then genetic optimisation can be employed to select which of the components to use for model building.

(4) Detailed Neural Model building, for preference using a constructive algorithm such as Cascade Correlation so that the user is freed from the burden of specifying the number of hidden units in advance.

(5) Model validation. A properly constructed model development cycle must support the generation of appropriate statistics on model performance, with comparisons between training, test and validation data. This is important for assessing the generalisation capability of models, comparing different models, and for assessing the financial returns which can be anticipated when the model is deployed.

Appropriate test statistics depend upon the kind of problem. If a real function approximator is required, it will be appropriate to consider correlation between actual output value and the model estimates, and also the Root-Mean-Square (RMS) error over an entire data set, and perhaps also within sub-range partitions of the output value range.

For a classification problem, classification rates (in-class, out-class and average), relative entropies and contingency tables are appropriate. Relative entropy is a measure of the mismatch between two data series. In this context one series is the set of {0,1} flags representing the actual class to which an input vector belongs, and the second series is the corresponding set of real numbers in [0,1] which are the neural network outputs. Relative entropy is a scaled version of the sum over all outputs (classes) of the sum over all data items of -d*log(y), where d is the class flag (0 or 1) and y is the model estimate. Low counts good.

Incidentally, it is possible to constrain the outputs of a classifier model to sum to unity. When this is done, and the objective function for parameter optimisation is the relative entropy function, and when the occurrence of the different classes in the training data is similar to that in the universe from which they are drawn, the numbers at the outputs of a Neural Network may properly be treated as input-conditional class probabilities.

(6) Model deployment. A useful model spends most of its life in recall mode. The software paraphernalia required for optimising the model and validating it are no longer required. Practical development environments allow the deployment of models outside the development environment, e.g. by generating "Flash code". This is a representation of the model in a programming language such as "C", Basic, Fortran etc., with the parameters as fixed at the end of training. The model can then easily be deployed in target applications, with cross-platform compilation if necessary.

Software environments which support such a modelling cycle are becoming commercially available. One example is NeuralWorks Predict, distributed in the UK by Scientific Computers Ltd. Such products can be viewed as productivity tools. They differ from older Neural Computing environments by offering only a restricted range of kinds of Neural Network, but are superior in that they deliver robust models with a minimum of trial and error from the user. The operations they perform on the data are essentially domain independent, and so it is unlikely that an optimal result will be achieved on a single pass through the modelling cycle. Domain knowledge may be added by reviewing, and where necessary overriding, the decisions taken on the first pass. In this way, the investigator and the software work together to evolve a good solution; typically five or six passes through the cycle might be required.

PRACTICAL APPLICATION OF NEURAL NETWORKS

In this section, we will concentrate on a practical application which demonstrates how Neural Networks may provide significant benefits in process based industry. As a side issue, the example also illustrates how Fuzzy Logic can be combined with Neural Networks to provide a robust solution. The application involves an industrial drying plant and was undertaken as part of ERA's Neuromonitor programme[1]. This project has involved the practical application of neural computing and associated technologies for energy efficient monitoring and control of industrial processes.

---

[1] The programme was set-up in 1994 and is part funded by the Department of the Environment (DOE)

The drying process is typical of the type of complex problem encountered in process engineering. A simple schematic of the process in question is shown in Figure 4.

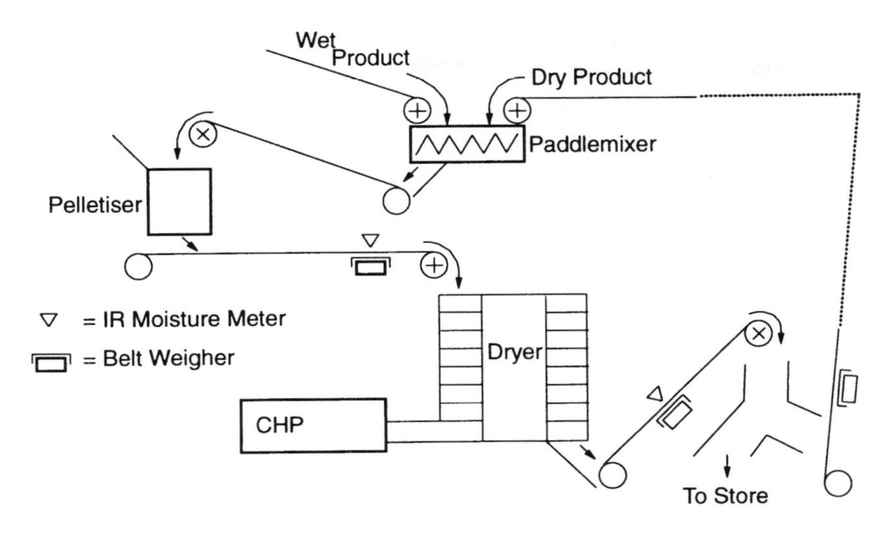

Figure 4: Industrial drying process

Wet and dry product are combined and delivered to the dryer on a continuous conveyor at a fixed speed. During the drying process the moisture content of the combined product is reduced from approximately 22% to a target of 11% (+/- 2%). A key factor in understanding the problem is that it takes approximately 40 minutes for the product to pass through the dryer and appear at the output. Heat is supplied to the dryer from a turbine with after burner. The supplied hot air is nominally at 600 degrees C but is manually adjusted by plant operators to maintain the desired target output moisture content. A conventional controller maintains the temperature of air supplied in conjunction with the exhaust air temperature by controlling an air damper to vary the volume of air supplied.

The main problem in this application is the variation of moisture content of the final product which can vary wildly. Control of this variable is made difficult because of the long 40 minute time lag through the dryer. The existing control method relied on the intuition of the operators to maintain the final moisture content by varying the heat supplied manually. Automatic control strategies for controlling the process have also been explored using largely classical methods but the long time lag between input and output, and difficulties in modelling a complex non-linear multi-parameter process has showed these to be unsuitable.

The drying process is an energy intensive process and there are significant savings to be made. Over drying the product is one source of excessive energy consumption while scrap or re-work of under dried product is another. This is an ideal application for Neural Networks. In particular, the self-learning and non-linear modelling abilities of a Neural Network were considered to be of great merit.

Neural Networks along with Fuzzy Logic technology have been used in this application to monitor and control the drying process in order to achieve tighter tolerances on the moisture content of the final product and subsequently reduce energy wastage. In the first instance this involves training a Neural Network to model the drying process and to predict dryer output moisture 40 minutes ahead, therefore enabling model-based predictive control and fault detection of the drying process. This arrangement is illustrated in Figure 5. The prediction of final output moisture is used within the fuzzy post-processing stage to determine the control setpoints for achieving minimum energy consumption while maintaining consistency of product quality. This element also alerts operators to fault conditions which are indicated by departure from normal process behaviour characterised by the Neural Network.

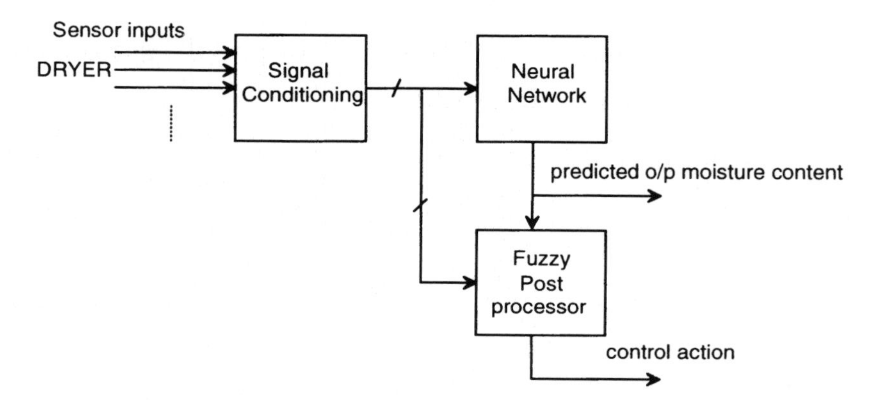

Figure 5: Neural Networks and Fuzzy Logic applied to dryer control

The LabView based Neuromonitor system underwent trials back in September at the trial site. Figure 6 shows a screen shot of the system while Figure 7 shows some of the results obtained during this trial. This shows the actual dryer output moisture plotted against that predicted by the Neural Network. The error was consistently within 1.5% in terms of moisture which was found to be perfectly adequate for this application since a +/- 2% quality tolerance is required.

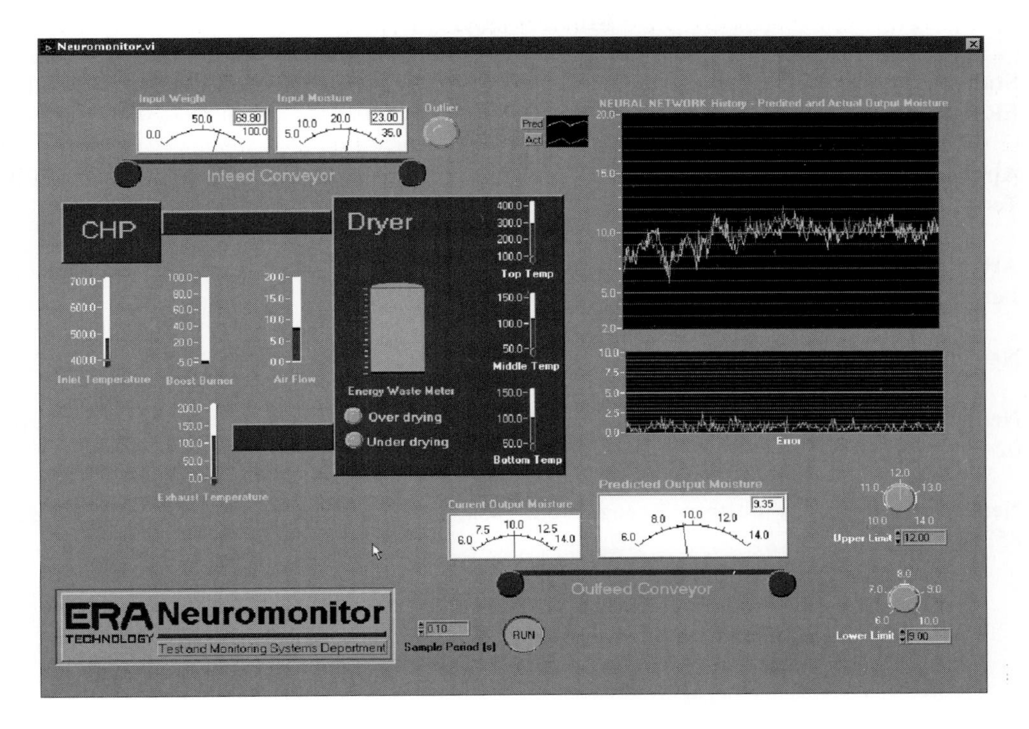

Figure 6: The Neuromonitor System in Action

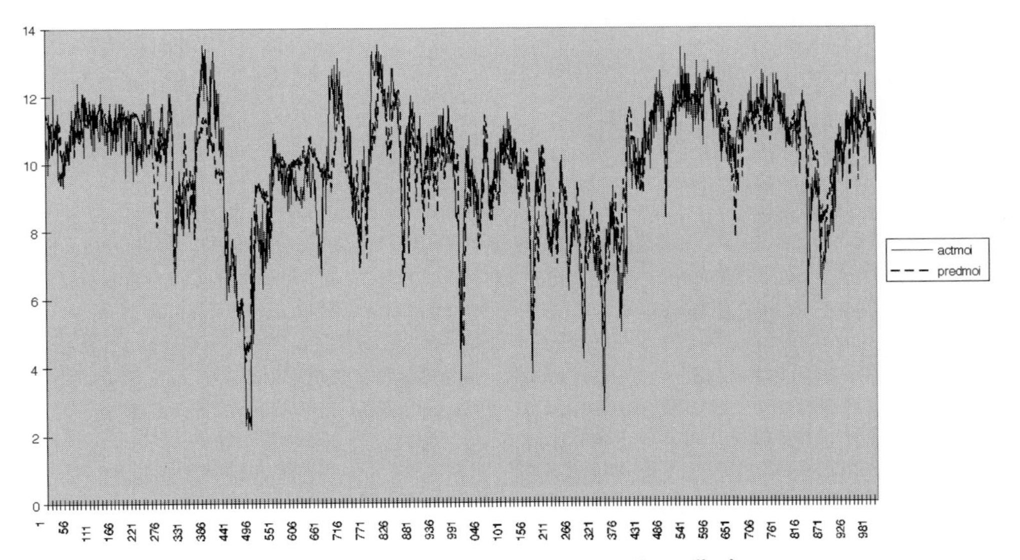

Figure 7: Trial results showing model prediction

BIBLIOGRAPHY

State-of-the-art Review in Neural Computing Research, Palmer, ERA Report 94-0418, ERA Technology Ltd, Leatherhead, 1994.

Applications of Artificial Neural Networks, Bjorkegren, ERA Report No.92-0871R, ERA Technology Ltd, Leatherhead, 1992.

Artificial Neural Networks - A Tutorial, Amin, IEEE Communications Conference, Jamaica, 1995.

Neural Computing: An Introduction, Beale and Jackson, IOP Press, ISBN 0-85274-262-2.

Neural Networks: A Comprehensive Foundation, Haykin, Macmillan Publishing, ISBN 0-02-352761-7.

Neural Networks for Pattern Recognition, Bishop, Oxford University Press, ISBN 0-19-853864-2.

Microlayer Coextrusion: Processing and Applications

C. Mueller, J. Kerns, T. Ebeling, S. Nazarenko, A. Hiltner and E. Baer
Department of Macromolecular Science and
Center for Applied Polymer Research
Case Western Reserve University
Cleveland, OH 44106-7202

ABSTRACT
New processing technologies are required for creating the engineered microstructures and nanostructures of the future. Numerous examples from the literature illustrate how the coextrusion of film with three or more polymeric layers is economically used to achieve a desirable mix of end-use characteristics. With layer-multiplying technology, two coextruded polymers can be multiplied into tens or thousands of alternating layers. In this way, polymers of widely dissimilar solid state morphologies and properties can be combined into unique layered and gradient structures. Two examples illustrate how microlayering can be used to obtain designed architectures by orientation and alignment of anisotropic particles in a polymeric matrix. The versatility of the microlayer process has recently been expanded by modification of the 2-component system to incorporate a third polymeric component at each interface. This is illustrated by insertion of a tie layer of maleated polypropylene between alternating layers of polypropylene and polyamide.

INTRODUCTION

Laminated steels have been known to exist as early as the first millennium B.C., but probably the most famous example is the Japanese sword. These blades were exceptionally strong in compression - that is, they were hard enough to retain a sharp cutting edge - but they were also tough enough to absorb blows in combat without breaking. The supertough and superplastic properties of contemporary laminated steels are achieved with laminates with as many as 2500 layers and layer thickness on the order of 1-5 microns [1,2]. Layered composites analogous to the laminated steels in terms of both the number of layers and the thickness of individual layers can also be fabricated from polymeric materials.

Continuous layer-muliplying coextrusion is a method by which two or more dissimilar polymers are combined into microlayer and nanolayer laminates with thousands of layers [3]. Recent studies of these unique materials have revealed dramatic improvements in toughness and impact resistance when the layers are ultra-thin due to the synergistic interactions of the key properties of the constituent components [4,5]. For example, a ten-fold enhancement in toughness and impact strength is achieved in microlayers of polycarbonate (PC) with brittle styrene-acrylonitrile copolymer

(SAN) when the size of the layers is reduced from the macroscale (tens of microns) to the microscale (several microns) [6]. This dramatic improvement in energy absorbing capability is attributed to fundamental changes in the micro-mechanical deformation modes in the composite that result from decreased layer thickness.

As the individual layer thickness is further decreased to tens of nanometers, the layer thickness is on the same size scale as the dimensions of the polymer molecules. In this unique instance, unusual structures result, especially if one of the components is crystallizable [7]. If the layers are thin enough, the interfacial regions totally dominate the bulk behavior. The large interface to volume ratio of nanolayered polymers facilitates fundamental studies of interfacial phenomena such as adhesion, interdiffusion, interfacial reactions, and surface-nucleated crystallization. Furthermore, new classes of materials with unique properties are possible with nanolayering. Because the coextrusion process requires stringent laminar flow conditions and short processing times, two miscible polymers can be brought into intimate contact with minimal mixing. Subsequently, gradient compositions can be produced by controlled interdiffusion [8-10].

The complex behavior of microlayered systems must be analyzed in terms of interactive hierarchical levels of structure. In particular, through analysis of the effects of layer thickness and adhesion between layers, further structure and property enhancement can be achieved. "Microprocessing" can also be the basis for creating "smart skins", film and sheet materials with multiple functions. The feasibility of creating tough, transparent layers that filter radiation of specific wavelengths and additionally possess barrier characteristics or unusual electronic properties seems possible. As the layer thickness is decreased further to the nanoscale, where the layer thickness is on the same size scale as the dimensions of the polymer molecule, new phenomena can be expected. Interesting effects may be achievable by "nanoprocessing" of polymers if the high interface to volume ratio is exploited. It is possible to imagine epitaxial effects especially if one or both polymers are crystallizable, high degrees of orientation not normally achievable during processing, and unusual interfacial properties if the coextruded polymers are partially miscible.

New opportunities for development and application of layer-multiplying technology are discussed in this paper. The process for microlayering two polymeric components is described, and examples of designed structures with oriented and aligned anisotropic particles that serve a mechanical or electrical function are presented. A recent modification to the layer-multiplying technology makes it possible to incorporate a third polymeric component as a thin "tie" layer at each interface. The potential of this new capability is illustrated by insertion of an adhesive layer between layers of two incompatible polymers.

COEXTRUSION OF POLYMER MICROLAYERS

Microlayers are comprised of alternating layers of two or more components with individual layer thicknesses ranging from the macroscale to the nanoscale. Typically, the total number of layers ranges from tens to thousands. By varying the melt feed ratio, the final sheet or film thickness, and the number of layers, the individual layer thicknesses can be precisely controlled. The coextrusion system used to prepare microlayered and nanolayered materials on an experimental scale consists of two 3/4 inch single screw extruders with melt pumps, a coextrusion block, a series of layer multiplier elements, and a tape or film die, Figure 1a. Metering pumps control the two melt streams that are combined in the feedblock as two parallel layers. From the feed block, the two layers flow through a series of layer multiplying elements; each element doubles the number of layers. In each element, the melt is first sliced vertically, then spread horizontally and finally recombined. An assembly of n multiplier elements produces an extrudate with the layer sequence $(AB)_x$ where x is equal to $(2)^n$, Figure 1b. The microlayer can be in the form of 1-2 mm thick tape, or film as thin as 25 μm. Microlayers with up to 4096 layers and individual layer thcknesses less than 100 nm have been successfully produced with this system.

As with any coextrusion, the quality of coextruded microlayers depends on the viscosity ratio of the components. The material with the lowest viscosity will tend to encapsulate the other so that it forms a slip film between the high viscosity component and the wall [11]. In addition to encapsulation, interfacial instabilities can occur as the melt streams come together if the viscosity mismatch is large [12]. To minimize these effects, the viscosities of the components should be as close as possible. Viscosity effects on layer structure were investigated by microlayering polycarbonate (PC) and poly(methylmethacrylate) (PMMA). The PC was Calibre 200-10 (Dow). Two PMMA resins were used: VM-100 (Ato-Haas) (PMMA1) with molecular weight 95,000 and V826-100 (Ato-Haas) (PMMA2) with molecular weight 135,000. Viscosity was determined as a function of temperature using a Kayeness Galaxy 1 melt flow indexer at a shear rate of 10 sec^{-1}. The melt flow rate was measured according to ASTM D1238-90B. Viscosity was calculated as the shear stress at the wall divided by the shear rate at the wall.

The viscosities, shown as a function of temperature in Figure 2, were used to identify temperatures where the materials would have the same viscosity. The viscosities of PMMA1 and PC matched at 241° and 292°C, respectively, a difference of about 50°C. For PMMA2 and PC, the viscosities matched at 271° and 292°C, respectively, a difference of only about 20°C. To coextrude microlayers of PC and PMMA, the extruder and melt pump temperatures were adjusted to these temperatures so that the viscosities matched as the melts entered the feedblock. The temperature of the feedblock, multiplying elements, and die was held at an intermediate temperature, 280°C. The feed ratio was 1:1 and 5 elements were used to produce microlayers with 64 layers.

To observe the layer structure, the samples were sectioned, polished with wet sandpaper on a metallurgical wheel, and viewed in the optical microscope. The layer structures of the two PC/PMMA microlayers are compared in Figure 3. These micrographs show a region of the cross-section near the midpoint in the width. Layer uniformity was much poorer in the microlayer with the lower molecular weight PMMA. Only 58 of the 64 layers could be distinguished in the PC/PMMA1 microlayer. Furthermore, with a feed ratio of 1:1, each layer should have had the same thickness; however, there were large variations in layer thickness. In contrast, all 64 layers could be identified in the PC/PMMA2 microlayer, and the layer-to-layer thickness was much more uniform than in the PC/PMMA1 microlayer.

The difference in microlayer quality was attributed to the smaller viscosity mismatch of PMMA2 compared to PMMA1. Although the temperatures in the extruders were adjusted so that both components would have the same viscosity when they entered the feedblock, it was anticipated that a viscosity mismatch would develop as the melts equilibrated to the temperature of the feedblock and multiplier elements. Thus the viscosity of the PC would increase and the viscosity of the PMMA would decrease. The magnitude of the mismatch was much larger for the lower molecular weight PMMA2, where there was about a 50°C temperature difference in the two melts as they entered the feedblock, than for PMMA1 where the temperature difference was only about 20°C. The higher viscosity PMMA1 would have had less of a tendency to flow to the edges during layer multiplying. This resulted in less variation in layer thickness, and also less of a tendency to encapsulate PC at the edges.

MICROLAYERS WITH DESIGNED ARCHITECTURES

Microlayering is an attractive approach for creating designed architectures from particulate-filled polymers. If the particles are anisotropic, for example platelets, flakes, tubes or short fibers, the geometric constraints imposed by layer multiplying ensure orientation of the particles in the plane of the layers. Furthermore, filled and unfilled layers can be alternated. For example, if the filler is added for stiffness, the toughness can also be enhanced by alternating a ductile layer to arrest cracks. A more sophisticated version of this approach is used by nature in the construction of sea shell nacre which consists of alternating nanolayers of high modulus inorganic and ductile organic [13]. If the objective is anisotropic electrical properties, conducting layers filled with metallic particles can be alternated with unfilled insulator layers. Both opportunities were investigated.

Design for Ductility

Microlayered structures were prepared with polypropylene Pro-Fax 6523 (Himont) as the matrix polymer and talc flakes (Jet Fil 500) as the particulate. The talc particles had a thickness of about 1 μm and an average aspect ratio of 10. To create structures with uniformly dispersed, aligned particles, talc-filled polypropylene (F) was used as both components in the coextrusion of microlayers with 256 layers. The talc content of these F-F microlayers was varied up to 40% by weight. For comparison, the talc-filled polypropylene was also compression molded at 190°C with approximately the same thickness (2mm) as the microlayered samples. Typically a filler is added for stiffness, and accordingly both compression molded and microlayered samples of talc-filled PP exhibited a linear increase in tensile modulus with increasing talc content. Relative to unfilled PP with the same process history, microlayering produced a larger increase in modulus than compression molding because of particle alignment.

The increased stiffness of talc-filled PP is usually achieved at the expense of toughness. This effect was demonstrated by compression molded, talc-filled PP. Without filler, PP was ductile in uniaxial tension and deformed by necking and stable neck propagation with strain hardening after the neck had propagated the length of the gauge section. However, with as little as 10% by weight talc, PP lost its ductility and fractured during formation of the neck, Figure 4. In contrast, microlayered PP with 10% by weight talc was almost as ductile as unfilled PP, and even with 20% talc a stable neck formed and propagated some distance along the gauge section before it fractured. For a filled polymer with poor adhesion, the ductility is determined by the amount of load-bearing matrix in the cross-section [14]. When this drops below the critical value needed to sustain the local stresses during drawing, the ductility is lost. The increased ductility achieved by microlayering resulted from particle alignment. In this configuration, the amount of the load-bearing PP in the cross-section was higher than if the particles were randomly oriented.

Another unique structure was created by alternating talc-filled PP layers (F) with unfilled PP layers (U) in a 1:1 ratio. The talc content of the F layer was maintained at 40% by weight, and the thickness of the alternating filled and unfilled layers was gradually decreased by increasing the number of layers from 16 to 1024. To observe the layer morphology, the microlayers were cryo-fractured through the cross-section, coated with 90Å of gold and examined in the JEOL JSM 840A scanning electron microscope. Two examples are shown in Figure 5. The alternating layers of talc-filled PP and PP were clearly discernable if the number of layers was 256 or less. The particles were uniformly distributed in the filled layers and showed preferential orientation in the plane of the layers. The quality of the particle orientation improved significantly as the layers were made thinner. The layers were difficult to differentiate in the 512 and 1024 layer microlayers because the particle thickness was comparable to the layer thickness.

The F-U microlayers had a net talc content of 20% by weight, which was the transitional composition for the F-F microlayers. By alternating a filled layer with an unfilled layer, PP with 20% talc was ductile, Figure 6. The fracture strain of the F-U microlayers gradually increased as the layers were made thinner to a maximum when the layers were in the range of 5-15 μm. A further decrease in the layer thickness to 1 μm resulted in decreased fracture strain. The optimum layer thickness was thought to occur when the unfilled layer was just thick enough to inhibit propagation of microcracks that originated in the talc-filled layers. With a low number of layers, poor particle orientation significantly increased the probability of crack initiation. Furthermore, in the thick layers the crack had the opportunity to grow to a critical size before it encountered a ductile PP layer. With a large number of very thin layers, particle agglomeration was observed. These agglomerates acted as stress concentratiors and accelerated crack initiation.

Design for Conductivity

The ability of microlayering to "organize" anisotropic particles was also used to obtain nickel-filled polypropylene with highly anisotropic electric characteristics. The conductive nickel flakes (Novamet) had an average thickness of 1 μm and an aspect ratio that ranged from 1 to 20. Blends of polypropylene with 5 to 25% by volume nickel flakes were prepared in a Haake Rheometrix 600 by mixing at 190°C for 8 min at 60 rpm under dry nitrogen. All blends were compression molded into 2 mm thick plaques at 190°C. Microlayers of polypropylene with 15% by volume nickel (F-F) were coextruded with 128 layers, and microlayers with alternating layers of polypropylene with 15% nickel and unfilled polypropylene (F-U) were coextruded with 64, 128 and 256 layers. To measure resistivity, contact faces were polished and a 500Å gold layer was deposited for good electrical contact. A resistivity test cell was equipped with a series of batteries so that the voltage could be varied up to 40V. The current through the specimen was measured as a function of applied voltage. All conductors were isolated with Teflon standoffs and the cell was electromagnetically shielded. Nonlinear behavior was observed in all specimens when the voltage was applied for the first time. The nonlinear effect was especially noticeable in blends with low nickel content. In these cases, the behavior resembled electrical break-down where the current changed very slowly with applied voltage until the electric field strength reached a critical level at which the current increased dramatically indicating a significant drop in resistivity [10]. In all cases, the current-voltage relationship followed Ohm's law in the second and succeeding applications of the voltage and the constant slope was used to calculate the resistance.

With low nickel content, the compression molded blends were virtually nonconductive. However, at a critical volume fraction of about 8%, the resistivity fell sharply to 10^2-10^3 Ohm-cm which corresponded to the resistivity of a good semiconductor, Figure 7. Increasing the nickel content further did not appreciably reduce the resistivity. Assuming that the critical volume fraction corresponded to a percolation threshold, a value of 8% was significantly lower than the 35% that was reported for spherical particles [11]. The greater specific surface area of platelet particles compared to spheres, which increased the probability of particle-particle contacts, could have accounted for the low percolation threshold. Particle-particle contacts were enhanced when particles were oriented by microlayering. Orientation decreased the in-plane resistivity of the F-F microlayers with 15% nickel by an order of magnitude compared to the compression molded blend with the same composition. The orientation also produced anisotropy in the resistivity. The cross-plane resistivity was higher than the in-plane resistivity, and matched that of the compression molded blend.

Anisotropy in the resistivity was magnified to ten orders of magnitude when layers of insulating PP were alternated with layers of nickel-filled polypropylene (U-F). The microlayer with 64 alternating layers and a total nickel content of 7.5% by volume was a semiconductor in-plane, with a resistivity comparable to nickel-filled PP above the percolation threshold, and cross-plane it was an insulator with resistivity similar to that of unfilled PP, Figure 7. No conductivity was detected for the U-F microlayers with 128 and 256 layers. It was possible that the more three-dimensional arrangement of particles in the the thicker layers facilitated formation of a continuous conductive network of particles within the filled layers.

MICROLAYERS WITH A TIE LAYER INCORPORATED

The versatility of the microlayer technology has been expanded with a second coextrusion system that adds a third polymer between layers of polymer A and polymer B. This component may be added for certain desirable properties such as barrier, strength, or adhesion. Insertion of the tie layer (T) at each interface is accomplished by extruding three polymers into a feed block that combines the melts into five layers with the sequence ATBTA. Typically, the thickness of the T layers is one tenth that of the A and B layers. The five layer melt is then fed into the layer-multiplying dies which repeatedly cut, spread and recombine the melt with a layer sequence $(ATBT)_xA$ where x is equal to $(2)^n$ for an assembly with n multiplier elements. An example of the 3-component microlayer is shown in Figure 8. The three components are all the same polyethylene resin but with color added to make the layers distinguishable. This example contains 33 layers (n = 3) and the ATBT ratio is 9:1:9:1.

The ability to insert a thin tie layer at each interface between dissimilar polymers A and B by microlayering creates opportunities for efficient incorporation of compatibilizing agents. When a compatibilizer is dispersed by the conventional method of melt blending, there is no guarantee that the compatibilizer will be efficiently distributed at the interface. This can be of particular concern because the compatibilizer is frequently costly. As a result, the molecular weight of the compatibilizer may be less than optimum so that it can diffuse easily to the interface. It is also the case that fundamental understanding of compatibilization is difficult to achieve with conventional melt blended materials when it is uncertain how much of the compatibilizer is actually at the interface. Furthermore, the adhesive strength can not be measured directly in melt blends.

The high surface to volume ratio attainable with microlayers makes them ideal for studying interfacial phenomena related to polymer blends. As an example, polyamide66 and polypropylene were microlayered with maleated polypropylene as the tie layer (T). This is a blend system of commercial importance because it combines the high ductility of polypropylene with the relatively high yield strength of polyamide. However, because PP and PA are immiscible and incompatible, the blends require a compatibilizer if good dispersion and good mechanical properties are to be achieved [17-20]. This is typically a maleated polypropylene which reacts with PA end groups *in situ* to form a block copolymer [21]. The resulting block copolymer acts as an emulsifier to decrease the interfacial tension and reduce the tendency of the dispersed particles to coalesce during processing. In the solid state, the graft copolymer promotes adhesion between the phases.

The PA was Zytel 42 polyamide 6,6 (Dupont) and the PP (Himont) had a melt flow index of 14. Maleated polypropylenes (PP-g-MA) were Polybond 3002, 3150, and 3200 (Uniroyal) with 0.2%, 0.5%, and 1.0% maleic anhydride by weight, respectively, and with molecular weights of 440,000, 330,000, and 110,000, respectively. Microlayers were prepared without a tie layer by shutting off the tie layer extruder. These had 33 layers (n = 4 without T layers) and a composition of 80:20 PP:PA. The three-component microlayers had 65 layers (n = 4) and a composition of 76:5:19 PP:PP-g-MA:PA. The average layer thicknesses were calculated to be 60 μm for PP, 15 μm for PA and 2 μm for the tie layer.

The interfacial strength was measured by the T-peel test (ASTM D1876). Specimens 15 to 25 mm wide were notched with a fresh razor blade and loaded at a rate of 2.0 mm/min. Some of the tests were interrupted and the crack tip region was sectioned with a low speed diamond saw. The sections were polished on a metallurgical wheel with wet sandpaper and alumina oxide aqueous suspensions, then examined in the optical microscope. The fracture surfaces were coated with 90Å of gold and examined in the scanning electron microscope.

Normalized load versus displacement curves showed an initial increase in the load as the specimen arms bent into the T-peel configuration. The load increased until a crack started to propagate and then remained constant as the crack grew. The peel strength varied by more than two orders of magnitude depending on the MA content of the tie layer, Figure 9. Microlayers without a tie layer fell apart without application of a peel force. With 0.2% MA in the tie layer, a small peel force of about 8N/m was required to propagate the crack. This increased to approximately 1000N/m when the tie layer contained 0.5% or 1.0% MA.

Matching fracture surfaces from the microlayer without compatibilizer indicate brittle fracture with no evidence of interfacial deformation before separation. The spherulitic morphology of PA is visible, Figure 10a. The matching PP surface contains an imprint of the PA morphology, which formed while PP was still in the melt, with additional features of the PP morphology, Figure 10b. Matching fracture surfaces from the microlayer with the 0.2% MA tie layer also showed brittle fracture although there were occasional holes at spherulite boundaries on the PA surface, Figure 10c. These appeared to correspond with fibers on the PP surface which were located at the imprints of the PA spherulitic boundaries, Figure 10d. Similar holes and matching fibers have been reported on PP joined to LLDPE, and were attributed to influx of the lower-melting material along the spherulitic boundaries as the higher melting material crystallizes and contracts [22]. Apparently the small amount of graft copolymer formed with the 0.2% MA tie layer resulted in some PP being pulled into the PA as it crystallized. This imparted some level of interfacial adhesion.

Fracture surfaces with 0.5% and 1.0% MA in the tie layer exhibited extensive plastic deformation. The PA surface contained many small fibrils, Figure 10e. The PP surface had the fibrillated texture characteristic of craze fracture, Figure 10f. Examination of the crack tip region showed extensive crazing in the PP layer ahead of the crack tip. Clearly the interfacial strength was high enough with these tie layers for crazes to initiate in the PP layer before the interface failed. The delamination crack propagated through the crazes, leaving short PP fibrils adhered to the PA surface. The craze fracture mode resulted in the very high peel strength measured for these microlayers.

SUMMARY

Microlayer coextrusion technology has applications in many areas. It is capable of producing laminates with up to thousands of layers with layer thicknesses in the micro- and nanoscales. The components in such structures often combine synergistically to produce novel composite materials with improved mechanical, optical, barrier and electrical properties. This technology also creates opportunities for fundamental research in the areas of adhesion, diffusion and crystallization. The symmetry and extremely large specific interfacial area of microlayers make them attractive as model systems for a broad range of problems in polymer science.

ACKNOWLEDGMENTS

This research was generously supported by the Army Research Office, grant DAAL03-92-G-0241 and the National Science Foundation, grant DMR-9400475.

REFERENCES CITED

1. O. D. Sherby and J. Wadsworth, "Damascus Steels", *Scientific American*, **252(2)**, 112 (1985).

2. J. Wadsworth, D. W. Kum and O. D. Sherby, "Welded Damascus Steels and a New Breed of Laminated Composites", *Metal Progress*, June, 1986, pp.61-67.

3. C. D. Mueller, S. Nazarenko, T. Ebeling, T. L. Schuman, A. Hiltner and E. Baer, "Novel Structures by Microlayer Coextrusion - Talc-Filled PP, PC/SAN, and HDPE/LLDPE", *Polym. Eng. Sci.*, **37**, 355 (1997).

4. M. Ma, K. Vijayan, J. Im, A. Hiltner and E. Baer, "Thickness Effects in Microlayer Composites of Polycarbonate and Poly(styrene-acrylonitrile)", *J. Mater. Sci.*, **25**, 2039 (1990).

5. K. Sung, D. Haderski, A. Hiltner and E. Baer, "Mechanisms of Interactive Crazing in PC/SAN Microlayer Composites", *J. Appl. Polym. Sci.*, **52**, 147 (1994).

6. J. Im, A. Hiltner and E. Baer, "Microlayer Composites" in *High Performance Polymers:* (E. Baer and A. Moet, eds.), Hanser, 1991, pp.175-198.

7. S. J. Pan, J. Im, M. J. Hill, A. Keller, A. Hiltner and E. Baer, "Structure of Ultrathin Polyethylene Layers in Multilayer Films", *J. Polym. Sci.: B: Polym. Phys.*, **28**, 1105 (1990).

8. G. Pollock, S. Nazarenko, A. Hiltner and E. Baer, "Interdiffusion in Microlayered Polymer Composites of Polycarbonate and a Copolyester", *J. Appl. Polym. Sci.*, **52**, 163 (1994).

9. D. Haderski, S. Nazarenko, C. Cheng, A. Hiltner and E. Baer, "Crystallization of a Copolyester in Microlayers and Blends with Polycarbonate", *Macromol. Chem. Phys.*, **196**, 2545 (1995).

10. S. Nazarenko, D. Haderski, A. Hiltner and E. Baer, "Concurrent Crystallization and Interdiffusion in Microlayers of Polycarbonate and a Copolyester", *Macromol. Chem. Phys.*, **196**, 2563 (1995).

11. W. Michaeli, *Extrusion Dies for Plastics and Rubber: Design and Engineering Computations*, Hanser Publishers, New York, 1992.

12. W.J. Schrenk and T. Alfrey, Jr., "Coextruded Multilayer Polymer Films and Sheets" in *Polymer Blends, Vol.2*, (D.R. Paul and S. Newman, eds.) Academic Press, 1978, pp.129-165.

13. M. Sarikaya, J. Liu and I.A. Aksay, "Nacre: Properties, Crystallography, Morphology, and Formation", in *Biomimetics: Design and Processing of Materials*, (M. Sarikaya and I.A. Aksay, eds.) AIP Press, 1995, pp. 35-90.

14. S. Bazhenov, J.X. Li, A. Hiltner and E. Baer, "Ductility of Filled Polymers", *J. Appl. Polym. Sci.*, **52**, 243 (1994).

15. A. Celzard, G. Furdin, J.F. Marêché, E. McRae, M. Dufort and C. Deleuze, "Anisotropic Percolation in an Epoxy-Graphite Disc Composite", *Solid State Comm.*, **92**, 377 (1994).

16. *Metal-Filled Polymers: Properties and Applications*, (S.K. Battacharya, ed.), Marcel Dekker, Inc., 1986.

17. J. Duvall, C. Sellitti, V. Topolkaraev, A. Hiltner, E. Baer and C. Myers, "Effect of Compatibilization on the Properties of Polyamide 66/Polypropylene (75/25 Wt/Wt) Blends", *Polymer*, **35**, 3948 (1994).

18. J. Duvall, C. Sellitti, C. Myers, A. Hiltner and E. Baer, "Effect of Compatibilization on the Properties of Polypropylene/Polyamide-66 (75/25 Wt/Wt) Blends", *J. Appl. Polym. Sci.*, **52**, 195 (1994).

19. Y. Takeda, H. Keskkula and D. R. Paul, "Effect of Polyamide Functionality on the Morphology and Toughness of Blends with a Functionalized Block Copolymer", *Polymer*, **33**, 3173 (1992).

20. R. Fayt, R. Jérôme, and Ph. Teyssié, "Interface Modification in Polymer Blends", in *Multiphase Polymers: Blends and Ionomers* (L. A. Utracki and R. A. Weiss, eds.), ACS Symposium Series 395, Washington DC, 1989, pp38-66.

21. F. Ide and A. Hasegawa, "Studies on Polymer Blend of Nylon 6 and Polypropylene or Nylon 6 and Polystyrene Using the Reaction of Polymer", *J. Appl. Polym. Sci.*, **18**, 963 (1974).

22. B.L. Yuan and R.P. Wool, "Strength Development at Incompatible Semicrystalline Polymer Interfaces", *Polym. Eng. Sci.*, **30**, 1454 (1990).

FIGURE CAPTIONS

Figure 1. Two component microlayer system: (a) Layout showing extruders, pumps, feedblock, multiplier dies and film die; and (b) schematic showing cutting, spreading and stacking of two components in the layer multiplying element.

Figure 2. Plot of PC and PMMA viscosities as a function of temperature with the melt temperatures and die temperature used during processing indicated.

Figure 3. Optical micrographs of PC/PMMA microlayers: (a) PC/PMMA1; and (b) PC/PMMA2.

Figure 4. Fracture strain as a function of talc content for compression molded talc-filled PP and F-F talc-filled PP microlayers showing the ductile-to-brittle transition.

Figure 5. SEM micrographs of F-U talc-filled PP/PP microlayers: (a) 16 layers; and (b) 128 layers.

Figure 6. Fracture strain as a function of average layer thickness for F-U talc-filled PP/PP microlayers.

Figure 7. Resistivity as a function of nickel volume content for compression molded nickel-filled PP and microlayers with nickel-filled PP.

Figure 8. Optical micrograph of a three-component microlayer with 33 layers comprised of red, clear and yellow polyethylenes.

Figure 9. Peel strength of three-component PP/PP-g-MA/PA microlayers as a function of anhydride content of the tie layer.

Figure 10. SEM micrographs of microlayer peel surfaces: (a) and (b) matching PA and PP surfaces without a tie layer; (c) and (d) matching PA and PP surfaces of the microlayer with 0.5%MA in the tie layer; and (e) and (f) matching PA and PP surfaces of the microlayer with 1.0% MA in the tie layer.

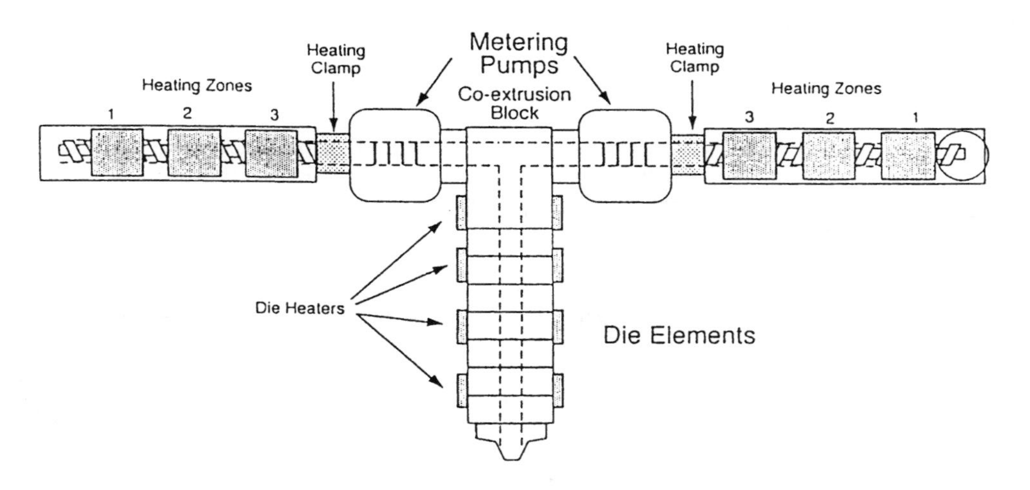
a
Heating Zones
1
2
3
Heating Clamp
Metering Pumps
Co-extrusion Block
Heating Clamp
Heating Zones
3
2
1
Die Heaters
Die Elements

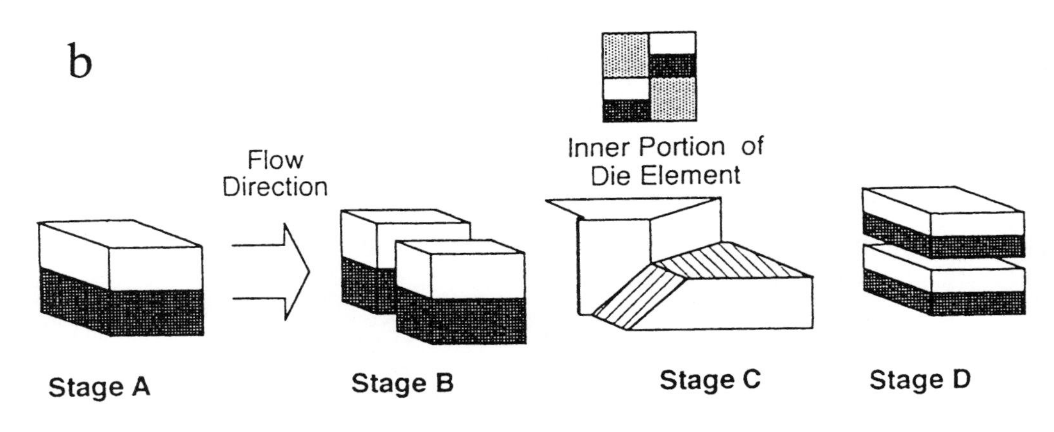
b
Flow Direction
Inner Portion of Die Element
Stage A
Stage B
Stage C
Stage D

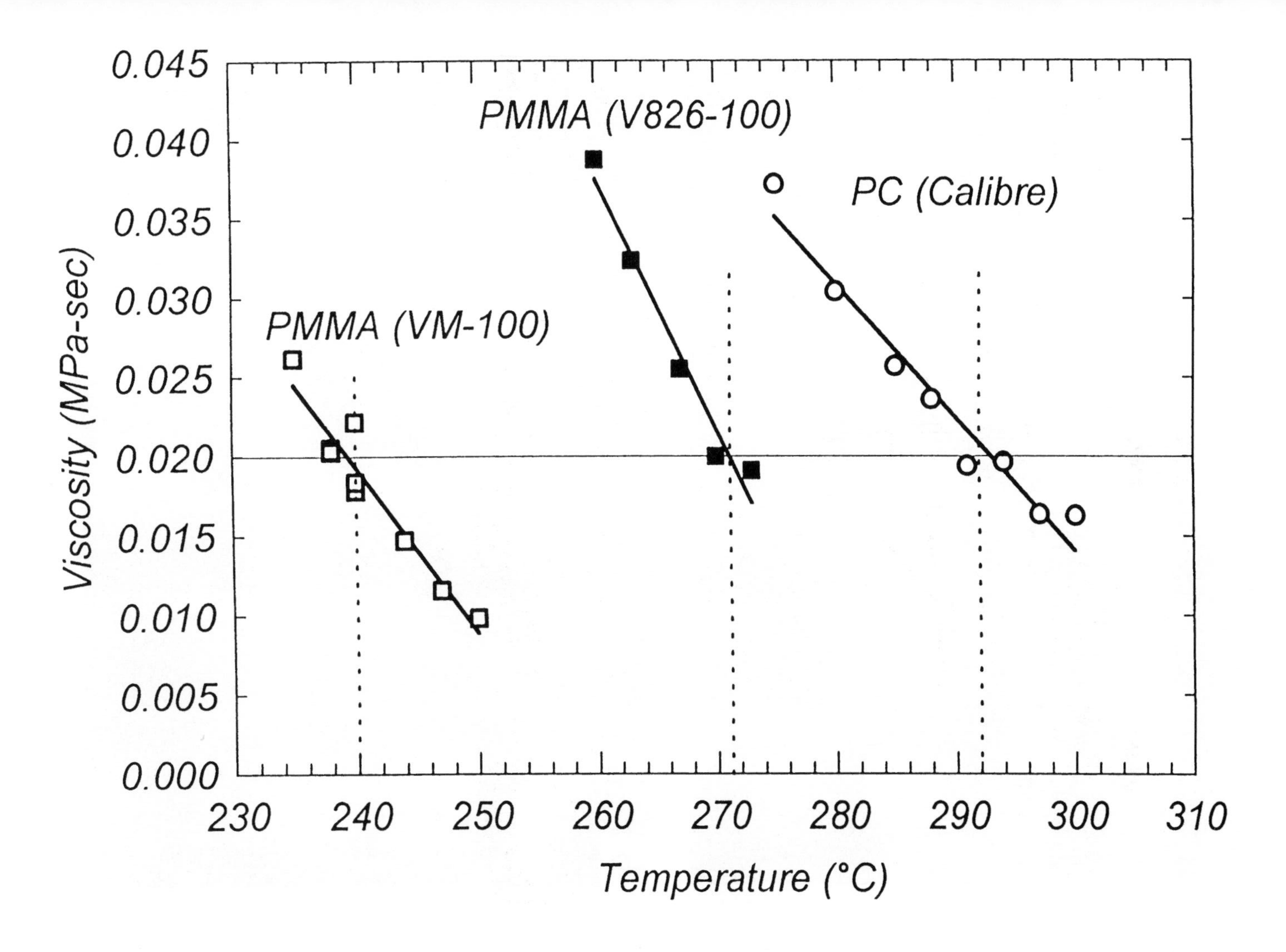
PMMA (V826-100)
PC (Calibre)
PMMA (VM-100)
Viscosity (MPa-sec)
0.045
0.040
0.035
0.030
0.025
0.020
0.015
0.010
0.005
0.000
230
240
250
260
270
280
290
300
310
Temperature (°C)

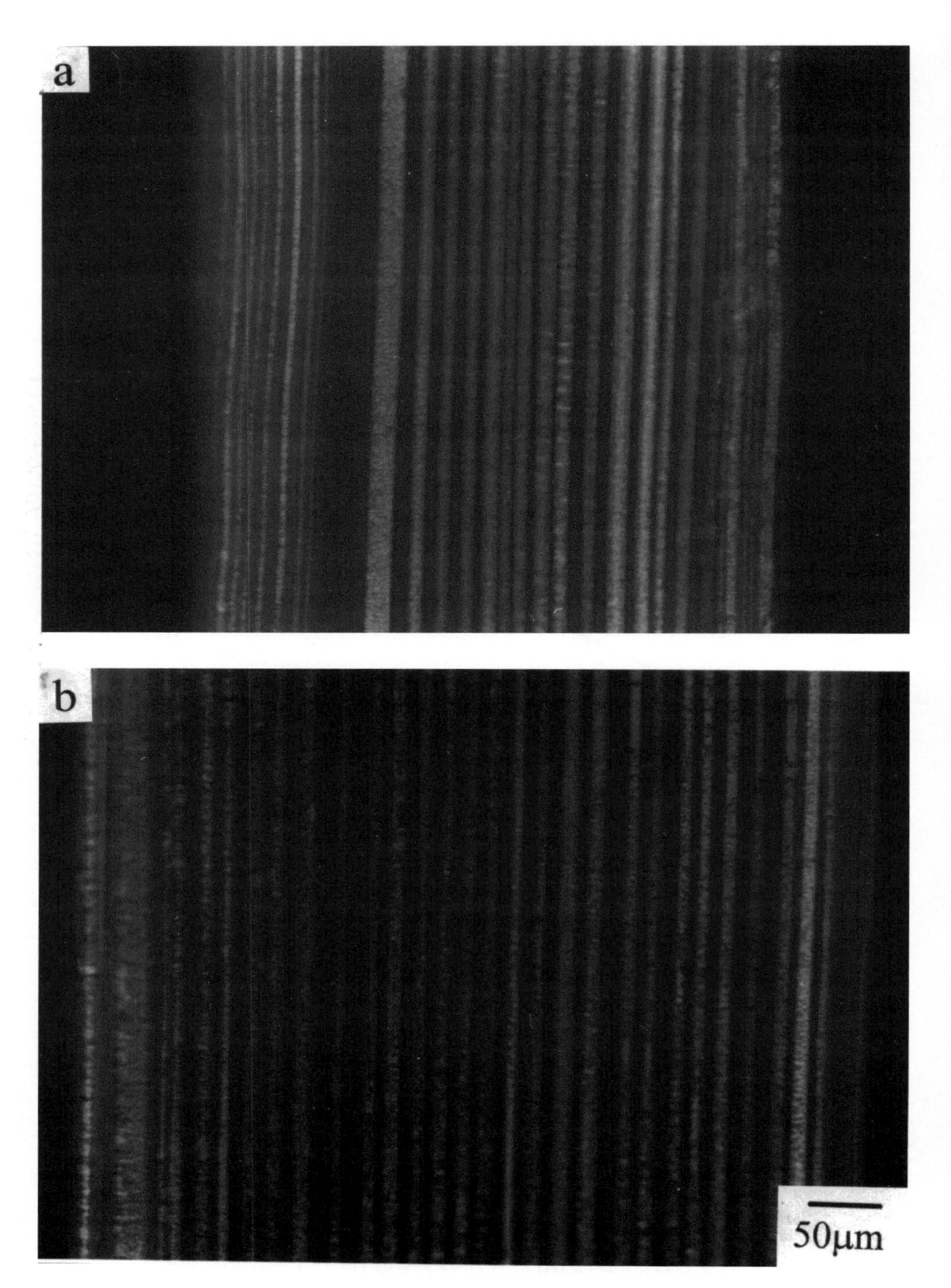
a
b
50μm

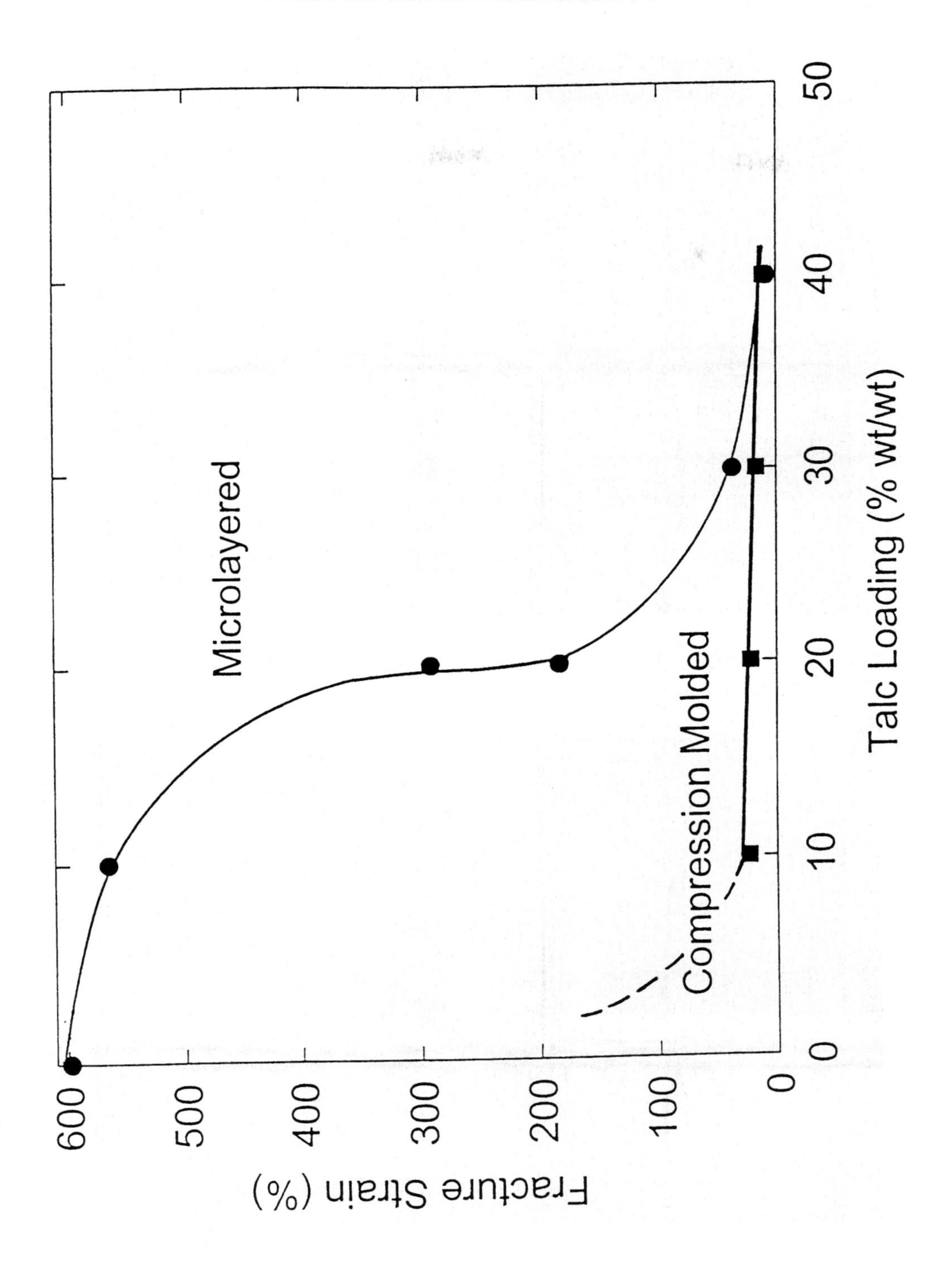
Microlayered
Compression Molded
Fracture Strain (%)
Talc Loading (% wt/wt)
600
500
400
300
200
100
0
0
10
20
30
40
50

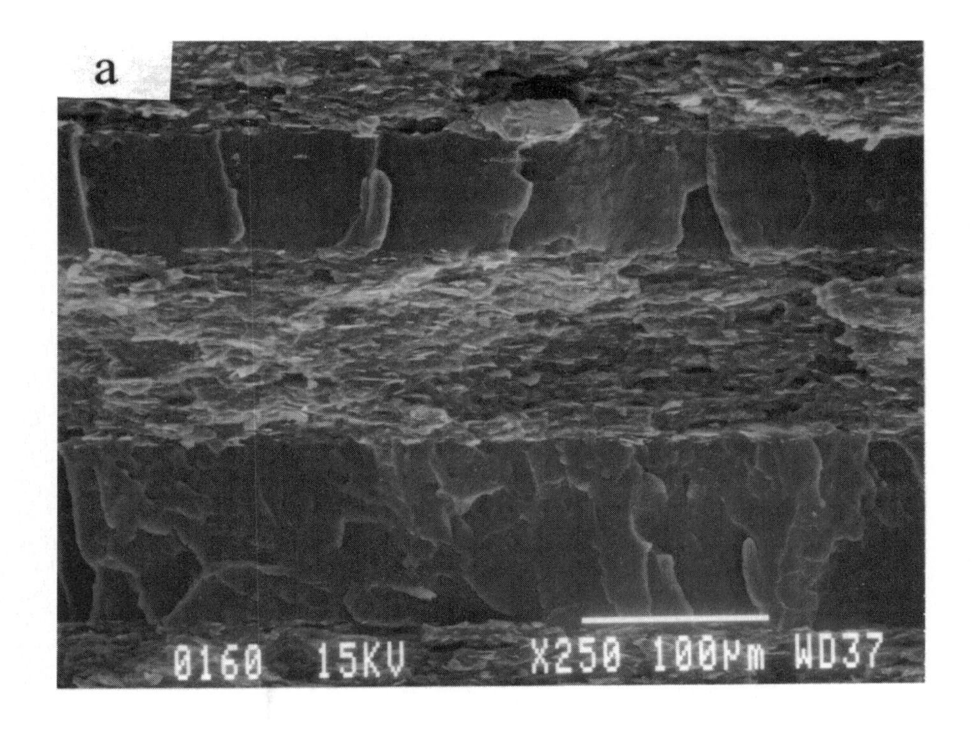
a
0160 15KV X250 100µm WD37

b
1280 15KV X250 100µm WD38

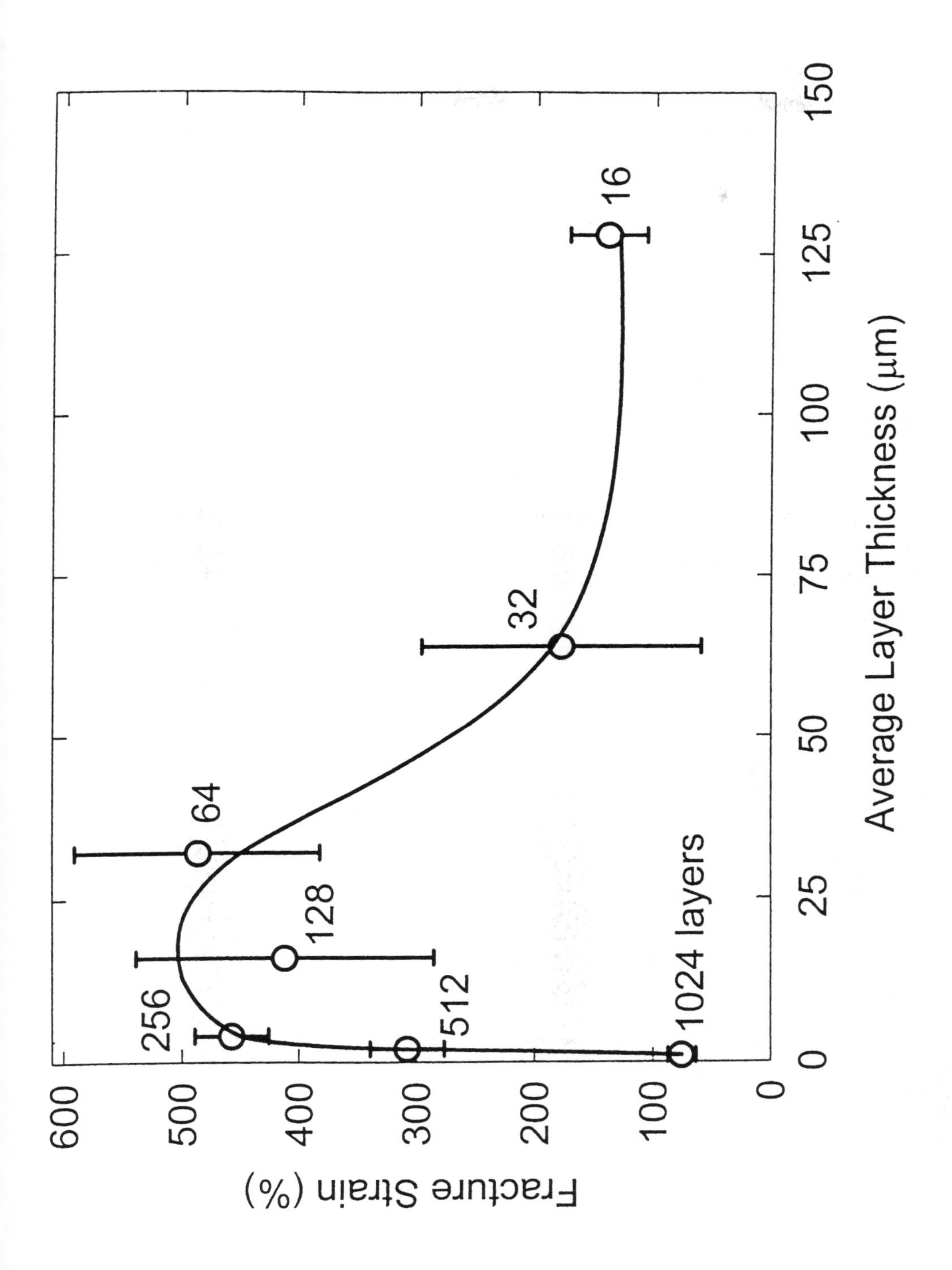
Fracture Strain (%)
600
500
400
300
200
100
0
0
25
50
75
100
125
150
Average Layer Thickness (μm)
1024 layers
512
256
128
64
32
16

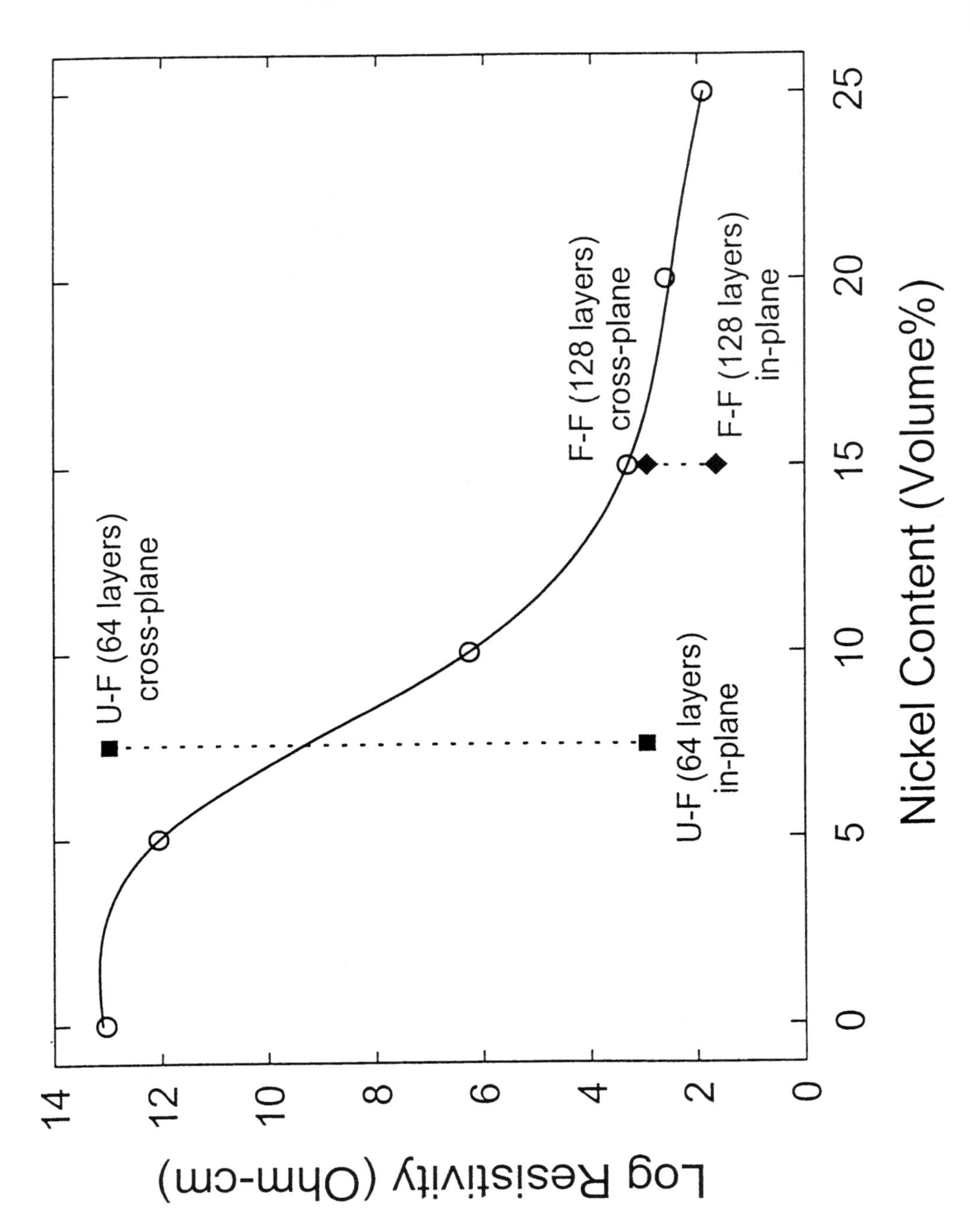
Log Resistivity (Ohm-cm)
Nickel Content (Volume%)
U-F (64 layers) cross-plane
U-F (64 layers) in-plane
F-F (128 layers) cross-plane
F-F (128 layers) in-plane
0
2
4
6
8
10
12
14
0
5
10
15
20
25

100µm

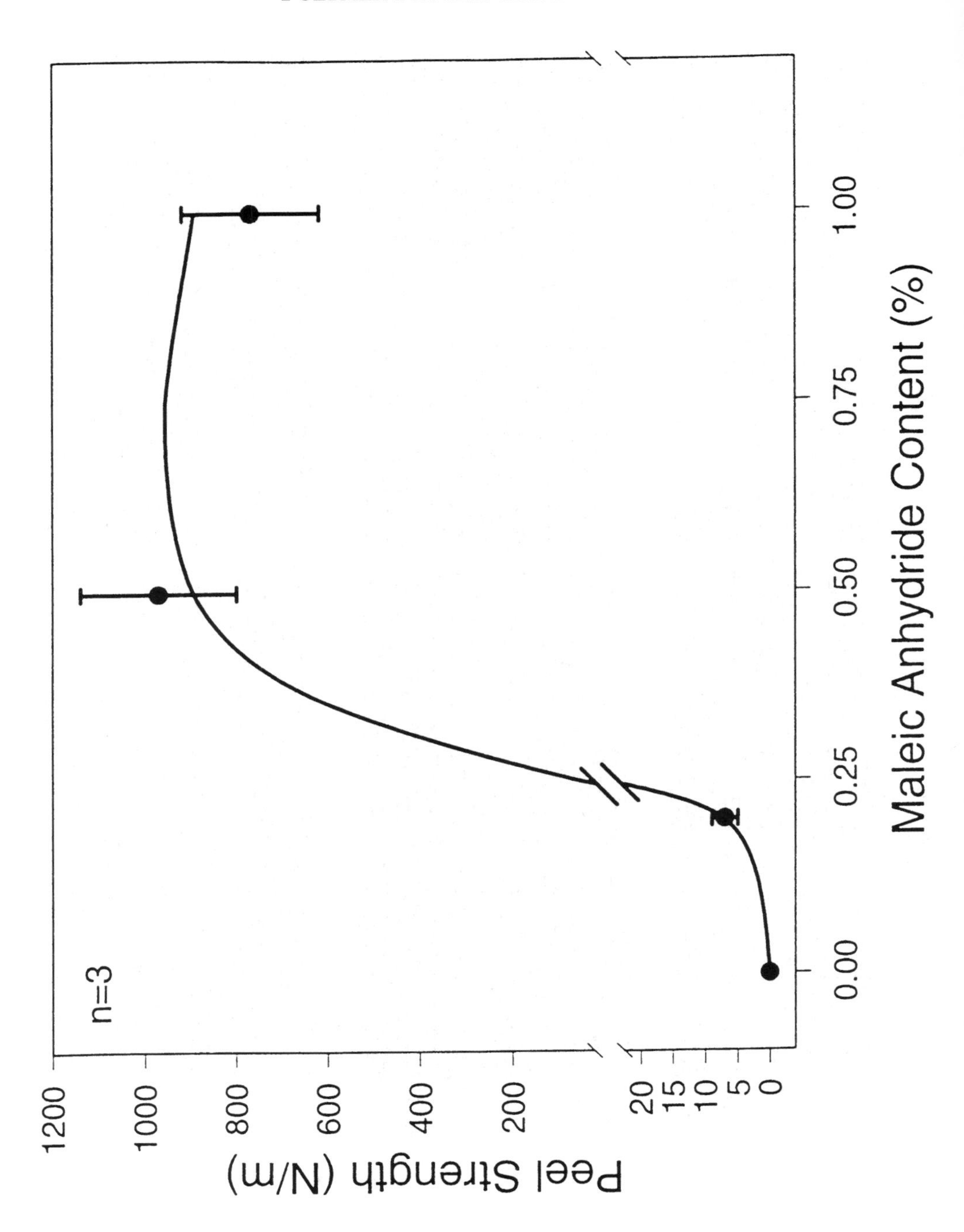
Peel Strength (N/m)
1200
1000
800
600
400
200
20
15
10
5
0
n=3
0.00
0.25
0.50
0.75
1.00
Maleic Anhydride Content (%)

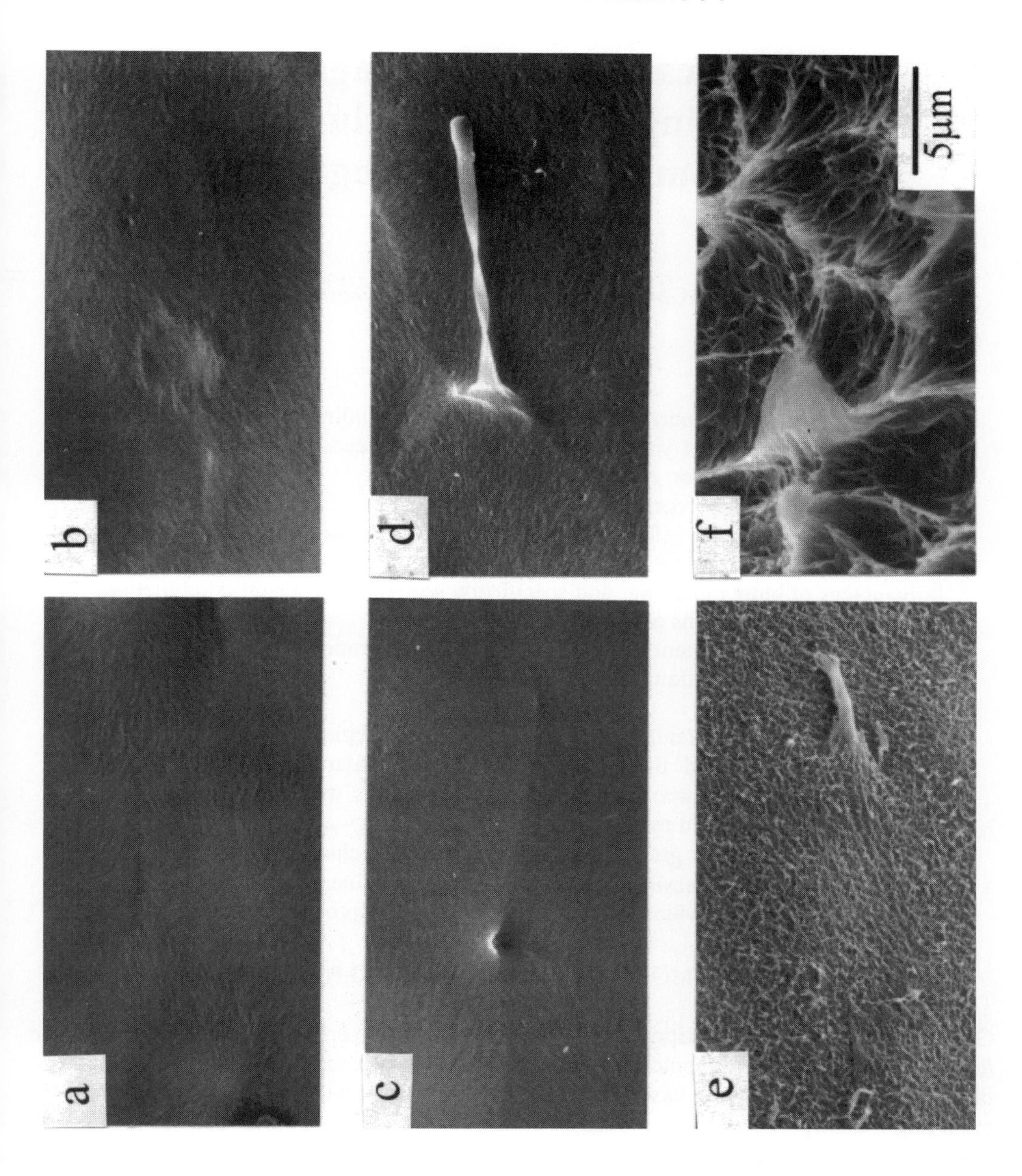
a
b
c
d
e
f
5μm

Technical Blow Mouldings from Engineering Polymers, including two component mouldings.

Peter W Deane
DuPont (UK) Ltd, Maylands Avenue, Hemel Hempstead, HP2 7DP

ABSTRACT
The use of engineering polymers in technical blow moulding has been growing in Europe over the last 3-4 years at a rate which far exceeds their growth in other processes such as injection moulding or extrusion.
This paper discusses recently developed technologies in the blow moulding industry, with particular reference to the fields of coextrusion and sequential blow moulding as applied to applications in the automotive area.
Advantages of blow moulding, and specifically sequential hard/soft combinations, are demonstrated in terms of:
- Rubber/metal replacement and multi-component integration
- Cost and weight reduction
- Recycling issues
Details are given of currently available material combinations and their respective properties, benefits and limitations. The use of materials in blow moulded automotive air ducts, turbo pipes and coolant pipes, as well as more traditional applications such as fluid reservoirs, is discussed.
Finally a summary is given of the basic polymer characteristics which are necessary for good processing in a blow moulding engineering resin, including a quantitative technique which was developed to compare processing performance.

WHY BLOW MOULD WITH ENGINEERING POLYMERS ?

Given the inherent characteristics of many engineering polymers which would tend to make them unsuitable for blow moulding - high processing temperatures, sharp melting points, rapid crystallisation etc. - it has been difficult to get these materials accepted in the blow moulding industry. However, three major factors have brought about a rapid change in the perception and use of engineering polymers in blow moulding.

Firstly, there has been the drive by car manufacturers to increase usage of plastics, particularly in under bonnet applications. Although this is mainly driven by cost (leading to requirements such as component integration and ease of assembly) there is also strong pressure to replace rubber (for recycling reasons) and metal (weight reduction and corrosion). By definition, because the environment conditions are particularly harsh - high temperatures, oils, fuels, road salt etc. - the material specifications invariably point towards engineering polymers such as Nylons, Polyesters, TPEs and various polymer alloys. Although injection mouldings are more normally associated with this trend, the argument applies equally well to hollow parts like air ducts, pipes and of course fluid reservoirs which are natural candidates for blow moulding.

Secondly, in recognition of the inevitable trend in demand for these high performance blow moulded components, the plastic raw material suppliers have responded by developing types and grades of engineering polymers which are tailored for blow moulding. The rheological characteristics which are required to meet the needs of this process are discussed in more detail later.

Finally, the technical innovation of blow moulding machine manufacturers has brought about significant new capabilities in blow moulding machines. Although originally developed mainly in Japan, we now have processes such as '3-D' and sequential blow moulding, where complex shapes in combination of 2 or more materials are possible.

BLOW MOULDING PROCESSES

The conventional blow moulding processes, which can be used for technical mouldings and which are well known and therefore not described in detail here, can be categorised as either

(a) Continuous extrusion - where the parison extrudes continuously and at a consistent rate from the die,
(b) Accumulator head - which involves a reservoir which is slowly filled and rapidly discharged with every moulding cycle
(c) Injection blow moulding where an injection moulded preform is subsequently pre-heated and then blown to its finished shape (a variation of this is the 'Pressblower' system where only the neck of the component is pre-moulded in the blow moulding machine).

A major disadvantage of these established processes is that they are not ideally suited to the blow moulding of long narrow components in 3 dimensions, such as air ducts, without producing excessive scrap and very long undesirable pinch-offs at the mould closing lines. This fact led to the development of the so-called '3-D' blow moulding process which essentially describes 3 different inventions, subsequently patented and commercialised by various machine manufacturers both in Japan and in Europe:

'Laydown' process

In this system [see Fig. 1] originally developed (with subtle differences) by EXCELL and PLACO CORPORATIONS in Japan, the parison is extruded vertically onto a horizontally fixed mould half, in such a way that the emerging parison is made to follow the path of the mould cavity by either moving the complete extruder head and die or alternatively moving the mould half (with extruder fixed). The parison is kept partially inflated with support air to prevent its collapse, until such time as the complete cavity has been filled, when the top half of the mould is closed over the lower half and the parison is fully inflated by inserting a blowing needle. The result is a moulding with virtually no scrap (except at each end) and no inherently weak "pinch-off".

The disadvantage of this process for engineering polymers is the relatively long time during which the parison is in contact with only one half of the mould, leading to premature freezing-off of the parison surface. However, this can be mainly overcome by use of high mould temperatures and specially developed resins which crystallise more slowly. There are several commercial applications for high temperature air ducts made with this process.

Movable Die Mould **Movable Die Head**

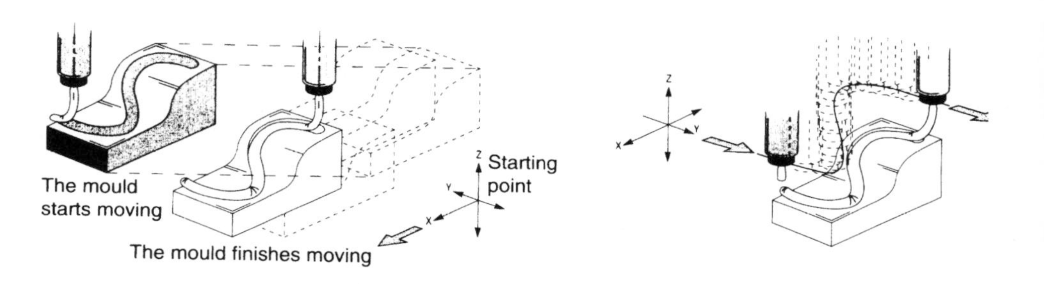

Figure 1. 3-D "Laydown" Process

"Parison Manipulation" Process

This technique [see Fig. 2] is a development of a conventional blow moulding (normally accumulator head) process, whereby the extruded parison is "manipulated" by a combination of robots and moving mould segments in order to make it conform to the 3-dimensional mould cavity. The parison is normally removed from the die by a robotic "gripper" which then positions it over the blow pin, or (in the case of subsequent needle-blowing) in one part of the multi-segment mould. Programmed movements of the robot arm and closing mould segments then position the parison in the cavity until finally the mould is completely closed and the parison is blown to produce the finished part. Although it may also exhibit some of the problems of the "laydown process" this process seems to be generally more suitable for more crystalline engineering polymers such as nylons.

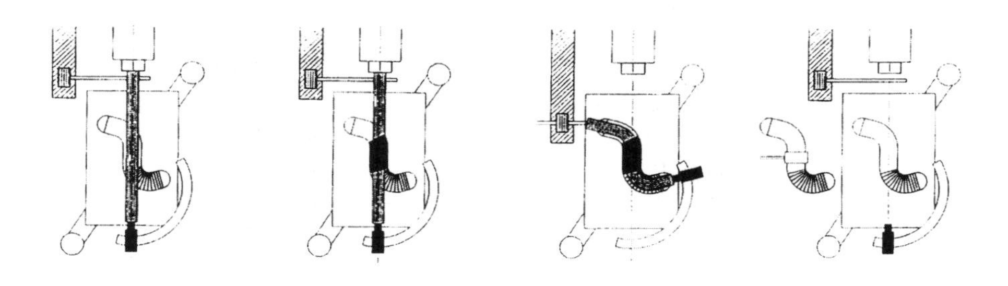

Figure 2. 3-D "Parison manipulation" Process

A variation on the "parison manipulation" theme is a machine which was developed anc patented by an Italian manufacturer, whereby the parison is extruded horizontally onto a conveyor belt, support air is used to maintain the parison shape, then after the complete parison is extruded the belt is stopped, the parison is "shaped" in two dimensions by special formers and finally the parison is dropped onto a horizontally fixed mould half (the conveyor having been automatically moved away). Gravity then forms it in the 3rd dimension, after which the top half of the mould is clamped over the lower half and the part is blown to its final shape. This machine also has the capability of sequential blow moulding to make hard/soft combinations. It would appear to be a relatively simple way to achieve parison manipulation without the complexity and associated cost of using a robot and segmented moulds.

3D "Suction" process

Here [Fig. 3] a basic accumulator head machine (with or without coextrusion/sequential facility) is used in combination with a specially designed mould having an air suction device connected to it.

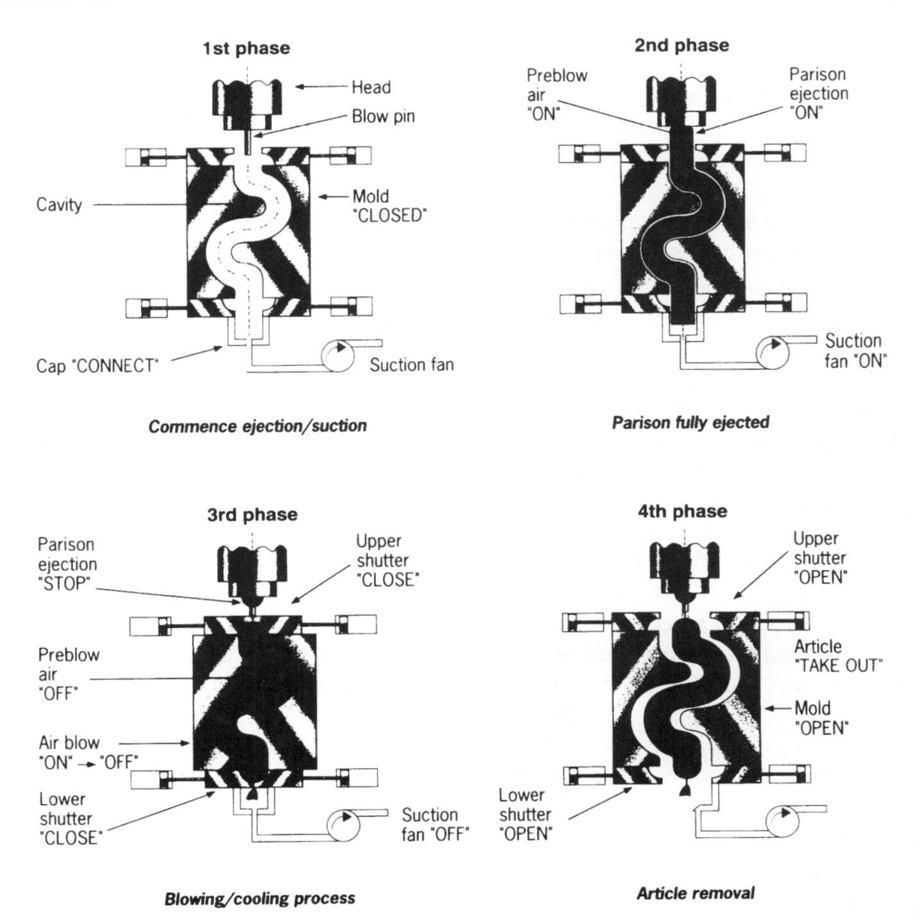

Figure 3. 3-D "Suction" Process

The process operates by extruding the parison into the cavity through an opening in the top of the closed mould, at the same time as a vacuum is applied to the lower end of the cavity. This suction and supporting air flow through the mould helps to pull and guide the parison until it reaches the lower end of the mould. At this point the parison is blown, either through a blow pin in the centre of the parison die, or by means of a needle which is inserted at some point in the parison.

This process is particularly suited to smaller diameter air ducts or pipes, especially where there is little change in diameter or cross section along the part length. However it does appear to have a practical limit to the angle and sharpness of bend that can be accommodated. In addition, it may be difficult to avoid some contact with the cavity wall which may affect its suitability for some engineering polymers.

ENGINEERING THERMOPLASTICS FOR BLOW MOULDING

Many types of stabilised engineering thermoplastics which are freely available for injection moulding have now been developed as blow mouldable grades. Of course there are many definitions of what exactly defines an "engineering thermoplastic" rather than a "commodity plastic". But for the purpose of this discussion, it is assumed that we are defining an engineering thermoplastic as one which can perform at a continuous temperature rating above 100°C and which can withstand a wide range of chemicals, oils, solvents etc. This generally implies a crystalline or semi-crystalline material, or an alloy with one of these materials.

Blow mouldable engineering plastics are therefore predominantly either Nylons, polyesters or certain thermoplastic elastomers.

Nylons

Several types of nylons (polyamides) and nylon alloys are now widely available in blow mouldable grades. For example it is possible to blow mould PA11, PA6 and PA6.6 types, as well as glass reinforced versions and certain flexible grades or alloys of these materials. In terms of temperature resistance the PA 6.6 and their alloys have the highest melting point and heat deflection (or VICAT) temperature and are generally better in high temperature ageing. However, PA 6 grades are more widely available and they also tend to have advantages in terms of slower crystallisation in moulding which can often give improved surface finish. PA 11 can be used where special chemical resistance (for example zinc chloride resistance) is needed. Flexible grades of both PA 11 and PA 6 are available which may be combined with their rigid equivalents to give "hard/soft" sequentially blow moulded components (see "Application of blow moulded engineering polymers".

Polyesters

Semi-crystalline polyesters for blow moulding are available in the form of PBT (sold as a toughened PBT under the Crastin® name by DuPont) and TEEE-copolymer elastomer (such as DuPont Hytrel®). Hytrel® has, in fact, been used for blow moulding for more than 15 years and is well established as a thermoplastic elastomer (see next section), used for automotive CVJ boots and other convoluted bellows.

Recent applications for both Hytrel® and Hytrel®/Crastin® combinations have been in air intake ducts, including turbo and intercooler pipes. The polyester elastomer is particularly useful where a flexible material is needed with a high level of oil resistance in combination with high and low temperature performance. PBT, being more rigid can withstand higher pressures, particularly at elevated temperatures and in this respect has similar performance to the nylons for applications such as turbo pipes. One significant advantage of PBT is its dimensional stability due to its much lower moisture pick up compared with nylon. However, due to the well known tendency for polyesters to suffer hydrolysis attack in the presence of hot water it is not possible to consider polyesters for applications like automotive coolant pipes.

Thermoplastic Elastomers (TPEs)

Using the previously defined criteria to define a true "engineering thermoplastic", there is a limited number of TPEs which can be classed as moderate or high performance materials. Amongst these the polyester elastomers are generally considered to have the best combination of both high and low temperature performance, as well as exceptional oil and grease resistance. Other flexible thermoplastic alloys (such as Santoprene®) are widely used, but since most are based on polyolefins they tend to have lower performance than the polyester elastomers. More recently, specially developed grades of Hytrel® and flexible nylon alloys have been developed, but there tends to be a trade off with either increased stiffness at low temperature or poor creep performance at high temperature. This latter point is important for flexible air ducts, where good sealing performance is normally required in order to maintain air-tight connections at elevated temperatures in the presence of internal pressure (often pulsating).

Material combinations for Hard/Soft mouldings - comparative properties

Table 1 below lists properties of various engineering polymer that can be combined in sequential blow moulding to give "hard/soft" properties at different positions in an air duct moulding. The best choice for a particular application will of course depend on the balance of properties and cost, taking into account flexibility, temperature resistance, general mechanical properties and fluid/chemical resistance.

Table 1

PROPERTY	MATERIAL COMBINATIONS						
	Hard Materials					Soft Materials	
	PA6 (Unrein)	PA6 15% glass	PA6 25% glass	PA6.6 (Tough unrein)	PA6.6 15% glass	PA11 (plasticised)	PA6 Flexible alloy
E-Modulus, Mpa (50% R.H. 23°C)	950-1500	2200-2500	3000-4000	800-1200	2500-3500	300	170
Melting Point, °C	220-225	220-225	220-225	260	260	180	220
Heat Defl. Temp °C (0.5 Mpa)	110-180	170-210	215-225	225	240-260	130	150

	PBT (toughened)	Polyester Elastomer (55D Hardness)	Polyester Elastomer Alloy Hytrel® HTR-8303
E-modulus, Mpa, 23°C	1500	150-170	400
Melting Point °C	225	95-210	224
Vicat Softening Point °C	125	175	210

	Polypropylene	TPO Alloy (40 shore D)
E-modulus, Mpa, 23°C	650-750	130-160
Melting Point °C	165-175	150-170
Vicat Softening Point °C	110	100-130

APPLICATION OF BLOW MOULDED ENGINEERING POLYMERS

The inherent properties of blow moulded engineering polymers has resulted in most of their applications being in automotive products. Their potential use in non-automotive areas is only likely where their value-in-use can be justified in terms of special requirements such as exceptional stiffness (glass reinforced nylons), or very dynamic performance (Hytrel®) or extremes of temperature.

Considering only automotive use therefore, we can list several key areas where engineering polymers are now widely used, or are being rapidly developed for new applications. These include:

- Air ducts - clean air or recycled air and turbo charge ducts
- Crancase venting hoses
- Coolant system pipes - all parts of the coolant system including oil cooler and radiator pipes
- Coolant and other fluid reservoirs
- Air conditioning (refrigerant) pipes
- CVJ boots, steering and suspension bellows/gaiters
- Fuel tanks and filler necks/pipes

Air Ducts

[Fig. 4] shows the basic layout of the various air ducts in a modern car engine.

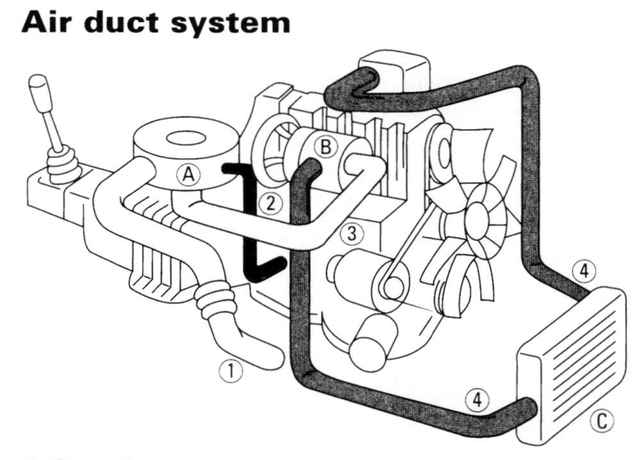

1) Clean air
2) Crankcase housing venting hose
3) Recycled air
4) Recycled charged air
A) Air filter
B) Charger
C) Heat exchanger for charged air

Figure 4

Upstream of the filter, clean air and usually relatively low temperatures predominate and in this area it may be possible to use polypropylene or lower performance TPEs. After the filter, where service temperatures may be higher (for example, closer to the engine) - typically above 120°C, it is normally necessary to consider materials such as nylons, polyesters or TEEE (Hytrel® types). This is particularly true for crancase vent hoses and ducts which contain so-called "dirty-air" which includes crancase gases. These gases are notoriously corrosive and will attack lower performing materials. [Fig. 5] gives an example of this type of air duct in Hytrel® polyester elastomer.

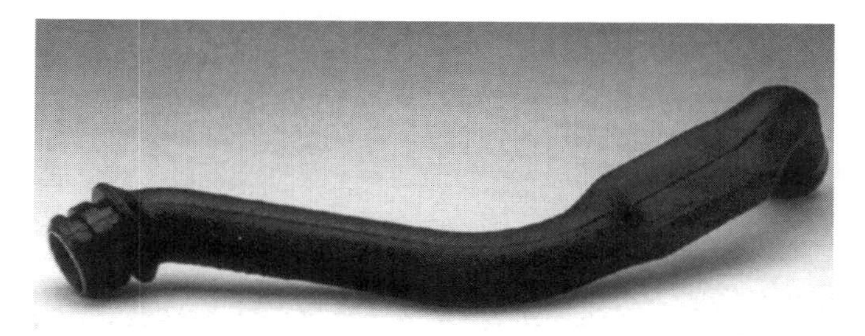

Figure 5. "Dirty Air" system duct

In turbo engines there is a rapid increase of the air temperature on the downstream side of the turbo and car manufacturers have actually measured running temperatures of between 150 and 180°C, with peaks to around 200°C (depending on car/engine design). The turbo discharge pipe therefore stretches the possibility of using engineering plastics to the limit and in some cases may go beyond the current limit, even for nylon 6.6 grades. This emphasises the need to check accurately the real-life temperature requirements before specifying an appropriate engineering thermoplastic. Fortunately, turbo engines also incorporate an intercooler which drops the air temperatures by perhaps 30-40°C after the intercooler, which usually brings conditions within the capabilities of a wider range on engineering polymers, including PA6, PBT and TEEE.

Coolant pipes

Traditionally, coolant pipes have been made from steel with flexible hoses from fabric reinforced rubber. As coolant systems [see Fig. 6] become more complex and pipe routings become more tortuous it becomes very costly to fabricate complex shapes in steel and in reinforced rubber with multiple branches and fixations. There is also the question of corrosion (of steel pipes) and recycling (of rubber/fabric hoses). For this reason, several companies are looking at rigid and flexible engineering thermoplastics to replace metal pipes and where possible eliminate, or at least minimise, the use of rubber hoses.

The service conditions for coolant pipes are usually well defined, maximum operating temperatures being in the range of 120-130°C, with internal pressure of around 2 bar. The fluid is, of course, a fairly standard mixture based on water/glycol, and the external temperatures can be typical underbonnet conditions - 100 to 150°C. For obvious reasons the chosen material must be very resistant to hydrolysis at elevated temperatures and for this reason special hydrolysis resistant unreinforced and glass reinforced grades of nylon have been developed. In general, due to higher service temperature capabilities and their proven performance record in injection moulded components such as radiator header tanks, nylon 6.6 materials are preferred.

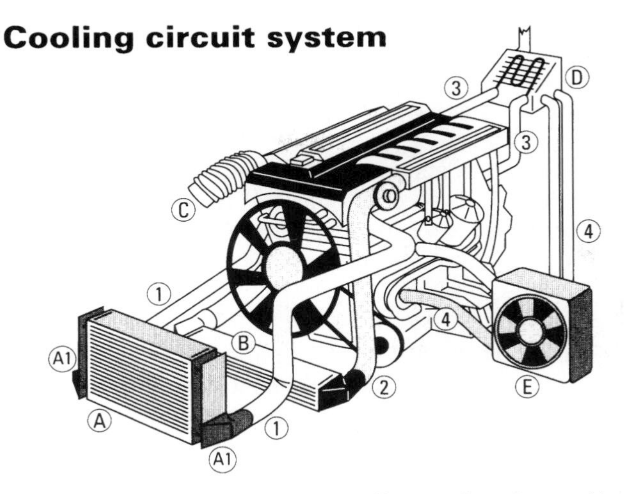

A) Coolant radiator
A1) Header tanks in ZYTEL
B) Oil cooler
C) Air duct
D) Heat exchanger for car interior
E) Air conditioner

1) Water cooling pipes and hoses
2) Oil circuit pipes and hoses
3) Car interior heating hoses
4) Air conditioner pipes and hoses

Figure 6.

Air conditioning pipes (refrigerant fluids)

This potential application for blow moulded pipes is relatively new, but the incentive is, generally, to replace metal pipes or reinforced rubber hoses, where cost saving and/or weight reduction is the expected benefit. The fluid involved in this case is a refrigerant liquid or vapour such as SUVA® (134A) which has replaced the old CFC fluids previously used.

The major consideration here, apart from temperature and pressure requirements, is also permeation (to prevent premature loss of fluid from the system). Taking into account all these factors, it is very likely that a rigid nylon, possibly glass reinforced, will fulfil the needs of this application. It will still be necessary however, for flexible hoses to be retained in parts of the system, and due to the pressures involved these will need to be textile reinforced. Conventional extruded and braided hose constructions in either rubber or flexible thermoplastics are probably more appropriate here.

CVJ boots and other bellows/gaiters

This is a well established product area for flexible thermoplastic elastomers (TEEE, TPO materials and thermoplastic urethanes predominate). It is not intended to discuss these applications further, except to say that - like in other areas - the performance limits are continually being stretched, particularly with respect to upper and lower temperature specifications.

Fuel tank and filler necks/pipes

There has been a step change in automotive permeation (emission) requirements in recent years. For example, current U.S. specifications call for all cars made in 1998 to meet the limit of 2 grams/24 hours total hydrocarbon missions from the complete vehicle. In Europe, the relevant specification for the future is Euro 2000, which is less severe than the U.S. requirement.

Consequently blow moulded plastic fuel tanks and filler necks have had to be developed with dramatically reduced levels of permeability. Without getting into too many details on this very specialised area, it is sufficient to say that major improvements have been made in fuel tanks by the use of (a) multi-layer coextruded fuel tanks, incorporating a barrier layer, and (b) Selar® modified HDPE where the Selar® (special resin and processing technology) provides overlapping "platelets" of barrier material throughout the tank wall. For engineering polymers however, which are generally too high priced to be considered on their own in fuel tanks, there is considerable potential for use in blow moulded filler necks, where low permeation properties and relatively small material content (compared with fuel tanks) may offset the higher cost of these materials.

PROCESSING CHARACTERISTICS NEEDED IN A BLOW MOULDED ENGINEERING PLASTIC RESIN

It has already been said that the development of engineering plastics for blow moulding has been a major challenge for material suppliers. The very special rheology and processing requirements in a blow moulding resin have had to be built into the resin without significantly compromising their inherent high performance in service.

Melt Strength and die swell

In blow moulding any polymer the prime requirement is of course high "melt strength". This relates to the performance of the molten resin during the formation of the extruded parison, as well as the blowing of this parison inside the mould cavity. Good melt strength means a high viscosity at very low shear rates (typically the only forces acting on the polymer after extrusion through the die will be gravity plus any residual internal stresses in the polymer itself caused by shear during extrusion). Another factor in melt strength is "melt elongation" which means the ability of the molten polymer parison to avoid breakage during the blowing stage, whilst hanging under its own weight, or during any parison manipulation process.

A second consideration which applies to all resins used for blow moulding is the "die swell" characteristics of the material. Die swell is essentially the increase in parison wall thickness or diameter relative to the extrusion die diameter and die gap. It can be defined as a simple ratio of parison diameter (or thickness) to die diameter (or thickness)and usually ranges from about 1.6 to 1.9 for polyethylene (diameter swell). The actual die swell achieved in practice can also be influenced by the die geometry and processing temperatures, but in general it is more dependent on the material itself. Since most tooling and parison programming has been established with polyolefins it is useful, though not essential, for engineering polymers to show similar die swell characteristics. A major influence on the die swell however, is the presence of glass fibre reinforcement in the resin. Due to the orientation effects of extruding glass reinforced resins, the die swell achieved with such resins tends to be very low - values of 0.9 to 1.2 are typical for both diameter and thickness swell.

An additional consideration here is when two resins are being sequentially blow moulded, for example as a "hard/soft" moulding. In this case it becomes especially important that the die swell of both materials are fairly well matched otherwise the parison control, especially in the transition from one material to the other, becomes very difficult. As a result of the key elements of parison "melt strength" and "die swell", several systems have been developed to quantify these characteristics in a blow moulding material. An example of an empirical method for making quantitative comparisons between different resins is described later.

Thermal Stability and moisture sensitivity

Because of the extreme sensitivity of the blow moulding process to material viscosity changes (in particular as it relates to melt strength), it is essential that the resin exhibits viscosity stability. This point refers not only to thermal effects (such as degradation or cross-linking) which might cause significant viscosity changes, especially as a result of machine stoppages or use of regrind, but also to any variations from lot-to-lot of delivered material. Of course there is a particular issue with some materials, such as nylons, where moisture content of the polymer granules entering the moulding machine can have a dramatic effect on the melt viscosity. For this reason it is essential to control tightly the upper limit of moisture in the feed materials - much more so than for injection moulding. Typical limits for moisture content when processing nylons are 0.05% and for polyesters 0.03%. The moisture issue is extremely important when handling regrind (scrap) material. Ideally, any trim materials and reground parts should be re-moulded within a few hours of first processing in order to minimise moisture pick-up and subsequent need for drying. In practice however, it will usually be necessary to allow for the drying of regrind material and also of virgin material which has been exposed to atmospheric moisture. In these extreme circumstances drying times may be up to 8 hours for nylons and perhaps 4 hours for polyesters.

Pinch-off strength and weldability

Since most blow moulded parts involve a "pinch -off" at some position on the moulding, the ability of the parison to effectively weld to itself in forming the pinch-off is important. The effectiveness of the pinch-off or "pinch weld" is determined partly by mould design (in the pinch area) and partly by processing conditions, but primarily by the inherent "weldability" of the resin. With engineering polymers special attention has to be paid during resin development to ensure that good pinch weld strength can be achieved in the finished mouldings.

The complexity of technical mouldings often requires the use of inserts - either injection moulded in similar material or machined metal inserts. In cases where plastic inserts are required to bond to the parison during the blowing process , it may be necessary to use an adhesive coating or mechanical "keying" of the insert to achieve the required bond strength.

With some less crystalline engineering polymers it may be possible (though not always) to achieve a pure weld to the insert.

Post moulding operations very often involve hot plate or other welding techniques where additional parts such as injection moulded nozzles or fixations have to be attached to the blow moulded component. For engineering plastics this means that the resin should have good welding characteristics and in particular is not easily degraded or affected by surface oxidation during the welding process.

Other Considerations

As in most manufacturing processes, there is a liability on behalf of the moulder and the resin supplier to provide reassurance to machine operators and others that there are no exceptional health or safety risks in processing the materials. Engineering thermoplastics, which are generally processed at higher temperatures than polyolefins, are inherently less thermally stable. This can result in decomposition if not handled correctly, causing in some cases potential problems with smoke and/or smell, or in extreme cases dangerous fumes. Material manufacturers make every effort to minimise these problems in the formulation of their products, but the safe operation of blow moulding machines and provision of adequate ventilation and extraction equipment is a responsibility of the moulder.

METHOD FOR CHARACTERISATION OF BLOW MOULDING POLYMERS

The method described below has been developed by DuPont for process characterisation of blow moulding engineering polymers and is believed to be particularly relevant for comparing materials in relation to their melt strength, die swell, pinch-weld performance and other features. Undoubtedly other material suppliers have their own techniques, which may or may not be similar.

The procedure is based on blow moulding of actual parts (a standard bottle shape is used) on a continuous extrusion blow moulding machine, having conventional features, including a suitable screw designed for use with engineering polymers. Whilst most of the resin characterisation is done without actually moulding the parts, it is necessary to make parts at some point in order to evaluate pinch weld strength. This latter test is done by cutting "test bars" from across the pinch weld in the moulding and them subjecting them to a standard tensile test. The die tooling and all machine settings (die gap, output rate as determined by screw speed etc.), are standardised for all materials - only the temperature settings are changed, according to the established optimum settings for each type of material.

The test is initiated by extruding the material to be evaluation through the die under standard conditions (and appropriate temperature for the material type). The molten parison is allowed to fall freely to the floor without contacting the mould throughout the procedure. Once conditions have stabilised the parison is cut at the die and a measuring system then records the position and time of the bottom of the new parison as it extrudes, under gravity, to the floor of the machine. A plot of the time/position of the "parison drop" is then used to analyse the parison behaviour of the resin and produce quantitative values which are identified as "Sag Index" and "Swell Index" for the resin.

[Fig. 7] below shows some typical curves which are obtained in this “parison drop test”.

"Parison Drop Test" for characterising processing behaviour of Blow Moulding resins

Resin identification	Swell Index	Sag Index	1200 mm Sag Time	Bottle Weight	Melt press.	Other Comments
Resin 1	11	4.5	35	72	410	
Resin 2	12	6	40	109	1100	
Resin 3	9	12	45	97	1650	
Resin 4	10	11	46	97	1420	

Parison Length (mm) / Time (sec)	mm:	0	100	200	300	400	500	600	700	800	900	1000	1100	1200
Resin 1	T	0	7	13	17.75	21.0	24	27	29	30.5	32	33.25	34.25	35.25
Resin 2	i	0	7.2	13	18.5	23	26.5	30	32.2	34.5	36.2	38	39	40
Resin 3	m	0	5	9.3	14.3	19.3	23.6	27.7	31	34.3	37.3	40	42.5	45
Resin 4	e	0	6	11	16	20.7	25.3	29.3	33	36.3	39.3	42	44.3	46.3

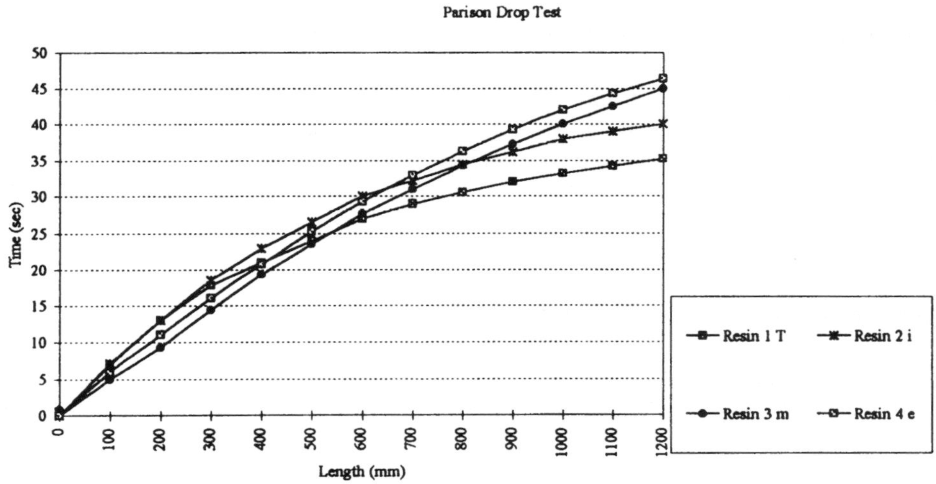

Parison Drop Test

Figure 7

From the shape of the curves it can be seen that the effect of gravity is to stretch the parison as the parison gets longer.

Ideally, of course, for minimum "Sag" characteristics, the curve should be a straight line. It will also be appreciated that a resin with high die swell will tend to give a curve which is more steep (at least in the initial part) than a material with low die swell - since more of the material is going into producing a larger diameter and thicker wall than is available for increasing the length of the parison (remember that the output rate is kept constant for all materials). In order to quantitatively compare several materials therefore, we need to analyse both the rate of curvature of the graph (representing parison sag) and the initial slope of the

curve (representing die swell). Mathematically of course, this implies the second and first derivatives of the curve respectively. However, in practice these functions are approximated, or simplified, by determining (a) the length of the parison at which point its rate of descent becomes twice that of its initial rate; this gives a number which is defined as the "Sag Index" and (b) the slope of the graph in the first section of the curve which gives a second number, which is defined as the "Swell Index".
It is clear that this method is very specific to the particular machine and processing set-up which is used for the test. A meaningful comparison of results from one machine and set of conditions, cannot be made with those from a different machine or conditions. However, as a general gauge of blow moulding characteristics and as a way of comparing materials in their process behaviour and of identifying subtle differences which may not be readily detectable otherwise, it has proven itself to be a very useful tool in developing resins and providing "bench mark" values against which new materials can be judged.

CONCLUSIONS

The use of engineering polymers in technical blow moulded parts is continuing to increase and will form a significant proportion of the use of these materials by the year 2000. Machine suppliers will continue to develop their machines to provide even more novel ways to produce useful parts from the blow moulding process. Simultaneously, resin producers will develop new grades and new combinations of materials with potential uses - particularly in the areas of coextrusion and sequential moulding processes. The overall result is that blow moulding will consolidate its position as a highly technical process for producing parts in engineering polymers which can offer unique solutions to the demands of the car industry.

® Santoprene is a registered trademark of A.E.S. Engineering Ltd
® Hytrel® is a registered trademark of DuPont de Nemours Inc.
® Crastin®is a registered trademark of DuPont de Nemours Inc.
® Suva® is a registered trademark of DuPont de Nemours Inc.
® Selar® is a registered trademark of DuPont de Nemours Inc.

MULTI-LAYER BLOW MOULDING OF FUEL TANKS

GRAHAM PICKWELL & NICK THACKER
British Polymer Training Association

Co-extruded plastics fuel tanks. There are many major issues related to the performance and production of blow moulded plastics fuel tanks and as regulations tighten, the criteria become more critical. This paper gives a broad overview of market requirements, the raw materials used, the design of equipment and the processing factors involved.

INTRODUCTION

Alternatives to metal. Over recent years, plastic fuel tanks have been replacing metal for both performance and cost effectiveness. However, mono layer versions are only partially effective in preventing fuel loss and, as emissions regulations become more stringent, manufacturers are faced with three alternatives; a) co-extrusion; b)fluorination; or c)use of alloys, as a means of improving the barrier properties of the tank. Although fluorination / sulphonation as an after treatment may be prohibitive on cost, some car manufacturers have taken this route. Some parts of Europe are using alloys but most other car producers have taken advantage of the co-extrusion process which is often more cost effective. There are disadvantages to co-extrusion such as the high cost of equipment (multi-extruders and dies), the complex control requirements and the need to produce large runs to reduce costs. However, the blow moulding of multi-wall fuel tanks (and other components) is set to increase.

MARKET PERFORMANCE

Plastic fuel tanks meet many important performance criteria. The following are important factors:

- High rigidity
- Good stress crack resistance
- Low fuel permeation
- Reduced weight
- High impact resistance (eg to resist burst when dropped from 6m at -40°C)
- Overall high mechanical strength
- Good performance over a broad temperature range
- Cost effectiveness

When comparing the overall strength of a mono layer tank versus a multi layer product there are no significant differences. The bulk of the polyolefin layers are dominant, any barrier layers are too thin to influence mechanical properties. The barrier layers are only used to restrict fuel emissions.

Emission Regulations

In the USA, overall testing of fuel tanks is more stringent. Fuel emissions are evaluated for the whole fuel system using the SHED test (Sealed Housing for Evaporative Determination). This specifies a threshold value of 2g in 2 hours. California (1995) has imposed other restrictions on mechanical performance and component lifetimes which are expected to be adopted by many other states. The EEC regulations are less critical. Regulation 34, Annex 5 allows a maximum weight loss at a temperature of 40°C over 8 weeks to be an average of 20g per day. Single layer high density polyethylene (HDPE) tanks usually comply with this EEC regulation, but not with the SHED test. Other manufacturers look at a fuel emission of 3ppm max. The problem of petrol permeation through plastic tanks becomes more critical as the level of polar additives, mainly oxygen containing chemicals such as methanol, is increased in respect to the non-polar constituents. The current maximum level of methanol allowed in petrol in Europe is 3%, but trends are moving towards increasing the amount of methanol used. All barrier materials, including fluorine, are more pervious to this methanol, but it has been shown that ethylene vinyl alcohol copolymer (EVOH) is 50% more effective as a barrier to methanol than polyamide (PA).

RAW MATERIAL COMBINATIONS

Support polymers. The major component of any plastic fuel tank is the material which gives the overall mechanical performance to the product, whilst still giving reasonable resistance to fuel permeation. The only real candidate which meets these requirements is high density polyethylene, HDPE.

High Density Polyethylene

The properties of HDPE are critical both in the performance of the resultant blow moulded fuel tanks and in the processing on the blow moulder itself. Some important parameters are outlined below:

- High crystallinity. Results in high tensile strength, high rigidity, good chemical resistance and high moisture barrier. Also gives low elongation and high stiffness, making the parison easier to control.
- High molecular weight. Use of low MFI gives high impact strength, enhances chemical resistance and barrier properties.
- Ease of processing.
- Relatively low permeation rate. The petrol permeation of HDPE is 20g/mm/m^2/24hrs@23°C. (Permeation is proportional to surface area, but inversely proportional to thickness).

It would be reasonable to assume that polypropylene (PP), especially random ethylene copolymers, would meet many of the mechanical requirements of fuel tanks. However, PP suffers from two major disadvantages; a) its very high crystallinity causes the tank to become brittle at low temperatures and it fails the drop test; and b) it has an appreciable absorption of petrol and potentially explosive vapour mixtures with air are formed around the tank.

Barrier materials. These are thermoplastics which provide good barriers to chemicals, solvents and gases. For use in multi layer petrol tanks the two polymers commonly used are PA and EVOH.

Polyamides

In general, polyamides are tough materials with high tensile strength. They form good barriers to solvents and chemicals (better than polyolefins) but absorb moisture readily, which impairs the barrier property. They must be protected by layers of good moisture barriers (eg polyolefins).

In blow moulding of co-extruded fuel tanks, high viscosity grades are used to ensure similar flow characteristics to the encapsulating HDPE, 6-6, 6 copolyamides are used as they process at lower melt temperatures than polyamide 6 (220°C vs 250°C), they can be extruded at higher outputs and form thinner layers. The incorporation of a polyamide layer only 2-5% of the total wall thickness of the tank can reduce the weight loss of petrol by up to 90%.

Polyamides need drying prior to extrusion to prevent hydrolytic degradation of the polymer at the elevated processing temperatures.

Ethylene Vinyl Alcohol Copolymer

EVOH is a random copolymer of ethylene and vinyl alcohol, characterised by its outstanding barrier properties and ease of processing. It has acceptable mechanical strength and processes similarly to polyethylene.

In common with PA, EVOH absorbs moisture which reduces its barrier properties and must also be protected by polyolefin layers. It must also be dried prior to extrusion.

Although EVOH is more expensive than PA, it is replacing PA as a barrier layer for both standard petrol mixes and high methanol content fuels due to its lower permeation rates.

Adhesives

In multi layer co-extrusions, polar barrier polymers such as PA or EVOH will not adhere to non-polar support materials such as HDPE. A specific tie-layer adhesive polymer is needed to cement the two layers together to prevent de-lamination. Tie-layer adhesives are made up of a two component system; 1) a non-functional modified polyolefin that will bond effectively to other polyolefin substrates; and 2) functional, reactive groups (often anhydrides), which are co-polymerised into the polyolefin backbone to enable it to bond to a variety of polar materials, eg PA or EVOH. Adhesion is dependent upon the thickness of the surrounding layers, the processing temperatures and the length of time the layers are in contact before cooling and solidifying from the melt. Grade selection is important in achieving similar flow and viscosity to the two layers to be joined. In co-extrusion blow moulding, the polyolefin base is usually an MDPE (medium density polyethylene) copolymerised with acrylic acid (EAA), or a 9% vinyl acetate content EVA (ethylene vinyl acetate copolymer).

Regrind

The use of co-extruded blow moulded fuel tanks has given producers the opportunity to use a percentage of regrind HDPE in the construction, thus reducing the overall cost without jeopardising the performance of the tank (eg drop test and weld strength). The regrind material can be sandwiched with layers of virgin polymer.

As manufacturers reduce the levels of barrier materials in the tanks, there is also the opportunity to recycle any below specification or scrap tanks, as they only contain approximately 4-8% nylon and tie-layer adhesives which is too small a level to cause any processing problems, the 90% plus HDPE is dominant.

Tank Construction

A typical 6-layer tank is composed of the following materials:
[interior] HDPE / TL / Barrier / TL / Regrind / HDPE [exterior]
(where TL = tie layer adhesive)

Ratios between the layers vary from manufacturer to manufacturer, but typically the inner and outer HDPE layers are in the region of 40% and 10% of the total thickness respectively. As the thickness of the barrier layer is reduced the regrind layer is increased accordingly.

THE MACHINERY USED

The design of blow moulders used to produce multi layer products is very complex and only some important features are considered here.

Extruders and screws. One of the main criteria in the selection of extruders in a co-ex machine is to ensure that each individual extruder is correctly sized for the application. If an extruder is to deliver 40% of the total output as a percentage of the layer thickness then it must be much larger than the extruder needed to produce the barrier layer at only 3% of the total construction. This maximises outputs, minimises residence times and allows the extruders to run efficiently.

Screw design is also critical. Radically different screw designs are required for PA to allow for the melting characteristics of this material. The HDPE extruders usually incorporate grooved feed sections, whereas the barrier materials usually run with smooth feeds.

EVOH is highly adhesive to metal surfaces and many parts of the machine must be chromed and streamlined to prevent any build up and subsequent degradation of the polymer which will affect its properties.

Continuous vs Accumulator Processes. There are two types of process used, both with a single parison as standard. These are continuous parison extrusion and the accumulator method, where a reservoir of material is built up in the head and then is extruded by piston ejection. With the accumulator principle barrier layer reduction is often more difficult to control and has a minimum value of 3-4% of the total construction. Continuous extrusion can result is down-gauging of the barrier to only 1% with a corresponding increase in regrind of up to 45%. With continuous parison extruder there must be one extruder per layer.

There are many designs of accumulator die heads and feed blocks; dependent upon the manufacturer. Some use a system of multiple pistons, others use a single piston. The laminate can be formed both prior or after the accumulator.

Die heads. Whichever process is used the co-ex die heads are very expensive, very complex and difficult to clean and maintain. They must be designed to give constant material flow and pressure to ensure consistent layer distribution and often incorporate sleeves through which the material flows to ensure this layer integrity.

Parison control. The individual layer thicknesses of each material are controlled by the respective extruder speed and pressure (ie output). The overall wall thickness/weight is controlled by a typical parison programming unit. When the die gap is adjusted it is accepted that all of the layers will change proportionally by the same amount, thus maintaining the layer distribution. Gravimetric control of extruder output is ideal.

Moulds. The mould parting lines is at 90% to the extruder axis for ease of withdrawal of the finished tanks. High clamping forces of 1000-1500KN are typical for the long welds and thick walls involved. Divergent tooling is normally used and the blow pins can either enter from below or from the side.

PROCESSING TECHNIQUES

Notwithstanding the correct choice of raw materials and the design of the equipment, there are many factors which influence the processing parameters that must be considered when producing blow moulded multi-layer articles. These include:

- Rheological compatibility between the polymers
- Polymers should have similar melt viscosities at their processing temperatures
- Any chemical or interfacial interaction between the layers
- Crystallinity effects (especially on cooling)
- Surface tensions must be similar at the processing temperatures to ensure good adhesion
- The shear compatibility between the components of the composite

The processing of polymers creates many potential problems, even with single layer articles. The number and frequency of these problems is increased proportionally as the number of layers increases in co-extruded products. Examples of the problems could be as follows:

- Non-uniformity of the wall thicknesses affects permeation
- Weak weld lines and weak pinch off welds may cause reductions in mechanical strength
- Holes in layers, thin corners and gels may create avenues for petrol permeation.

It is critical that comprehensive training in the production of multi-walled extruded products is given to help minimise all of these factors and problems.

Intermittent Encapsulation of Plastics and other Techniques for Discontinuous Extrusion

Mr Simon Dominey
Killion Extruders, Division of Davis-Standard
200 Commerce Road, Cedar Grove, New Jersey 07009 USA

INTRODUCTION

The established principle of plastics extrusion is that it is a continuous process, that is polymers should flow continuously from extruder, through a forming die, and be transported at a constant rate whilst being cooled.

Most effort in extrusion technology has been expended in reducing variables of the process to try and ensure absolute continuity. However, there has been some work done in changing these principles to non-continuous, variable output from the extrusion process. The objective of this work was to produce a continuous extrusion that varied in shape/cross section or composition/structure. This paper looks at some of the techniques for non-regular, non-continuous extrusion.

Typical Problems

There are several problems in developing a system for intermittently stopping and starting the flow of one polymer within another. If one attempts to simply "stop" one extruder and turn another one "on", such attempts fail because the walls of the die create a drag on the polymer and a relatively long purge of the original material at the walls will follow. Furthermore, since extruders take many minutes to stabilize it is generally not advisable to start and stop them while trying to make a consistent product. The viscoelastic nature of most polymers further complicates the situation.

Due to the viscoelastic properties, polymers flowing through channels are subject to wall drag. FIG. 1. Common visual evidence of the effect of the local velocity gradient occurs in extruded sheet when one colored plastic has changed to another. The new color flows intrusively through the system and is seen as a parabola (or flattened parabola) having a leading edge exiting from the center of the die. Because of the locally slower velocity at the outside edges of the sheet die, the new colored extrudate takes a long time to completely purge the die of the original material. It is common for sheet dies to take 30 to 60 minutes to completely purge. However, the core of the extrudate often purges in seconds it is not dragged by the walls.

As a material flows through a die, the walls of the die resist the flow. A boundary layer in the fluid material moves much more slowly at the walls than the material away from the walls. The boundary thickness depends upon the power law index and it has a definite thickness. It is important to notice that the effect of the drag is exerted at the boundaries. Because of the above, it would be preferable to stop and start the output without an equivalent stop start in flow!

An Experimental Die System

In this special die (Fig. 2) an A-B rod is produced and flows through channel 2. The B stream is surrounded by the A stream. A separate stream flows through channel 1 composed completely of A material. The two streams are connected by a transposition plate that is a passageway for the A-B stream, another passageway for the A stream and an unused passageway 3.

The transposition plate can be moved in transition to where an A-B segment is between the two channels. The other portion of the valve holds the A segment for later use. If the plate moves very quickly, only a momentary interruption to the flow takes place. This interruption is too short to allow pressure to build within the channels or in the extruders supplying the materials.

At the end of movement the entire A-B segment has moved into channel 1 where it will be pushed by the A stream of material until it exist the die. The previously unused passageway 3 now lines up with channel 2.

If the transposition plate is moved back to the original position, the A segment that was removed (and held in the plate) from channel 1 will now be back in the same location in channel 1 but surrounded by new material. The new slice of A in channel 1 is at the same time moved to channel 2. This means that from this point on in the cycle, channel 2 will be extruding A-B sections surrounded by A streams.

As the A-B and A sections are in the molten state, the flows will fuse together (in particular the identical A resin). As the cycle of transportation plate movement is continued, sections of B resin totally encapsulated in A resin will be produced in a continuous extrusion.

Note the earlier comments on boundary layer flow. If we plot velocity in a die or channel for a Newtonian fluid then we will see a parabolic shape with high velocities in the center of the channel and low velocities at the wall. As the power low index moves toward zero then the wall drag effects penetrate less distance into the flow channel.

Referring to Figure 3 (and assuming a polymer being processed with power low index of say 0.2) it can be seen that the wall effects are limited to the outer sections of the flow channel or the areas of Resin A only. This leaves the encapsulated B resin intact.

In actual fact, the interchanging sections of Resin A are being deformed by wall drag but because they are the same material we cannot discern the effect of drag flow. In practice, adjusting the distance between flows and the flow velocity enables us to accommodate resins who temperature, etc., can be adjusted so as to exhibit a power low index from 0 to 0.3. Outside of this range very low extrusion rates must be used so as to prevent wall drag effecting the B resin.

Variations of the above transposition plate mechanism can be produced encapsulated multiple layer components.

Variable Pull Rate Control

An alternative to die control for adjusting cross section is to vary haul off rate in some intermittent manner. As long ago as 1958 attempts were made to adjust the haul off rate of tube extruders to create a tube in varying diameter. This is now commonly known as bubble, bump, pulse or taper tube. The basic principle is that if die flow velocity is constant, then speeding up pulling rate will reduce product diameter, slowing pulling rate will decrease product diameter.

The Stop Start Concept

The easiest way to understand this technique is to imagine that a clutch brake has been fitted between the belt puller drive motor and the belt transport. A rotary pulse encoder would be mounted on the puller belt to obtain a measurement of forward movement. The series of pulses, is used as input into a preset counter. The operator would set a specified length on this counter, which is equivalent to the length of the small diameter. When the present value is reached, an output signal is generated which disables the clutch and activates the brake causing the belts to stop. At this point a timer can be used to initiate a delay preset for a specific time. When this preset is reached the brake is disabled and the clutch is engaged bringing the belts back to the preset speed.

A Practical System

In practice the rapid changes of velocity need rapid changes of air volume within the tube (so the tube does not collapse) a controlled air supply system is therefore necessary. When the belts are engaged a specific volume of internal air is necessary to size the tube. This volume needs to be adjusted during the stoppage to create the bubble. Normally for tubes with more than one bore separate air supply regulators are necessary, each capable of being individually adjusted. It is also necessary to electrically shift these regulators in sequence with the activation of the clutch-brake. Delays may also be necessary to control the exact profile of the tube.

Completely stopping the movement of the tube as suggested in the above clutch-brake system gives several problems. A more realistic approach is to operate at a "normal" speed and a "slow" speed.

If the air pressure (and flow) is adjusted in a more exaggerated manner; then the internal pressure in the tube can be increased sufficiently to inflate the tube by differential pressure. This inflation will take place at the point of lowest wall tension. For most plastics this occurs in the molten state. Typically this occurs at the point the melt exists the die , before it contacts the cooling medium.

This air inflation technique requires materials that have good melt strength and extensibility but little elastic recovery, further we must be able to cool the product sufficiently rapidly to "freeze in" the shape change caused by inflation. Figure 4 shows the control computer screen from a typical system. The exact shape of the tube is controlled by adjusting (synchronizing) the pull velocity an air pressure relative to each other in relation to a length base line, until the desired large and small diameters are produced. Until recently the exact profile or shape of the approach transition and exit transition were a function of the fixed dynamics of the puller mechanism and air control solenoids.

For the past 18 months we have been using advances in electronic servo control to accurately adjust all stages of these various processes. By employing high speed control loops and a high acceleration servo motor for driving the pulling device in conjunction with a special ultra low inertia servo air valve, it has been possible to break the process down into small discrete (digital) steps. The most advanced of these control units was capable of defining 256 separate points along any given length of each control axis. This enables extremely accurate profiling of the transitions. This increased level of precision has enabled the adoption of this technology in areas outside of conventional (plain) tubes, it has now been successfully used for tubes with up to five different bores (lumens) with differing cross sections.

Detecting Variations from In-Line Forming

Corrugated Tube

For many years a semi discontinuous extrusion process has been in common use. This is the well known, in line tube corrugator that forms a continuous "set" of bellows in a tube wall. This is accomplished by fitting segmented die formers on a caterpillar track and thermoforming/blow-molding the shape in the die cavities as the tube passes between them. It is possible to extend this technique by fitting different shape dies in sequence on the caterpillar track. This system has been developed in the automotive industry to improve productivity and reduce stock level when a wide range of flexible tube configurations are required. The disadvantage of this process is that the continuously extruded tube must now be cut off line to split it into its individual configurations.

Attempts to cut this type of product in line have failed due to lack of accuracy in determining the cut point. Using time or length for the product has not worked due to its flexibility (elasticity and compressibility) and inconsistencies from forming mold to forming mold. For several years we had used optical detection techniques with some limited success. This photo eye system was used as part of a compute r program counting the number of corrugations prior to each cut. Fig. 5. By using an encoder to measure length we could also cut between corrugations by linking the two parts of the program. Fig. 6.

A series of tests were run to find a small size photo detector that could be adjusted and aligned over a relatively large gap. Eventually a coherent beam infeed type sensor was found to have the best configuration of parameters. Appendix. The fast response time and small focus detection beam, whilst being necessary for resolution and accuracy of cut, gave problems in the form of "noise". This noise was generated by the vibration in the product that is the bellows pushing together and pulling apart. By using a noise averaging algorithm most of this could be eliminated. A further problem was introduced when it became necessary to make cuts not just between corrugations but on a raised cuff located between corrugations. Fig. 7. Because the raised cuff is typically less than 50% of the height of the primary corrugation, the noise filtration algorithm would frequently rejects it as noise! It was necessary to establish a simple method of detecting a genuine cuff and rejecting variations in corrugation. Some attempts were made using fuzzy logic with some success on single configuration products. It will be remembered that single configurations were not the driving force for this project. When multiple in line configurations were run the response time of the fuzzy logic circuit gave drift in the controller continually as it adjusted its learning pattern.

The solution was to generate a set of patterns for any given sequence of product configuration. The controller could then use simple pattern recognition techniques to identify that the next detection had a high probability of being the cuff and not noise. Statistical tests were run to determine the spread of error due to any coincidental noise/high probability or deletion errors/low probability coincidence.

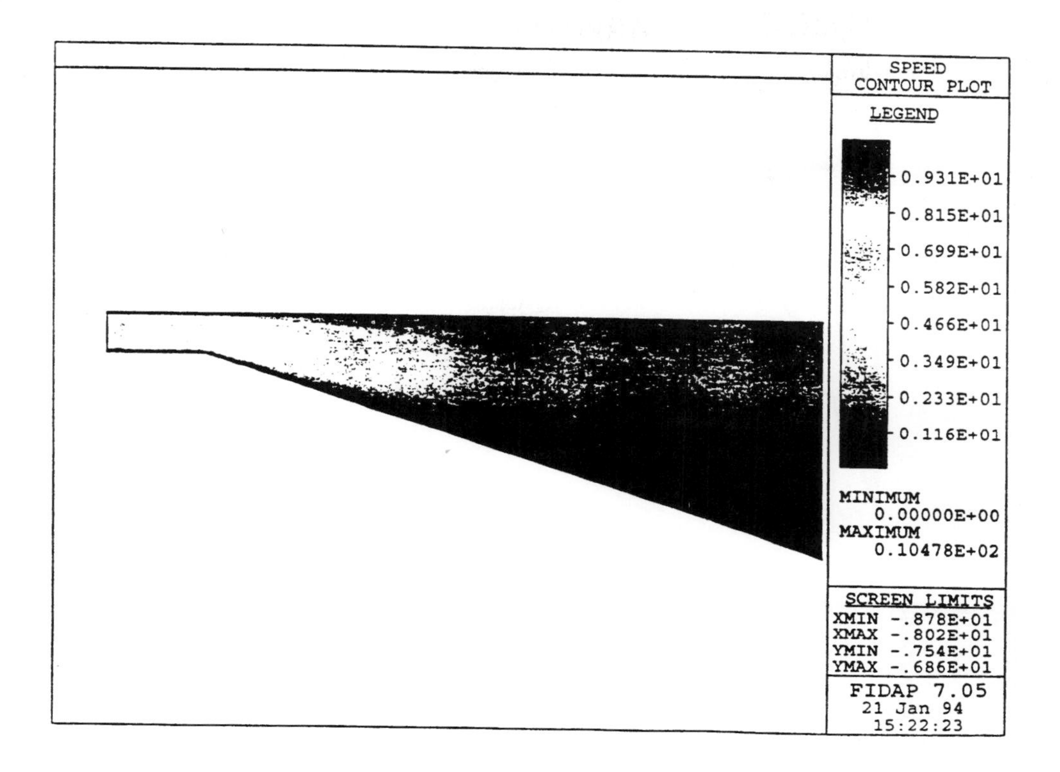

Figure 1

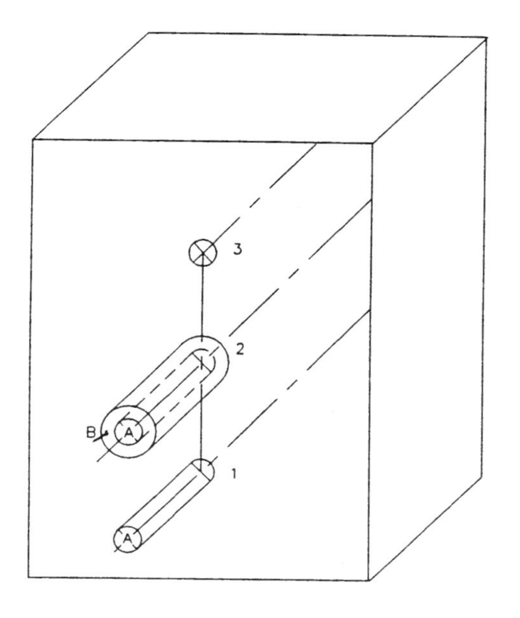

Figure 2

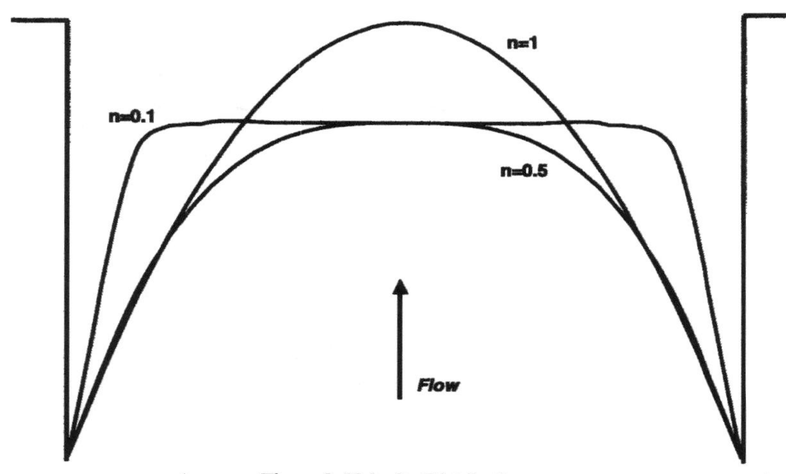

Figure 3: Velocity Distribution

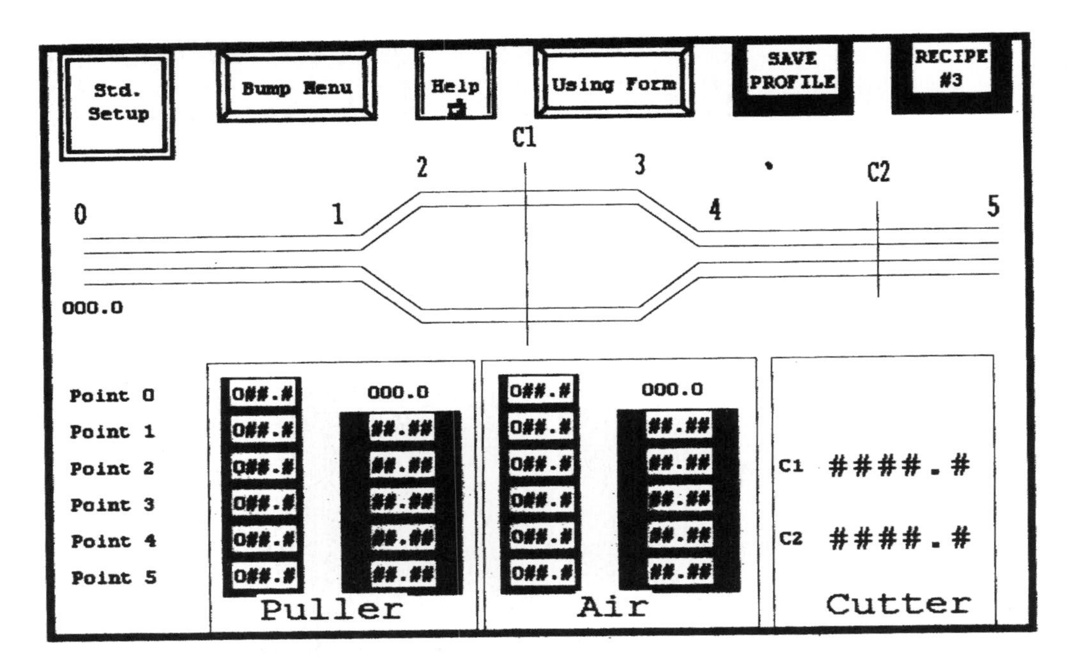

Figure 4

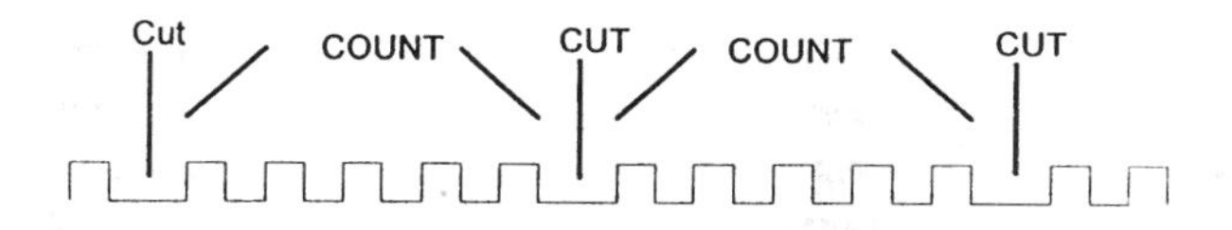

Figure 5

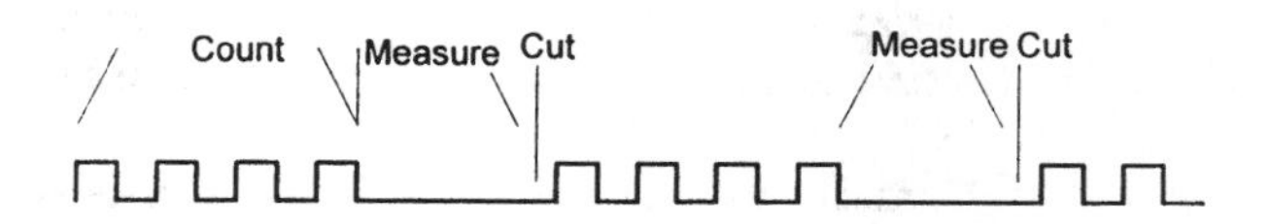

Figure 6

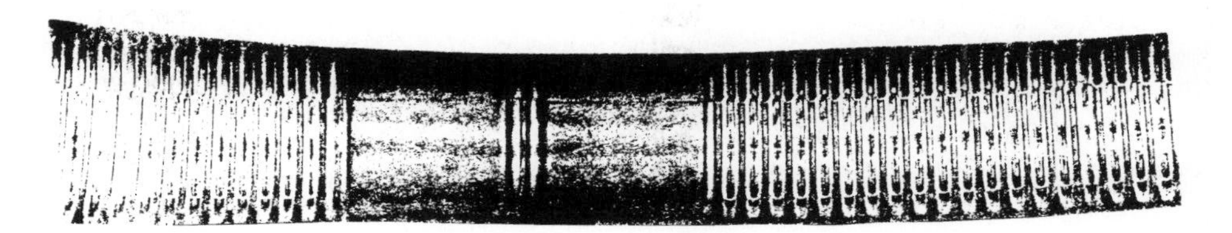

cuff

Figure 7

APPENDIX

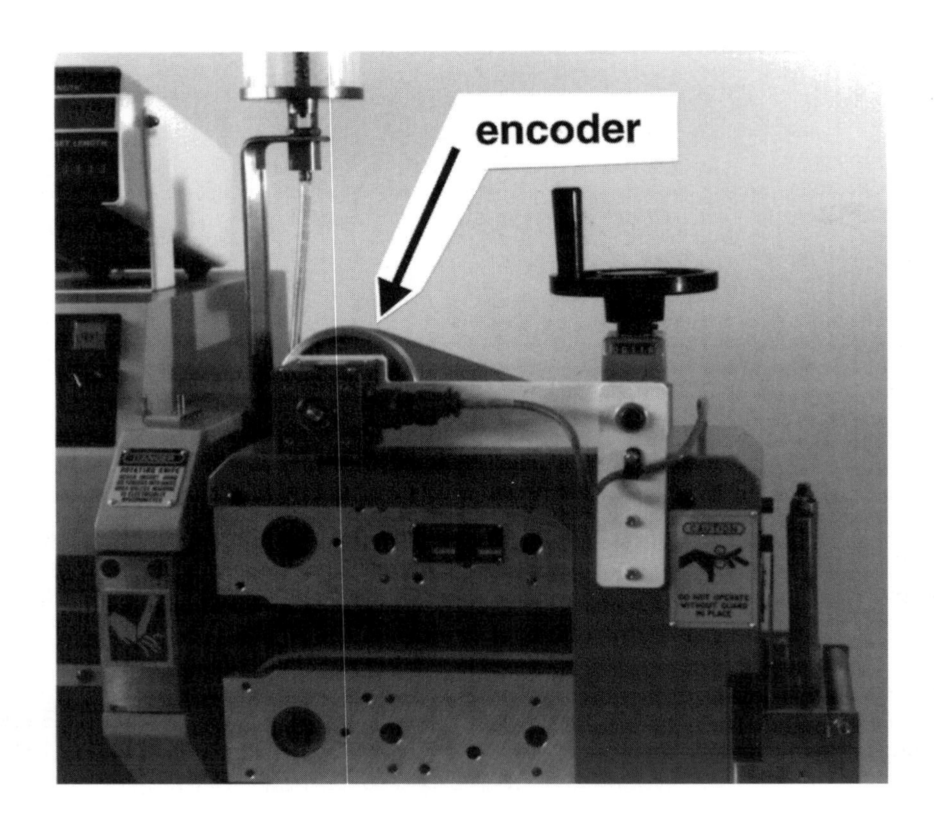

Scale up of Extruders, Practicalities and Pitfalls

J.A. COLBERT, TECHNICAL DIRECTOR
Betol Machinery Ltd

ABSTRACT
Most material and product development is carried out on a small scale, initially in Company Laboratories, Research Centres, or Universities. The successful developments are then put into large scale commercial production. However, some of these are doomed to failure at worst, or costly modifications, because insufficient attention has been paid to the task of scaling up.

This paper stresses the importance of understanding scale up especially the energy balance involved, not just at the time of designing production plant but also when carrying out the initial research. Two different approaches are used, one for single screw and one for twin screw extrusion.

SINGLE SCREW SCALE UP

This is not a new subject by any means, but despite the fact that some of the references used, and the basis for the approach advocated here, were published over 20 years ago, it is still poorly understood. Far too often processes are developed at a laboratory scale without due consideration being given to the longer term large scale production.

The extrusion process involves the transfer of energy from electrical power via viscous shear to an increase in the enthalpy of the polymer. This is then followed by a process whereby the enthalpy is reduced by transferring energy to a cooling media such as water or air. Hence it can be seen that, to fully understand the process and then achieve successful scale up, we must understand the heat transfer for processes taking place and the overall energy balance.

Energy Balance

An energy balance for a single screw extruder is illustrated in Figs 1 and 2. Essentially there are four main components:

Mechanical Work (E). The main drive motor, less any transmission losses, transfers its energy via viscous shear to the polymer.

Barrel Heater/Cooling (H). Heat transfer to and from the polymer is limited by the internal surface area of the barrel which is an important factor when considering scale up.

Enthalpy Change of Polymer (I). If the temperature of the feedstock and the extrudate can be measured then, by reference to enthalpy vs temperature data, available for most polymers, the energy input to the polymer can be found.

Heat Losses (S). These are difficult to calculate and are often considered as the balancing term in the equation. However, from work carried out on a range of extruders it has been possible to compile a table showing estimates for heat loss on a range of sizes and barrel temperatures. (Fig 3). These should only be used as a guide as obviously different guarding or insulation will make large differences to the heat loss.

These four components can be simply illustrated on a plot of Energy vs Screw Speed as shown in Fig 2. As output is broadly proportional to screw speed then, for similar temperatures, both the Mechanical Work input to the polymer, and the increase in enthalpy of the polymer will also be proportional to screw speed. With the barrel temperature set to a constant figure then the heat losses should be constant regardless of screw speed and at zero speed heat losses will be balanced by the barrel heaters. It can then be seen that, as screw speed is increased, the amount of barrel heating required to balance the equation decreases until an adiabatic condition is reached. Above this speed barrel cooling is then utilised more and more, increasing with speed. Obviously this is a simplification utilising a global view of the process, and really each barrel zone could be viewed separately. However, there is a range of screw speed over which the process will be running close to the adiabatic condition.

At this condition the process is almost independent of heat transfer from the inner surface of the barrel to the polymer or vice versa. This is an important consideration from the point of view of scaling the process. If the process is reliant upon heat transfer then the limiting factor becomes the surface area of contact between barrel and polymer. This will only scale up by the ratio of extruder diameters to the power of 2. The volume within the extruder, on the other hand, will increase by the ratio of extruder diameters to the power of 3. Hence, by operating at or near to adiabatic conditions we are giving ourselves the potential to scale up by a larger factor.

Thermal Scale Up

The thermal considerations of scale up were recognised by Schenkel [2] as early as 1963 where he utilised a "thermodynamic parameter" in his suggested rules for scale up. This parameter is defined as the ratio of, the increase in temperature due to barrel heating at the start of the process, over the total increase in temperature due to both barrel heating and viscous shear effects once melting has commenced.

$$\text{Thermodynamic parameter} \quad \psi = \frac{T_1 - T_o}{T_2 - T_o}$$

Although 'T_o' the feedstock temperature, and 'T_2' the final melt temperature were easy to measure, the real problem lay in establishing the value of 'T_1'. However, as the value of ψ must lay between O and 1 and it was possible to set rough limits on T_1, it was found that the thermodynamic parameter ψ normally lay between 0.3 and 0.7, although at the true adiabatic condition it would be equal to zero.

This approach worked fairly well for large, slow running polyethylene densification extruders where barrel temperature had a significant influence but tended to become inaccurate with faster running extruders.

DIMENSIONLESS NUMBER SCALE UP

Dimensionless numbers are the basis for scale up in a wide variety of applications utilising as they do critical dimensionless factors relevant to the particular process. One that most people will be aware of is

$$\text{Mach No.} = \frac{\text{Velocity of Body}}{\text{Velocity of Sound}}$$

where Mach 1 is the speed of sound, or the point at which a body travelling at that speed is said to "break the sound barrier".

Another that is used extensively in the analysis of fluid flow is

$$\text{Reynolds No. (Re)} = \frac{\text{Inertia Forces}}{\text{Viscous Forces}} = \frac{\rho\sigma d}{\mu}$$

This is said to get critical at Re = 2×10^3. Above this value the flow is turbulent, and below this flow is laminar. Reynolds No. for flow in an extruder channel is around 10^{-1} to 10^{+1} which is why we treat polymer melts as laminar flow.

To utilise dimensionless numbers however we need to define which numbers are relevant to our particular process. Work was carried out on this in the early 1970's and various papers on the subject were presented in particular some by Prof. Pearson [6]. These identified Graetz and Griffiths numbers, both dealing with thermal considerations, and Q/Wbh, a volumetric efficiency, as being the three numbers most relevant to the process.

Graetz No. (Gz) = $\frac{\text{Convection of heat from end to end of channel}}{\text{Conduction of heat to/from walls}} = \frac{H^2 V_z}{\alpha L}$

Where

V_z	=	Downstream velocity of Polymer	= $\frac{Q}{bh}$
h	=	Channel depth	
L	=	Length of section	
α	=	Coefficient of Thermal Diffusivity	= $\frac{K}{\rho Cp}$

As with Mach and Reynolds No there are significant levels of Graetz No.

Gz < 1	Temperature of polymer controlled by wall temperature. The flow will be fully developed.
1 < Gz < 10	The local wall temperature has less effect than does the previous history of the material.
10 < Gz	Wall temperature has no effect. All temperature changes will be caused by viscous heat generation.

Griffiths No. - Gf (or Nahme No. - Na)

= $\frac{\text{Temp. difference across channel generated by viscous polymer flow}}{\text{Temp. sensitivity of viscosity}} = \frac{b_o \mu V^2}{K}$

Where

V	=	Peripheral velocity of screw π ND
K	=	Thermal conductivity
b_o	=	Temperature increase needed to decrease viscosity by $\frac{1}{e}$

μ = Viscosity

The significant levels for Griffiths No are

Gf < 1	Insignificant generation of cross channel temperature gradients. Isothermal analysis will be valid.
1 < Gf < 10	Large cross channel temperature differences are generated. There will be errors in isothermal analysis.
10 < Gf	Isothermal analysis no longer relevant. Flow pattern dominated by generated temperature effects.

Volumetric Efficiency (Q/Wbh)

This number is obtained by taking the output equation for the metering section of an extruder, as derived from the Navier stokes equation, and making it dimensionless.

$$\text{Throughput (Q)} = \text{Drag Flow}\left(\frac{Wbh}{2}\right) - \text{Pressure Flow}\left(\frac{h^3 b}{12\mu}\frac{dp}{dz}\right)$$

$$\text{Hence } \frac{Q}{Wbh} = 1/2 - \frac{h^2}{12\mu W}\frac{dp}{dz}$$

Where Q = Volumetric Output
W = Peripheral Speed of Screw
b = Channel Width
h = Channel Depth

Significant values are

Q/Wbh	=	0.33	Maximum Pressure Generation
	=	0.5	Constant Pressure
	>	0.5	Pressure Drop Towards Die

By keeping these three numbers constant, and utilising the power low equation for viscosity, a set of scale factors can be established that are just dependent upon the power law index 'n' in the following equation;

Viscosity = Constant * (Shear Rate) $^{(n-1)}$

The scale factors, where D = ratio of screw diameter, are as follows;

	Newtonian	Non-Newtonian
Channel Depth	$D^{1/2}$	$D\left(\frac{n+1}{3n+1}\right)$
Screw Length	D	D
Channel Width	D	D

Screw Speed	D^{-1}	$D\left(\frac{-2n-2}{3n+1}\right)$
Throughput	$D^{3/2}$	$D\left(\frac{5n+1}{3n+1}\right)$
Power	$D^{3/2}$	$D\left(\frac{5n+1}{3n+1}\right)$
Specific Output Q/N	$D^{5/2}$	$D\left(\frac{7n+3}{3n+1}\right)$

SCALE UP IN PRACTICE

There are a variety of pitfalls one can find when scaling the single screw extruder, some o which are listed below;

a) Heat transfer - As mentioned earlier if the original process is dependent upon heat transfer to or from the extruder barrel then scale up is limited by the surface area. Unfortunately this is often the case with laboratory machines as the ratio of surface area to process volume is very high, and this ratio decreases significantly with production machines. It is therefore imperative that laboratory development should pay attention to the energy balance, and how close the operation is to the adiabatic condition.

b) Basic Design Factors - There are other dimensionless numbers related to the design of an extruder that also need to be kept constant. There are really simple ratios relating certain measurements to the diameter, such as

Length is expressed in diameters	L/D
Flight clearance is normally	D/1000
Flight width is normally	D/10

c) Screw Speed - This must be scaled according to the factors already set out. Too often laboratory extruders are run very fast and it is then not possible to achieve the scaled speed on the production machine.

d) Un-Scaled Items - The feedstock is obviously unchanged so feed throat design is more critical with smaller machines. This can mean that a laboratory machine may be feed limited whereas the larger machine is not. This can alter the nature of the process so attention must be paid to the possibility of a feed limitation.

e) Pressure Similarity - If an extruded product is being produced, such as a tube of certain diameter, then for a larger extruder, with a higher output, the pressure will also increase. This will again change the process considerably. For true scale up the pressure at the end of the screw must be close to the original pressure.

With compounding this is easily achieved by increasing the number of strands and keeping the output per die hole constant. With sheet a wider die can be used, or blown film a larger diameter die.

With tube however, where a fixed diameter is required, the pressure will change and hence the melt temperature of the product. This will then cause problems with sizing the product and downstream cooling.

f) Universal Screw - Different polymers do not scale in the same proportions hence a screw that works well on two polymers at one size of operation may scale up to two different designs at a larger diameter. This is illustrated in Fig. 4 where a 4.5 inch screw running polypropylene and nylon was scaled up to a 6 inch design. Even though this is only a small increase it can be seen that two different designs are arrived at. Eventually only one 6 inch screw was manufactured but the final design was a matter of using experience to produce a compromise design.

SCALE UP OF TWIN SCREW EXTRUDERS

This was the subject of a previous paper at this conference [7] and although dimensionless numbers or design ratios were used, the approach was more empirical. The critical design ratios to keep constant were;

Centre Line Ratio - The ratio of the distance between the centres of the shafts divided by the radius of the barrel.

Barrel Length L/D - Length of the barrel expressed in terms of the number of diameters.

Screw Clearance - This is expressed as a fraction of the diameter and is normally of the order D/100

As co-rotating twin screw extruders are segmental in design there is a lot of flexibility in screw design. Hence, although screw design should be similar, it is relatively easy to change the design slightly to modify the process. Also with the process being starve fed there is an extra variable that can be changed, feed rate, again to modify performance. This means that accurate scale up is not so critical as with single screw, but it still needs to be understood, and the pitfalls can be just as painful.

Essentially the output potential and power required will scale up under adiabatic conditions by the diameter ratio to the power of 3. If however, there is a need for some heat transfer the scale up factor will reduce to around 2.5. In practice the majority of co-rotating twin screw extruders scale up by a factor between 2.65 and 3.0.

CONCLUSION

Scale up should not be left until the production machine is required but should be included in the experimental planning to ensure that full consideration of scale up limitations is included in the development plan.

An understanding of the energy balance is critical to future scale up as thermal considerations form the basis of the key dimensionless numbers in the scale up technique. A knowledge of the viscosity/shear rate characteristics of the polymer, and hence the power law factor 'n' is required for scale up of the single screw extruder.

Co-rotating twin screw extruders can also be scaled up but with their increased design flexibility a more empirical approach can be adopted.

REFERENCES

1. Design Formulae for Plastics Engineers by Natti S. Rao, published by Hanser (1991).
2. Plastics Extrusion Technology and Theory by G. Schenkel, published by Iliffe (1963).
3. Polymer Extrusion by Chris Rauwendaal, published by Hanser (1986).
4. Polymer Melt Rheology by F. N. Cogswell, published by George Godwin Ltd., and PRI.
5. Flow properties of Polymer Melts by J. A. Brydson, published by George Godwin.
6. Pearson, J. R. A. Reports of University of Cambridge, Polymer Processing Research Centre (1969).
7. Scale up of Twin Screw Extruders by J. A Colbert, Polymer Process Engineering Conference July 1995 at Bradford University.

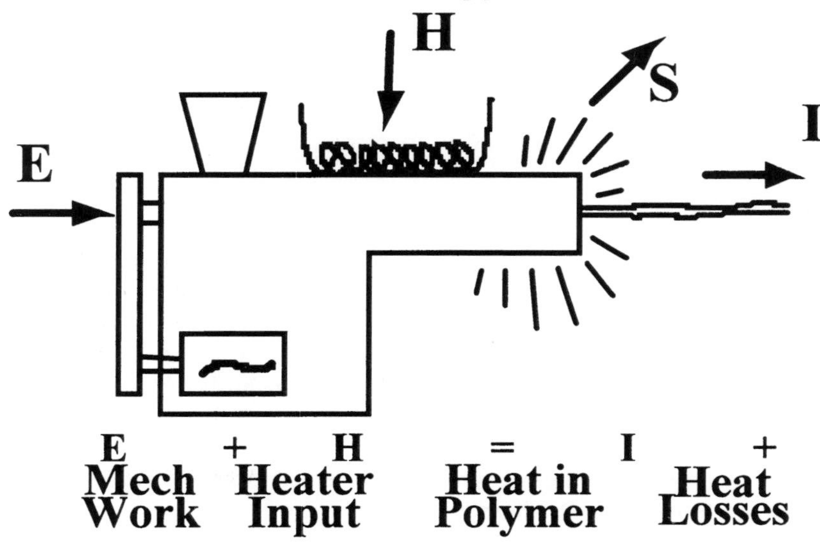
Extruder Energy Balance
H
S
I
E
E + H = I +
Mech Work
Heater Input
Heat in Polymer
Heat Losses

FIG.1

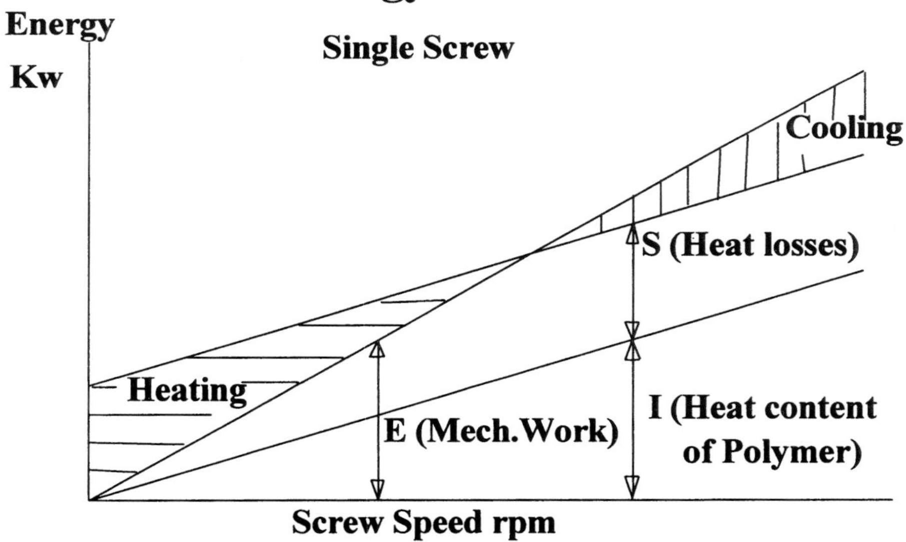
Energy Balance
Energy
Kw
Single Screw
Cooling
S (Heat losses)
Heating
E (Mech.Work)
I (Heat content of Polymer)
Screw Speed rpm

FIG.2

HEAT LOSS FOR SINGLE & TWIN SCREW EXTRUDERS

Barrel Temp Deg.C	Barrel Dia. inches	Barrel Dia. mm	Radiative Heat loss W/in2	Convective Heat loss W/in2		Heat loss factor		Heat loss KW	
				Single	Twin	Single	Twin	**30D single**	**35D twin**
150	2.0	50.8	1.20	1.12	0.73	0.029	0.037	**3.48**	**5.13**
250	2.0	50.8	3.16	2.28	1.48	0.068	0.088	**8.16**	**12.35**
350	2.0	50.8	6.51	3.58	2.33	0.126	0.168	**15.14**	**23.52**
150	4.5	114.3	1.20	0.73	0.70	0.024	0.036	**14.64**	**25.59**
250	4.5	114.3	3.16	1.48	1.43	0.058	0.087	**35.27**	**61.78**
350	4.5	114.3	6.51	2.33	2.24	0.111	0.166	**67.13**	**117.85**
150	6.0	152.4	1.20	0.70	0.70	0.024	0.036	**25.65**	**45.49**
250	6.0	152.4	3.16	1.43	1.43	0.057	0.087	**61.93**	**109.82**
350	6.0	152.4	6.51	2.24	2.24	0.109	0.166	**118.14**	**209.50**

FIG.3

SCALE UP using dimensionless number approach

ORIGINAL SCREW								SCALED UP DESIGN						
Screw Dia. (ins)	Channel depth			Output Kg/hr	Screw Speed rpm	Scale up factor "n"	Polymer	Screw Dia. (ins)	Channel depth			Output Kg/hr	Screw Speed rpm	Specific Output Kg/hr/rpm
	Feed (ins)	Inter. (ins)	Meter (ins)						Feed (ins)	Inter. (ins)	Meter (ins)			
4.5	0.600	0.410	0.205	400	90	0.90	Nylon	6.0	0.696	0.475	0.238	613	67	9.2
4.5	0.600	0.410	0.205	400	90	0.70	Nylon	6.0	0.703	0.480	0.240	607	66	9.3
4.5	0.600	0.410	0.205	400	135	0.50	PP	6.0	0.713	0.487	0.244	598	96	6.3
4.5	0.600	0.410	0.205	400	135	0.40	PP	6.0	0.721	0.492	0.246	592	94	6.3

FIG.4

CHEMICAL ASSIST FOAMING AND THE ROLE OF SUPERCRITICAL FLUIDS IN EXTRUSION

Michael E. Reedy
Reedy International Corporation (RIC)

ABSTRACT

Processing and properties of polymers are the two most important criteria that are many times in conflict with each other. The processor looks for the largest processing window while the designer looks for good surfaces, optimum density reduction, excellent thermal properties and great impact behavior. Now, chemical foam assist (CFA) technology enables both commodity and engineering polymers to process more easily and with improved properties for a wide variety of extrusion processes.

Carbon dioxide is used extensively as a physical blowing agent in the production of thermoplastic foams. For example, in the extrusion of foam sheet, carbon dioxide alone or in conjunction with other gases can produce light weight products with excellent physical properties. Recent studies have shown that CO_2 has unique low pressure solubility. In most polymers, CO_2 when pressurized to 1,700 psi becomes a supercritical fluid and acts as a solvent resulting in a lowering of the glass transition temperature and improved polymer melt flow characteristics.

During transition between a supercritical fluid and a gas, the CO_2 will absorb heat energy, improve melt cooling and consequently facilitate increased extrusion rates. The improved melt flow characteristics and heat energy absorption results in improved physical properties of extruded board and sheet materials.

GAS AND POLYMER SYSTEMS

A great deal of progress has been made in producing polymeric foams by introducing a gas into the polymer melt stream. Exothermic foaming agents, such as azodicarbonamide, generate N_2 when subject to temperatures above their decomposition temperature. Although economical, N_2 is a highly insoluble gas which can create problems with surface roughness as the gas breaks through the skin. Subsequently, gas counterpressure technologies were developed to pressurize the mold, assuring good skin formation and a class A surface. Direct gas injection systems are also used to introduce a gaseous or liquid blowing agent directly into the polymer melt stream. While very economical to use, they require a substantial capital investment and a properly designed screw. The next development is CFA technology which offers easier processing, better product quality, and the advantage of a drop-in additive. CFA is the process of using endothermic nucleating and blowing agents that decompose at process temperatures to create CO_2. CO_2 acts as a nucleant and foaming agent and creates a microcellular structure with a smooth solid skin around a fine cellular core. Particle size, distribution purity, and controlled gas release are tailored to provide many, very small nucleation sites that create this fine and uniform microcellular structure.

The permeation of the gas is also dependent on the polymer structure. Crystalline polymers exhibit much greater resistance to gas diffusion because the crystalline structure is tightly packed in lamellae crystals. The gaseous molecule is too large to penetrate these crystals. The interstitial space between the amorphous polymer molecules is large enough to allow the gas molecules to diffuse.

PROCESS IMPROVEMENT USING CO_2

CFA technology can improve melt flow characteristics, ensure uniform cell formation and facilitate increased extrusion rates. In many cases, the endothermic agent is not used to make foams, but will only be used to improve the melt behavior of the polymer as a processing aid. When most endothermic agents are added to the material blend, a solid pellet or powder is first converted to gaseous CO_2 through decomposition of the additive. Below 1,100 p.s.i., CO_2 acts as a lubricant improving melt flow. At higher pressures, above 1,700 p.s.i., CO_2 is converted from a liquid gas to a supercritical fluid. In this state, the CO_2 is more readily absorbed into the polymer melt and dramatically lowers the viscosity of most polymers. By lowering the viscosity of the polymer melt, one has the choice of reducing processing temperature or utilizing the improved melt flow at the same temperature. Improved melt flow can mean improved extrusion rates, less energy consumed and reduced burning through shear heat.

Probably the most important advantage of lowering viscosity is the possibility to achieve equivalent flow characteristics at lower melt temperatures. When high concentrations of gas molecules exist in the polymer matrix, there is significant plasticization of the polymer as evidenced by a decrease in the glass transition temperature (Tg). For specific microcellular PVC foams, the Tg drops from 78.2°C for virgin PVC to an estimated 14°C upon saturation. Bubble nucleation and foam growth are possible below the Tg of the original polymer through CO_2 saturation. When CO_2 changes phase from a super critical fluid to a gaseous bubble due to the drop in pressure, heat is absorbed. Similar to a refrigeration system, the changes in phase results in an endothermic process of heat absorption.

The saturated CO_2 exerts an internal pressure in the product that results in improved material flow characteristics which enables the processor to feed polymeric material very consistently. This internal pressure is similar to the pressure inside a soda bottle. Once the cap is removed, the pressure drops to atmospheric and the CO_2 begins to escape the liquid. Since the product is still saturated with CO_2 and is significantly plasticized, lower processing stresses are evident in the product. The material has more time for the molecular chains to relax and for the stresses to dissipate. Eventually, the gas will permeate through the polymer and escape to the atmosphere.

USING CHEMICAL FOAM ASSIST TECHNOLOGY TO ADD CO_2

For extrusion processing, we expect to see a major growth in the applications of sophisticated endothermic nucleating and foaming agents. This technology enables designers to produce closed cell sheet and board products with increased processing windows and with additional reductions in part weight while maintaining important physical properties, such as, rigidity and strength. To produce extruded CFA products it is necessary to add only .1% to .8% active level of an endothermic blowing agent. Endothermics are often formulated with carrier resins that also improve the melt strength and melt tension and carbon dioxide solubility. In considering the process design, you need to remember that the flow of the material will have been improved when using CFA technology, due to reduced viscosity of the polymer.

A critical element to successfully utilizing CFA technology for processing extruded materials is the die design and especially the die-lip design. For proper foaming, the die land needs to be as short as possible without adversely affecting surface appearance of the skin through melt fracture. A large die land will cause premature foaming, that is, foaming inside the die. This is because a large land allows the melt/gas pressure to drop below equilibrium and the polymer begins to nucleate and a cellular structure develops. The surface in a foam sheet will degrade because the newly formed cells will move to the outside and rupture open, throwing molten plastic back on the die face creating die fuzz. Common techniques, other than a new die, involve increasing extrusion output rate to increase melt pressure using a gear pump and/or lowering the melt temperature. Remember, using an endothermic nucleating agent will lower the viscosity and hence the equilibrium level of the polymer/gas mixture.

Closed cell foams have many discrete voids contained by polymeric struts and walls. Many of the unique properties of closed cell foams relate to the cellular structure and to the proportion of solid volume to the void volume, or density. Also since polymeric foams are composed of a solid phase and a gas phase, both phases will contribute to the behavior of the foamed material's physical properties. For example, during flexing, bending, and bucking of foamed materials, the cell walls contribute most during lower deformations, while the Boyles Law behavior of the trapped gas begins to dominate during increased deformation. It is the strength and integrity of the struts which give the higher density foams their high compressive and flexural strength and toughness. Likewise the pneumatic behavior of the trapped gas, as well as the resiliency and flexibility of the solid struts and walls, give low density foams excellent cushioning properties and flexibility. Also, the shape of the cells and the extrusion direction are two factors which play important parts in determining the physical properties. For example, in PET foams, the cell size and orientation directly affect crystallization and crystallization rate in thermoforming. In foam plank, the compressive strengths vary when tested in different directions. The density of the foam also determines how the foam will perform under compression. The lower the density, the less polymer contained in the foam and thus the contribution to compressive strength is less from deformation of the cell's walls.

CFA technology is applicable for processing polymeric closed cell foams for thermal insulation. Since foams have a small volume fraction of solid and a large volume fraction of gas, they will have a lower thermal conductivity than a solid body made of the same material. As you are aware, the heat transfer in foams takes place by conduction through the solid cell walls and struts, conduction through the gas-filled cell interiors and radiation throughout the foam. By combining CO_2 nucleation with CO_2 gas in combination with other improved lower thermal conductive gases, we can dramatically improve thermal insulation while decreasing V.O.C. (volatile organic compounds) and O.D. (ozone depletion) emissions.

To produce good surface finish extrusion with a good distribution of foam, it is necessary to observe both design and processing recommendations. CFA technology cannot only affect the mechanical and thermal properties of foams, it also has an impact on other foam properties, such as, surface texture and surface appearance. These characteristics are affected by cell size. It is recognized that small cells are generally more difficult to produce. The smaller the cell size, the more rapid the foam expands during cell formation and therefore, the greater the tendency for wrinkling or corrugation.

As foam is extruded, several seconds are required to reach the mandrel or vacuum sizers located a short distance from the die. These help apply a thin skin and control the dimensions and density. If a foam of two different cell sizes is being produced, the smaller size cell will expand first and will be stretched or elongated to form wrinkles or corrugations. CFA ensures that foams of identical cell sizes are formed and provides the processor with the best surface finishes.

From the processing aspect, most standard extrusion machines are capable of producing good quality CFA foams. The extrusion rate, melt temperature and the amount of nucleating agent will need to be adjusted to reflect the desired change in the viscosity of the polymer melt. When establishing processing conditions, it is important to change only one variable at a time. Remember that increasing the extrusion rate, usually increases shear heat so lowering the melt temperature as well works as a counter process change. A middle ground set of parameters can be quickly established. The die temperature control is also critical in maintaining consistent melt temperature throughout the die. In all cases, machine operators will find that the processing window for each polymer will be variably widened by using CFA technology.

CONCLUSION

What we have described here is the first state of chemical foam assist technology and the effect of CO_2 on the flow behavior of a polymer. Over the last ten years, a number of processes have been developed to produce extruded board and sheet materials with improved physical properties and process economics. Now with increased environmental pressure and a new emphasis on the use of CO_2, new opportunities are available to the processor. In some applications, CFA technology complements traditional direct gas injection applications. While the high pressure injected gas acts as the accelerator for the polymer melt, the CO_2 polymer solution improves melt flow characteristics which leads to improvements in overall performance.

As people begin to appreciate the benefits of chemical gas assist as a processing aid and not just as a way of solving die problems or reducing density of board and sheet products, I am sure endothermic chemical blowing agents will continue to have tremendous growth in the extrusion industry. I hope this paper has shown the benefits of CFA and endothermic blowing agents and their contribution to foamed materials.

ACKNOWLEDGMENTS

The author would like to acknowledge the following people and companies for their assistance.

Barry Penney, Pentex Sales
Walt Harfmann, Harfmann Technologies
Mike Caropresso, Caropresso & Associates
Larry Currie, Mack Molding Company

REFERENCES

Clark, Christopher and Williams, Rick, "Gas Assist Injection Molding: Controlling the Flow," Proceedings from the 23 Annual Conference, Structural Plastics Division, Boston, MA, April 2-5, 1995,

Currie, Larry "Challenge of Building and Injection Mold." Proceedings from the 23 Annual Conference, Structural Plastics Division, Boston, MA, April 2-5, 1995.

D'Orazio, Lawrence, "Technology and Applications of Closed-Cell Foams" Proceedings from Technology and Applications of Closed-Cell Foams, Lancaster, PA

Kumar, Vippin, "Microcellular Polymers: Novel Materials for the 21st Century," Cellular Polymers, 1992, Paper 6, pp. 1-7.

Goel, Satish K. and Beckman, Eric J., "Generation of Microcellular Polymers using Supercritical CO_2", Cellular Polymers , 1993, Paper 5, pp. 1-11.

Greele, Peter F. "Solid vs. Gas vs. Foam: Who Has The Best Ribs in Town?," Proceedings from the 23 Annual Conference, Structural Plastics Division, Boston, MA, April 2-5, 1995.

McLaughlin, Thomas, "Copolymer and Ionomer Foams" Proceedings from Technology and Applications of Closed-Cell Foams, Lancaster, PA

Wessling, M. et. al., "Carbon Dioxide Foaming of Glassy Polymers" Journal of Applied Polymer Science, Vol. 53, 1994, pp. 1947-1512.

Integrated Compounding Technology for the Preparation of Polymer Composites containing Waste Materials

C.E. Bream, E. Hinrichsen, P.R. Hornsby, K. Tarverdi and K.S. Williams
Wolfson Centre for Materials Processing, Brunel University, Uxbridge, Middlesex, UK

Abstract

A novel twin-screw extrusion compounding process is described for the preparation of polymer composites containing low cost reinforcing additives derived from waste products. The technology is exemplified using natural fibre reinforcements, from agricultural sources and fibre-reinforced thermoset scrap, added to a polypropylene (PP) matrix. Central to the method is the integration of a preparation step, which through comminution controls the physical size and morphology of the additive component.This is combined with a treatment stage, which allows the surface chemistry of the filler to be modified, thereby promoting interaction with polymer during subsequent melt blending. A significant enhancement in modulus and tensile strength of PP can be achieved by this method, particularly using well bonded linseed flax and comminuted woven glass-fibre reinforced thermoset. Results are discussed in terms of the influence of the compounding route on the microstructure and properties of the composites produced.

1. INTRODUCTION

In general, the modification of polymer properties through incorporation of additives relies on the use of intensive mixing machinery capable of imparting the desired homogeneity into the ultimate composition. To this end, processing plant based on variants of screw extruders and internal mixers have been developed, which can combine a diverse range of polymer matrices and additive feedstocks. Intermeshing co-rotating twin-screw extruders are particularly versatile in this respect, and are in widespread use, for example, to prepare mineral filled and fibre-reinforced thermoplastics compounds [1].

To achieve maximum performance from these additives however, it is necessary to optimise their structural form in the composite by maximising the extent of filler dispersion, or by maintaining fibrous species with a high aspect ratio. Effective stress transfer between polymer and reinforcing phase is also of paramount importance and can be promoted by the presence of chemical treatment at the additive/matrix interface. This may function as a chemical coupling agent between the composite phases, or simply aid wet-out and surface compatibility between organic polymer and inorganic species [2]. Conventional practice with such additives generally necessitates their preparation in a form close to that which is required in the ultimate polymer composition. Hence particulates with previously defined morphology and size are commonly surface treated prior to use in the compounding process. Similarly, continuous glass fibre filaments are normally sized and surface coated immediately after manufacture, then introduced into the polymer during compounding, either as a chopped strand or as a continuous roving, which is subsequently fragmented by the intensive nature of the mixing procedure. To minimise excessive breakage yet ensure good wet-out by the polymer, it is advantageous to delay fibre addition until the polymer has fully melted, by downstream addition into the compounder. Surface modification may also be applied through a masterbatch technique, which relies on the treatment diffusing to the interface between additive and matrix during melt compounding, but is generally considered to be less effective than filler pretreatment.

It will be evident, therefore, that established polymer compounding procedures normally separate the stages of additive preparation and combination with the polymer. Whilst this facilitates control and consistency of compound quality, additional costs are incurred.

This presentation describes an alternative compounding strategy, which combines stages of additive preparation and melt blending within a single process unit [3]. A key advantage of this integrated approach is its ability to utilize non-uniform forms of particulate or fibrous feedstock derived from waste sources and present them in a condition suitable for subsequent combination with host polymer matrices. This may involve, for example, size reduction, surface treatment, or moisture removal from the additive at minimal additional cost, thereby maximising the functional benefits of utilising waste products and increasing their potential for adding value to polymers.

The general principles of the technology will be explained, then exemplified by reference to the preparation of thermoplastics composites containing thermosetting plastics recyclate and natural fibres derived from arable waste.

2. PRINCIPLES OF INTEGRATED COMPOUNDING TECHNOLOGY

The method is based on modified twin-screw extrusion technology requiring assembly of well-defined process steps, which condition the additive phase, prior to its incorporation into the polymeric matrix. Hence in the early stages of the process only the additive is present, which is then combined with the polymer further downstream. As was mentioned earlier, this concept is the inverse of conventional extrusion compounding procedures.

Generalised functional stages of the process are shown schematically in Figure 1, and its operating characteristics are considered more fully below.

2.1 Feeding and shear modification of the additive feedstock

Depending on the physical form and transport behaviour of the additive feedstock, some limited pre-processing may be necessary prior to its introduction into the compounder. This is especially so when handling waste materials of the type considered later in this paper, which may require a preliminary shredding, or size reducing operation, to facilitate handling and enable consistent metering into the extruder. During the early stages of processing within the compounder, imposed shear within and between the screw elements can be applied to structure the feedstock further. This may include fracture and comminution of brittle materials, deformation, cutting and crumbing of ductile thermoplastics or rubbers, or pulping of natural fibre cellulosic products. In this context, the application of twin-screw extrusion procedures for the physical preparation of cellulosic pulp, derived from wood sources for use in the paper industry, is well established industrially.

Mechanical pulping is achieved by defibrillation and cutting of fibres between intermeshing screw elements thereby reducing their size to simplify handling in conventional bulk paper making operations. These principles can be applied to a variety of annual plant crops, including cotton, hemp and flax [4]. Combined chemical and mechanical pulping operations are also feasible using an optimum screw configuration and the introduction of active liquids. For example, alkaline peroxide bleached pulps have been produced by additions of sodium hydroxide, hydrogen peroxide and sodium silicate during the pulping process.

Twin-screw extruders can also be configured to effect disintegration of other solid materials, including plastics waste, through shear deformation in the solid state [5,6]. This provides a continuous pulverisation method utilising pressures and temperatures close to the melting point of the material, where energy is required to form a thin film of polymer over the screw flights. When this film is intensively cooled and accompanied by high shear, particle disintegration ensues. This action originates from the principles of high shear deformation of metals under the action of pressure and shear, which were established many years ago [7] and have been applied more recently to pulverisation of thermoplastics and elastomeric polymer waste [5,6]. Mechanical attrition of rigid thermosetting plastics can also be achieved by this approach, as will be demonstrated later. However, with such materials, the inclusion of process lubricants are beneficial to facilitate comminution by reducing energy input and minimising screw wear. Preferably, these processing additives are selected with a view to their inclusion in the ultimate composition.

2.2 Volatile extraction

A consequence of the size reducing stage is the internal generation of frictional heat, which if augmented by externally applied heat through the barrel wall, can aid removal of adsorbed moisture, or other volatile matter, present within the recyclate additive. The rate of volatile removal from within the process chamber is also greatly increased using a reduced surrounding pressure created by a vacuum extraction unit fitted a barrel exit port. Since material in the extruder is constantly being re-distributed during transport and broken down due to the applied shear, new surfaces are constantly being exposed, which also enhances the rate of volatile removal.

2.3 Application of surface treatment

As mentioned previously, promotion of adhesion between reinforcing and binder phases is generally considered a necessary requirement to achieve optimum properties. However, there are major difficulties in applying conventional treatment technology to recyclate materials, which may have irregular form or may undergo further breakdown during subsequent mixing operations, thereby exposing uncoated additive surfaces.

These problems can be ameliorated if treatment is introduced during the comminution stage, when only additive is present, and if necessary, during blending with the binder phase. Choice of the treatment system is critical as will be demonstrated later, and will strongly influence the ultimate mechanical performance of the composites produced.

2.4 Combination with binder phase

After judicious preparation of the additive reinforcement, it can be combined with binder material entering the primary extruder through a downstream delivery point. With molten thermoplastics or elastomers, this necessitates the use of a secondary screw extruder, in order to melt and transport this material. Resinous thermosetting polymer binder can be more conveniently introduced using a liquid displacement pump. Opportunities also exist for formulating the matrix phase, for example, by introducing pigments, prior to combining it with the recyclate. Co-rotating, intermeshing, twin-screw extruders are highly efficient and versatile mixing devices, which through optimum configuration of screws and mixing elements, together with appropriate use of operating conditions, can ensure effective wet-out and dispersion of additives within resinous or molten polymer binder. In this regard, their design and operating principles are well documented [7]. A further devolatilisation zone operating under reduced pressure, may also be incorporated during this stage to extract air or unwanted volatiles.

2.5 Die forming

The output from this process comprises natural fibre or thermoset recyclate reinforcement well blended with either thermoset resin or viscous thermoplastics melt. Pressurisation of the mixed composition permits extrusion into profile, including strand for subsequent pelletisation; sheet; or resinous dough for moulding into the final product form.

3. SPECIFIC APPLICATIONS OF INTEGRATED COMPOUNDING TECHNOLOGY

3.1 Preparation of thermoplastics composites containing natural fibre reinforcement

Natural fibres derived from renewable crops, such as wheat and flax, offer a potentially useful source of reinforcement for PP and other plastics matrices. In this regard, the inherent strength and stiffness of the fibres, their ultimate length after compounding and chemical affinity with the polymer, determine their effect on the mechanical properties of the composite. Since these reinforcing additives are thermally sensitive, it is also necessary to consider the effects of melt processing at temperatures close to 200°C on their structural and mechanical integrity. Previous work has considered the structural and compositional characterisation of these materials, using micro-mechanical testing and thermo-analytical techniques [9].

It has also been shown that the high pressures and shear forces encountered during melt compounding, exert damage to both fibre types, but in particular with wheat straw, where severe fibre attrition is evident [10]. This results in a significant reduction in fibre length and compaction of the cellular structure, causing some densification of the reinforcing phase. It has also been demonstrated that natural fibres, of the type under discussion, generally have no apparent chemical interaction with thermoplastics such as polypropylene. Chemical treatment of the fibre is a necessary requirement, therefore, to ensure good fibre-matrix bonding for optimisation of composite mechanical properties.

Linseed flax fibres are known to offer greater potential than wheat straw for the enhancement of thermoplastics mechanical properties. Indeed it has been demonstrated that polypropylene modified with linseed flax, has significantly higher stiffness and strength than equivalent compositions produced using wheat straw [10]

An important benefit of the integrated compounding technology considered in this paper is that low quality linseed flax may be upgraded for use as a polymer reinforcement during the initial stages of processing. Figure 2 shows how the pulping action of the twin-screws separates the fibres and releases debris associated with the outer epidermis and internode material. Although the quality of this pulp is inferior to continuous flax fibres, grown for textile purposes, processing costs are significantly reduced and less demands are placed on perfection of the crop species. A similar approach can be adopted to the preparation of wheat straw fibres, which after pulping, are reduced in length and become less cellular.

The need for fibre surface treatment to enhance interfacial bonding with the matrix is apparent in Figure 3, which demonstrates the lack of affinity between wheat and linseed flax straw in combination with polypropylene. Despite this deficiency, modulus of composites containing these fibres can be significantly enhanced, whilst tensile and Charpy impact strengths are maintained (Table 1). In this regard, as was inferred earlier, flax is a more effective reinforcement than wheat straw. It can be seen that promotion of interfacial bonding between fibres and polymer through the use of chemical modifiers, can result in further improvements to mechanical properties relative to neat polymer.

Since it is known that further fibre fragmentation occurs during the process of blending with polymer, particularly in the case of wheat straw [11], introduction of surface treatment *during* the compounding process increases the likelihood that coverage of newly exposed fibre surfaces is achieved.

3.2 Comminution and re-use of thermosetting plastics waste

The development of integrated polymer compounding technology has been instrumental in establishing an effective strategy for gaining value from thermoset waste. Emphasis has been given to fibre-reinforced polyester and phenolic scrap, derived from industrial sources, from which the principal aim has been to assess the reinforcing potential of such materials after comminution and subsequent inclusion into selected polymer matrices [12].

The successful re-use of such materials in this way depends on the need to reduce the size of the cured thermoset to a level where it is suitable as a functional filler. In this regard, the composition of the recyclate, particularly its glass fibre content, particle morphology and the length of residual fibre species, are critical factors in determining its reinforcing efficiency.

Since thermoset recyclate is a mixture of inorganic and organic components, as with naturally occurring fibres, the thermal stability of the resinous phase becomes an issue when the material is to be subjected to reprocessing at elevated temperatures, which is necessary for combination with thermoplastics or elastomers. As Figures 4 and 5 demonstrate, polyester based materials have less resistance to thermal breakdown than phenolic containing recyclate, although stability is both time- and temperature- dependent.

In common with alternative fibre-reinforced polymer composites, it is necessary to obtain effective stress transfer between fibre and matrix by enhancing interfacial bonding. As will be demonstrated below, this can be satisfactorily achieved by chemical modification of the filler phase during polymer compounding. Treatment can be applied during the comminution stage and if required, during subsequent additive blending with the polymer in molten or resinous states.

Uniform coating of comminuted thermoset prior to introduction into the extruder is problematic, due to the irregular and non-free flowing nature of this material. Furthermore, additional break-up of this filler continues to occur during melt mixing with polymer due to the high shear stresses encountered in this viscous medium. Creation of new surfaces in the filler at this stage would result in exposure of untreated reinforcement at the polymer interface. In such complex multiphase additive systems, selection of the optimum treatment package is critical, since ideally it must react with several different chemical species present.

Mechanical properties of polypropylene composites containing 30% by weight of thermoset recyclate are shown in Table 2. It should be noted that the DMC recyclate used in these compositions contained only 15% by weight of chopped glass fibres, in addition to 55% by weight of mineral fillers, with the remainder being cured polyester resin. The phenolic material originated from industrial scrap made from 80% woven glass impregnated with phenolic resin. These results reflect both the differences in glass level within these recyclate variants, the form of glass fibre present and the influence of interfacial bonding between additive and matrix phases. The performance of PP composites containing treated phenolic recyclate with high glass fibre content is most significant. Stiffness, strength and toughness greatly exceed polymer filled with calcium carbonate, at an equivalent loading, and approach the properties of commercial glass fibre-reinforced PP material. The exceptional impact strength of the phenolic recyclate composite is attributed to the presence of woven glass mat fragments which effectively inhibit crack propagation, whereas particulates of DMC recyclate are inherently weak and offer little resistance to crack development.

Pelletized compound has been successfully injection moulded into complex automotive parts as illustrated in Figure 6.

4. CONCLUSIONS

Specialised polymer compounding technology has been developed which enables waste materials, such as scrap or residue derived from industrial or agricultural sources to be physically and chemically modified, then combined with polymer within a single process unit. The technology integrates the additive preparation and combination stages in a more cost effective manner than is possible by separate unit operations and in addition, permits greater control of microstructure in the ultimate composite. By this means, the effectiveness of potentially reinforcing additives can be enhanced.

This process technology has been successfully applied to the re-use of polyester or phenolic based thermoset scrap and natural fibres, originating from annual crops such as linseed flax and wheat straw. In particular, the early stages of the process alter the physical form of the additive phase, either by controlled fracture and size reduction of brittle thermoset composites or by pulping of natural fibre species. Surface treatment can be applied during this stage and/or later, whilst blending additive and polymer matrix components. A key benefit of this approach is that treatment is constantly available to coat new surfaces which are inevitably created during the compounding process.

Application of this technology to the preparation of polypropylene composites reinforced with waste materials has demonstrated that marked increases in mechanical properties can be obtained when good fibre-matrix interfacial bonding is achieved. In this paper particularly, good strength, stiffness and toughness values have been reported for polypropylene containing suitably prepared linseed flax fibres and a woven glass-reinforced phenolic recyclate. It has been demonstrated that the use of waste products of this type have the potential to add value to polymers, through increasing mechanical performance, at minimal additional cost.

REFERENCES

1. Chapter 2 in Plastics Extrusion Technology, F. Hensen (ed), p26, Hanser ,Munich (1988).
2. Chapter 4 in Particulate-filled Polymer Composites, R.N. Rothon (ed), p123, Longman, Harlow (1995).
3. P.R. Hornsby and K. Tarverdi , UK Patent Application, (1996)
4. G. van Roekel, Paper 26, Proceedings from Conference on 'Uses for non-wood fibres - commercial and practical issues for papermaking', Peterborough, UK, 29-30 October, (1996), Pira International.
5. A. Riahi, J. Li, H. Arastoopour, G. Ivanov and F. Shutov, Society of Plastics Engineers Annual Technical Conference, ANTEC '93, New Orleans, May 9-13 (1993), p891.
6. H. Arastoopour, G. Ivanov and F. Shutov, 2nd International Conference on Cellular Polymers, Herriot-Watt University, Edinburgh, 23-25 March (1993) Paper 18.
7. P.W. Bridgeman, Physical Review, 48, 825, (1935).
8. Chapter 10 in Polymer Extrusion (2nd Edn), C. Rauwendaal, Hanser, Munich (1990).
9. P.R. Hornsby, E. Hinrichsen and K. Tarverdi, J. Mat. Sci., 32, 443-449, (1997).
10. P.R. Hornsby, E. Hinrichsen and K. Tarverdi, J. Mat. Sci., 32, 1009-1015, (1997).
11. E. Hinrichsen, M. Phil. Thesis, Brunel University, UK (1994).
12. M.J. Bevis, C. Bream, P.R. Hornsby, K. Tarverdi and K. Williams, Proceedings from 20th International BPF Composites Congress '96, Hinckley, UK, British Plastics Federation.

ACKNOWLEDGEMENTS

The authors are grateful to the following organisations for their support of the work reported in this paper through two EPSRC/DTI LINK research programmes: Cookson Plantpak (Crops for Industrial Use initiative), Alcan Chemicals, DSM Resins (UK), Balmoral Group, BIP Plastics, British Plastics Federation, Cray Valley, Croxton & Garry, Dow Europe, Filon Products, Ford Motor Company, Jaguar Cars, GKN Technology, Laminated Profiles, Owens Corning Fiberglas, Permali, Rover Group and Scott Bader (Structural Composites initiative). The collaboration with University of Nottingham in the last mentioned programme is also acknowledged.

Table 1. Mechanical properties of polypropylene/natural fibre composites (at 23°C)

Composition	Flexural Modulus (GPa)	Tensile Stress at break (MPa)	Charpy Impact Strength (kJm^{-2})
Unfilled PP	1.2 (0.01)	32.9 (0.09)	2.46 (0.26)
PP+ WS(a)	2.35 (0.01)	29.3 (0.3)	2.22 (0.17)
PP+ WS(b)	2.34 (0.03)	29.5 (0.09)	2.19 (0.04)
PP+ WS(c)	2.3 (0.11)	29.03 (0.19)	2.28 (0.21)
PP+ WS(d)	2.34 (0.04)	32.23 (0.2)	2.21 (0.13)
PP+ LF(a)	3.01 (0.04)	36.13 (0.1)	3.63 (0.22)
PP+ LF(b)	3.0 (0.04)	36.3 (0.74)	3.35 (0.2)
PP+ LF(c)	2.93 (0.05)	35.48 (0.3)	3.27 (0.09)
PP+ LF(d)	3.01 (0.06)	39.01 (0.2)	3.03 (0.39)

PP - polypropylene, WS - wheat straw, LF - linseed flax
(a) No fibre treatment, (b) with vinyltrimethoxysilane treatment, (c) with glycidoxypropyltrimethoxysilane treatment, (d) Including maleic anhydride functionalised PP
( ) Standard deviations
All fibre loadings 25% by weight

Table 2. Mechanical properties of polypropylene/thermoset recyclate composites (at 23°C)

Composition	Wt% Glass in Composite	Tensile Strength (MPa)	Tensile Modulus (GPa)	Notched Charpy Impact Strength (J/mm^2)
PP	-	26.5 (0.3)	1.8 (0.1)	7.7 (0.7)
PP/$CaCO_3$ (1)	-	20.4 (0.1)	2.1 (0.2)	3.2 (0.5)
PP/Glass Fibre	30	72.3 (0.6)	7.7 (0.4)	5.4 (0.2)
PP/DMC: Untreated	5	18.4 (0.9)	2.7 (0.3)	2.4 (0.2)
PP/DMC: Treated	5	29.1 (0.1)	3.1 (0.1)	2.4 (0.2)
PP/GWP: Untreated	24	28.9 (0.2)	5.2 (0.6)	4.7 (0.2)
PP/GWP: Treated	24	62.1 (0.4)	5.1 (0.2)	7.4 (0.2)

(1) 30% by weight filler
PP - Polypropylene, DMC - Dough Moulding Compound Recyclate, GWP - Glass (Woven Fabric) Phenolic Recyclate , ( ) Standard deviation

Figure 1. Integrated twin-screw extrusion compounding technology.

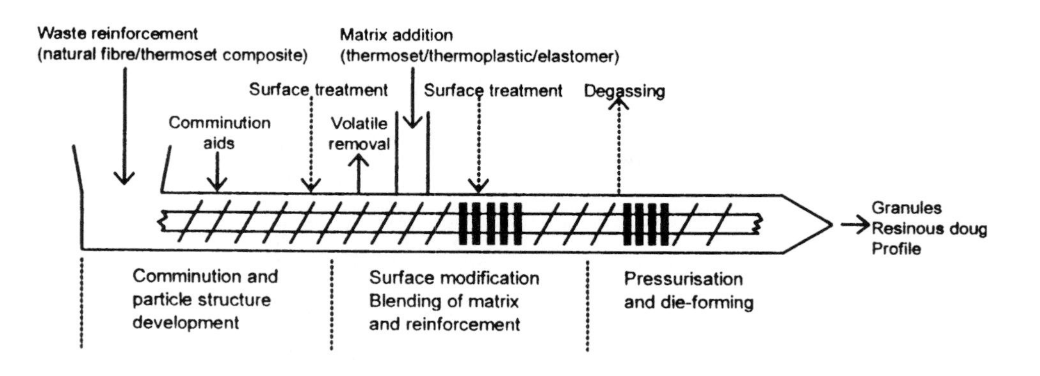

Figure 2. Linseed flax fibres and associated debris after extrusion pulping

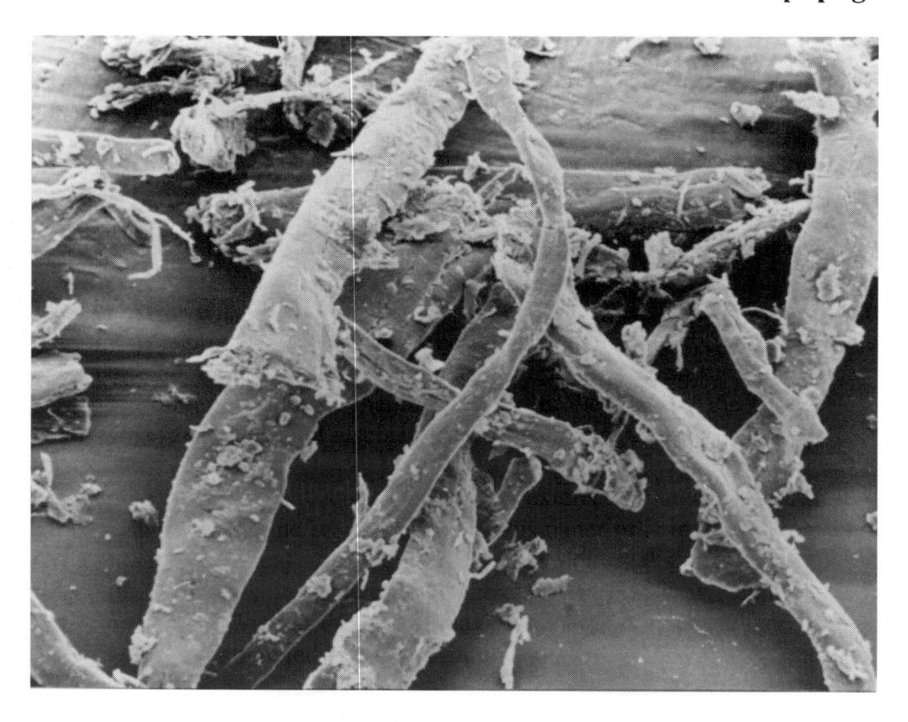

Figure 3. **Polypropylene compounds containing (a) linseed flax fibres and (b) wheat straw fibres. In both cases fibres were untreated**

(a)

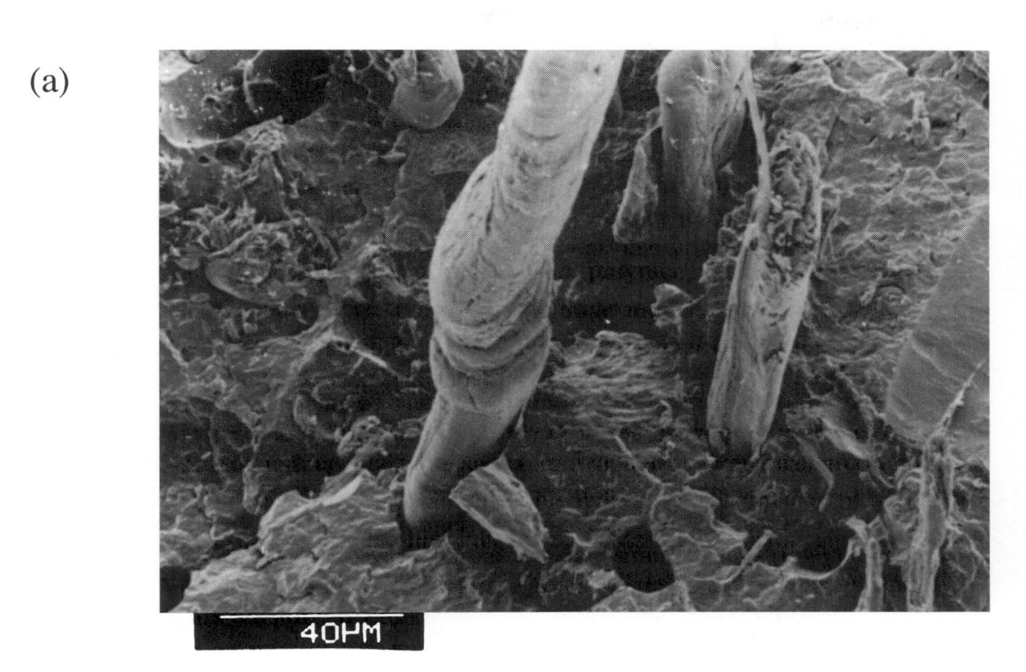

(b)

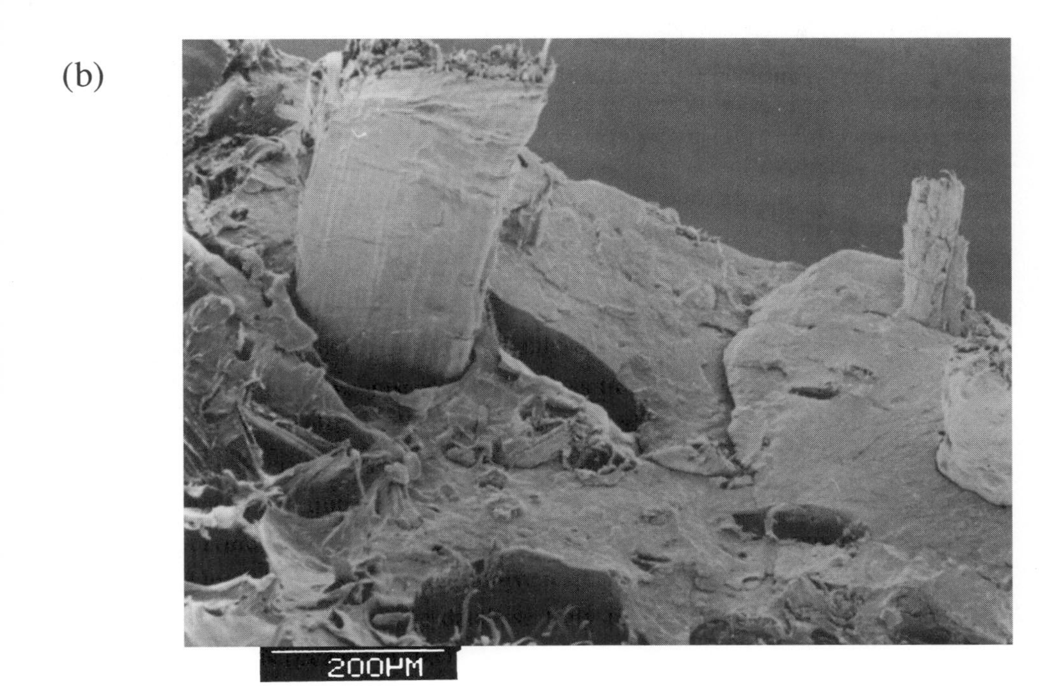

Figure 4. **Thermal Stability of DMC Recyclate: TGA isothermals (in air)**

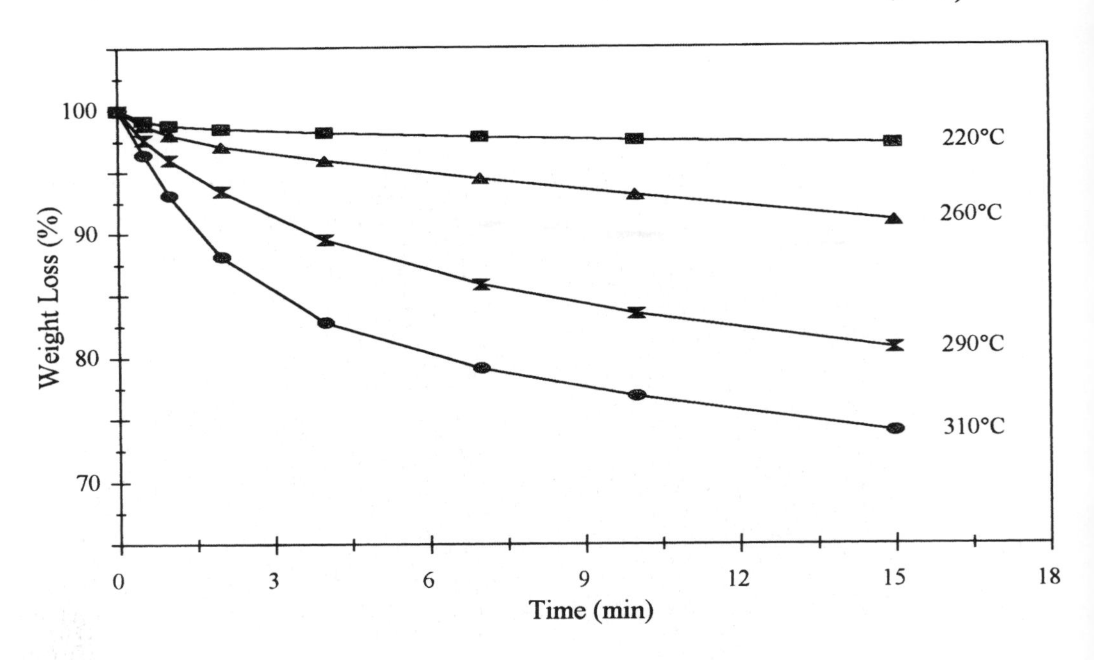

Figure 5. **Thermal Stability of GWP Recyclate at 290°C: TGA isothermal (in air)**

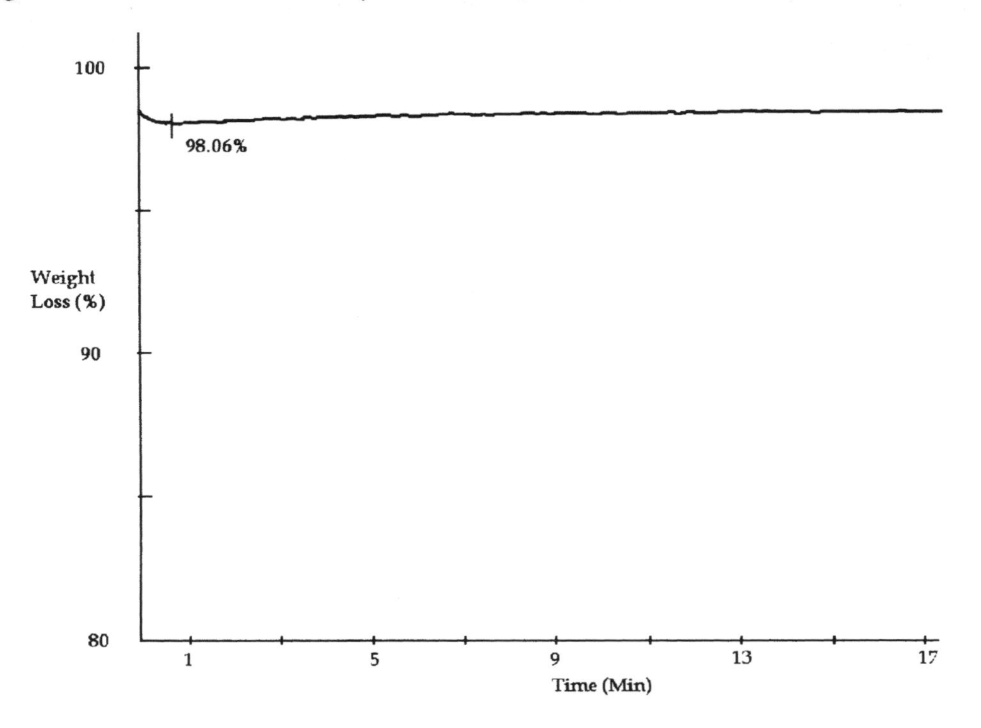

Figure 6. Automotive demonstrator component:

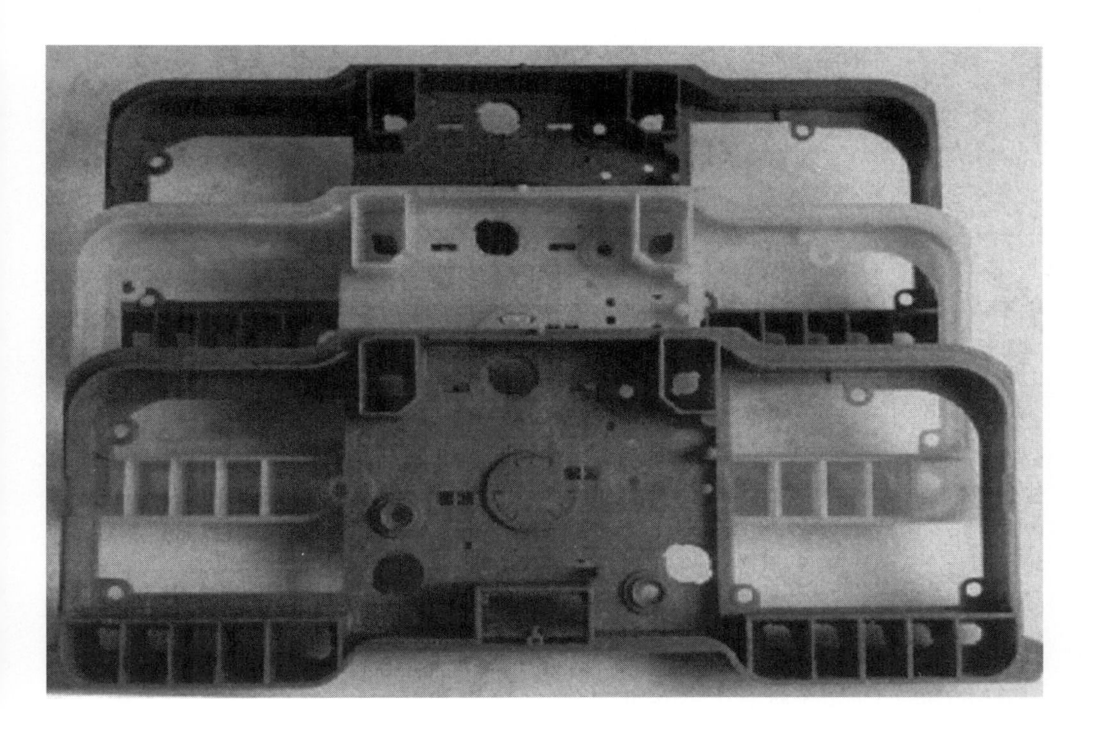

PP/DMC (front), PP/Glass fibre (middle), PP/GWP (back)

Simulation of the Injection Moulding Process - A Practical Validation

P A Glendenning
PERA Technology, Melton Mowbray, Leicestershire, UK

ABSTRACT
The work presented in this paper was part of the BPF Design Data Initiative Project 2/3 entitled 'Effects of Processing on Materials Properties and Final Product Dimensions'. The work was started in 1991 and concluded in 1994. A consortium of industrial companies sponsored the project, and additional funding was provided by the Department of Trade and Industry.

The first phase of the project reviewed the existing science base with regard to pre-mould analysis, prediction of the influence of various factors on shrinkage, warpage, & internal stresses in moulded parts, and the effects of processing on material properties. The second phase of the project involved practical moulding trials to validate mould filling software, and a comparison and practical validation of shrinkage & warpage prediction software. This was supplemented by an investigation into the effects of current moulding practices on the end properties of plastic components and a survey of the perceived usefulness of predictive software in industry.

In general, the packages tested showed reasonable agreement on fill patterns and were good for predicting qualitative trends in pressures and shrinkage/warpage. Prediction of quantitative values was more variable.

A lack of appreciation in industry of the factors which affect polymers prior to injection was identified together with a relatively poor understanding of the mechanisms which affect polymers at a molecular level. It was found that industry has a good appreciation of the usefulness of the software for mould filling simulation.

INTRODUCTION

The work presented in this paper was part of the BPF Design Data Initiative Project 2/3 entitled 'Effects of Processing on Materials Properties and Final Product Dimensions'. The work was started in 1991 and concluded in 1994. A consortium of industrial companies sponsored the project, and additional funding was provided by the Department of Trade and Industry.

The evaluation of injection moulding simulation packages for the prediction of mould filling, and for the prediction of shrinkage and warpage, was an important part of the project. Guidelines for injection moulded component design (especially with fibre filled polymers) were also produced, and conditions prior to injection and during the moulding process were reviewed with regard to their effect on final product quality. An assessment of current industrial practice and the level of industrial knowledge in the areas of injection moulding simulation and pre-injection process parameters was carried out and seminars were held to disseminate the results.

This paper aims to give an overview of the results from the project. With regard to the simulation packages mentioned, it is important to note that these are under continual development, and that along with advances in computer technology, the results available now are likely to be superior to those that were obtained at the time of the project, which itself covered a three year time span.

The Injection Moulding Process

In the injection moulding process, thermoplastic granules are fed via a hopper onto a plasticising screw which rotates to feed the plastic towards a mould. The granules are melted in the screw flights by a combination of conductive and shear heating and the screw moves back as a charge of molten polymer builds up in front of it.

At a set point the screw stops rotating and moves forward to inject the polymer melt into the mould. This forces the melt along runners and into the mould cavities where it cools and solidifies. During the cooling process, shrinkage of the polymer occurs and positive pressure (packing or holding pressure) is maintained on the melt to counteract the shrinkage by packing more polymer into the mould.

The Approach to Simulation

Filling analyses are reasonably well developed, and the majority of moulding situations can be analysed to predict whether specific moulding conditions will fill the mould, including where weld lines are likely to be located.

Prediction of shrinkage and warpage requires additional information such as the orientation of fibres or molecules, and details of flow history. To predict warpage, residual stress calculations have to be carried out. It is the distribution of residual stresses within the product which will determine whether warpage occurs and to what extent.

Mould Filling

During mould filling, material at the flow front advances in what is described as 'fountain flow' where material fed up the centre rolls back and hits the mould walls. Behind the flow front, the structure of the flow is more conventional. Flow conditions are highly non-isothermal and transient as material is continually cooling due to the low mould wall temperature and the flow length is continually changing.

A solid layer of frozen material builds up along the mould walls. However, significant forced convection and frictional heating also occur due to the flow velocity and the viscous nature of the material.

The viscoelasticity of the polymer varies according to the conditions experienced during the moulding cycle. In general, flow is modelled as viscous and under shear deformation only. Extremes of viscous and elastic behaviour occur during the injection and packing phases respectively, and in the transition regions complex constitutive equations need to be used to model the viscoelasticity.

In all of the packages studied, flow is regarded as two-dimensional because the majority of components are very thin compared to their length and width.

Other than viscosity, significant material properties that need to be accounted for during filling include density, specific heat capacity, and thermal diffusivity or conductivity.

Filling has an important influence on the shrinkage that occurs in the moulded polymer. This is because of two factors - fibre/molecular orientation and thermal/pressure history. The orientation of fibres and molecules is largely determined by the flow of the melt and it is this orientation which causes the anisotropy in moulded parts that results in differential shrinkage and directional variations in mechanical properties. The thermal and pressure history of the material influences its crystalline morphology (in the case of semi-crystalline materials) which also results in anisotopic shrinkage.

The Packing and Cooling Phases

The variation of density with temperature and pressure is fundamental to a simulation of this part of the process. The pressure-volume-temperature (PVT) relationship determines the thermal strains and residual stress distribution, which can result in warpage or distortion.

Therefore, the accuracy of PVT data is critical. It should be noted that there are certain limitations in measuring PVT data that can effect the simulation accuracy. For example, the cooling rate in the injection moulding process is much faster than can be used in measuring PVT data.

Software packages used

The software packages used for the mould filling simulation work were C-Flow 3.1 (AC Technology, USA),Fillcalc V (Rapra Technology, UK), Mefisto 4.0 (SIMCON, Germany) and MFL release 8 (Moldflow Pty Limited, Australia). For the shrinkage and warpage simulations, two leading commercial packages, C-Mold and Moldflow, were used. These are produced by AC Technology and Moldflow respectively. With the C-Mold package a layer 3 integrated analysis was run using the modules C-Pack/W version 3.2 and C-cool version 3.2. For the Moldflow analysis the following modules were used: MF/FLOW 8.0, MF/Cool 3.1 and MF/Warp 2.0.

SIMULATIONS & PRACTICAL VALIDATION

Methodology

Practical trials and software simulations were carried out using identical sets of processing conditions. The filling simulation results were evaluated practically by pressure measurements taken during moulding and by short shots used to show the mould filling pattern. The shrinkage and warpage results were evaluated by dimensional measurements and measurement of displacement along the walls of the component.

The simulations were run at Sintef SI in Oslo, Norway, and the practical validation work was carried out at PERA's Technology Centre in Melton Mowbray.

The mould tool used for the analysis was for a simple box component. Typical features likely to be of interest to a component designer were included, such as ribs, a boss and a slot to generate a weld line. Two geometries were defined for the tool, producing similar mouldings but with nominal wall thicknesses of 1.6mm and 2.2mm. A direct sprue feed was used into the centre of the box top. The geometry of the box is shown in Figure 1.

The materials to be used were chosen on the basis that they were fully characterised in all of the software programs to be evaluated and that they were readily available. The materials selected for the filling analysis were a grade of acrylonitrile butadiene styrene (ABS); Novodur P2H-AT, and a grade of 30% glass fibre filled polyamide 6,6 (nylon 6,6); Zytel 70G 30. For the shrinkage and warpage analysis a grade of polypropylene, Noblen W501, was used.

Filling analysis

Eight different sets of moulding conditions were used for each material. The parameters varied were fill time (injection speed), melt temperature, mould temperature and wall thickness. Details of the conditions used are given in Table 1 and Table 2.

Table 1 Moulding conditions used for comparisons between software and experiments for Zytel 70G30

	Fill time (s)	Nominal thickness (mm)	Melt temperature (°C)	Mould temperature (°C)
Z1	1.5	1.6	280	80
Z2	1.5	2.2	310	120
Z3	2.7	1.6	280	120
Z4	2.7	2.2	310	80
Z5	4.0	1.6	310	80
Z6	4.0	2.2	280	120
Z7	2.7	1.6	310	120
Z8	2.7	2.2	280	80

Table 2. Moulding conditions used for comparisons between software and experiments for Novodur P2H-AT

	Fill time (s)	Nominal thickness (mm)	Melt temperature (°C)	Mould temperature (°C)
N1	1.5	1.6	220	50
N2	1.5	2.2	250	90
N3	2.7	1.6	220	90
N4	2.7	2.2	250	50
N5	4.0	1.6	250	50
N6	4.0	2.2	220	90
N7	2.7	1.6	250	90
N8	2.7	2.2	220	50

For each material and set of conditions, simulations were carried out by using C-Flow, Fillcalc V (FCV), Mefisto and Moldflow (MFL) packages.

Pressure transducers were positioned in the tool and linked to a software program which would provide a graphical pressure profile. Details of the transducer locations are given in Figure 2. One transducer was placed in the nozzle of the injection moulding machine. Four other transducers were placed in the mould cavity. These were positioned such as to give pressure readings in the middle of each of the box side walls (one of which would contain the weld line), in the middle of one end wall, and in a location on the upper surface of the box.

Warpage Analysis

Eight sets of processing conditions were used for the shrinkage & warpage analysis. The conditions were produced by Sintef SI using a reduced factorial design within a processing window defined by PERA. It should be noted that the conditions used were extreme moulding conditions which were used to obtain particularly good and bad polypropylene mouldings. Many of these conditions, especially those with long cooling times or exceptionally high packing pressures, are unlikely to be used in a volume production situation. In this respect, the software packages have been tested to extremes as they are designed to predict shrinkage and warpage within the window of typical production processing conditions.

The Moldflow package includes warnings which are given when the processing conditions are outside its normal limits. Warnings occurred for each moulding trial where the packing pressure was exceptionally high.

Table 3 Moulding conditions used in shrink and warp trials

	Melt temp. [°C]	Mould temp. [°C]	Fill time [s]	Pack. time [s]	Pack. press. [MPa]	Cooling time [s]
W1	220	20	0.4	10	140	15
W2	250	20	0.4	5	40	15
W3	220	50	0.4	5	140	2
W4	250	50	0.4	10	40	2
W5	220	20	1.0	10	40	2
W6	250	20	1.0	5	140	2
W7	220	50	1.0	5	40	15
W8	250	50	1.0	10	140	15

Results from short shot mouldings

The short shot mouldings carried out clearly showed that different fill patterns were obtained when moulding ABS as opposed to those obtained when moulding glass reinforced nylon. The last point of fill for the ABS mouldings was consistently on the end of the box whereas for nylon it was consistently in the middle of the long side.

These differences can be attributed to differences in the rheological characteristics of the materials. Nylon has a low melt viscosity whereas ABS is a relatively stiff material in its melt state. However, a factor which may have a more significant effect is the glass content of the nylon. The glass fibres become aligned with the flow of the melt and they tend to become caught and reoriented at various points in the cavity where geometrical changes occur. This has an important influence on the flow of the melt.

Figure 3 illustrates typical short shots obtained for the glass filled nylon and Figure 4 shows simulated fill patterns for the last stage of filling. If the simulations are compared with the short shots, it can be seen that the simulations provide a good representation of the actual result. The work carried out indicated that reasonable agreement between the programs is usually obtained with regard to predicting fill patterns although slight discrepancies do occur. These can be attributed to differences in approach in the theoretical calculations.

Pressure measurements

An example of a pressure versus time plot obtained from the mould transducers is given for one of the trials (Z1) in figure 5. The actual fill time initiates at a point shortly after the nozzle pressure starts to rise and can be taken to be complete at the point of peak cavity pressure. The holding pressure plateau can be clearly seen in the nozzle pressure trace. Some problems with transducer drift were initially experienced but these were overcome by resetting the transducers after each set of pressure readings. The fact that the transducer 1 pressure does not drop back to zero in Figure 5 can be attributed to this.

The predictions of cavity pressure predicted by Mefisto were consistently higher than those predicted by the other packages. On average, it was found that the C-Flow and Moldflow pressure measurements were closer to the measured values than the other two packages.

There were many instances where significant differences occurred between measured and simulated pressures, as a result of which the pressure values obtained were somewhat inconclusive. A number of reasons can be suggested for this which in themselves relate to the way in which the packages were used and the ways in which the validation work was carried out.

It is possible for the user of C-Flow or Moldflow to specify a switch over time from injection pressure to holding pressure and to specify speed changes based on the screw position inside the barrel during filling. In this instance, the simulation packages were set to switch from constant flow rate to constant entrance pressure at a specified percentage of total fill. The differences between simulated and measured pressures can be affected by differences in approach relating to the switch over from injection to packing pressures.

Differences in fill times can also have a significant effect on pressure predictions because the specified fill time determines the initial flow rate into the cavity. Any variation in this flow rate will cause changes in melt temperature, viscosity, frozen layer thickness and pressure. Ideally constant flow rate would have been used as a measure for the trials rather than fill time. However, this would have caused practical problems in measuring the flow rate.

INDUSTRIAL VALIDATION

The questionnaire results showed that there is a general appreciation in industry of the usefulness of computer software to predict mould filling and to predict shrinkage and warpage of plastic injection mouldings. Areas in which the software is perceived to be useful include the selection of materials, positioning of gates to avoid weld lines, specification of wall thickness, weight reduction, avoidance of stress concentrations, and avoidance of sink marks. It is generally appreciated that the software assists in solving these problems before tools are made and consequently reduces cost and time to production.

It was found that often the software is expected to give information which is qualitatively accurate and in many cases it is accepted that quantitative accuracy is difficult to achieve.

In terms of the number of companies who use the software, the indication at the time of the project was that it is used by about 50% of large trade moulders. This does not necessarily include shrinkage and warpage software. In fact, the most widely used software is predictive software for mould filling. Not surprisingly, smaller companies do not use software as much as the larger companies. The percentage of companies using software in the UK would appear to be in-line with the situation in Europe as a whole.

There is a growing trend for large companies involved with injection moulding (for example, in the automotive industry) to require the use of software by their suppliers. Therefore, injection moulding companies who use predictive software are in a better situation to obtain major contracts.

The responsibility for providing software input to the design and production cycle is seen as involving all parties including material suppliers, end-users, moulding companies and toolmakers.

CONCLUSIONS

In terms of predicting filling patterns, the simulation packages provide a very useful tool with which costs can be reduced by eliminating the need to carry out expensive mould modifications at a later date. Filling pattern simulations were found to be in generally good agreement across the packages tested.

The accuracy of mould filling simulation packages is good enough to provide a good qualitative indication of when high or low gate pressures are to be experienced. Over the trials carried out, no single mould filling simulation package showed a consistently greater accuracy in terms of quantitatively predicting cavity pressures although Moldflow and C-Flow produced results on average closer to the practical pressures than Fillcalc V or Mefisto.

Differences in approach can occur in some instances between simulations and practical moulding methods. These differences can affect the measured and simulated results and it is important to bear these factors in mind both during set-up of simulations and set-up of moulding equipment. An example is the methods used to determine the switch over point from injection pressure to packing pressure.

C-Mold was found to under-predict shrinkage although good qualitative trends were shown with variations in processing conditions. Warpage was also under-predicted by C-Mold.

Moldflow was found to provide a more accurate prediction of quantitative shrinkage and warpage than C-Mold when normal industrial processing conditions are used (i.e. within the processing limits for which the package is intended).

Industry, in general, has a high level of appreciation for the usefulness of predictive software, mostly for mould filling simulation. The use of such software by large trade moulders is on the increase and companies who use the software are in a better position to obtain major contracts from large companies because there is a growing trend from such companies to require the use of simulation software by their suppliers.

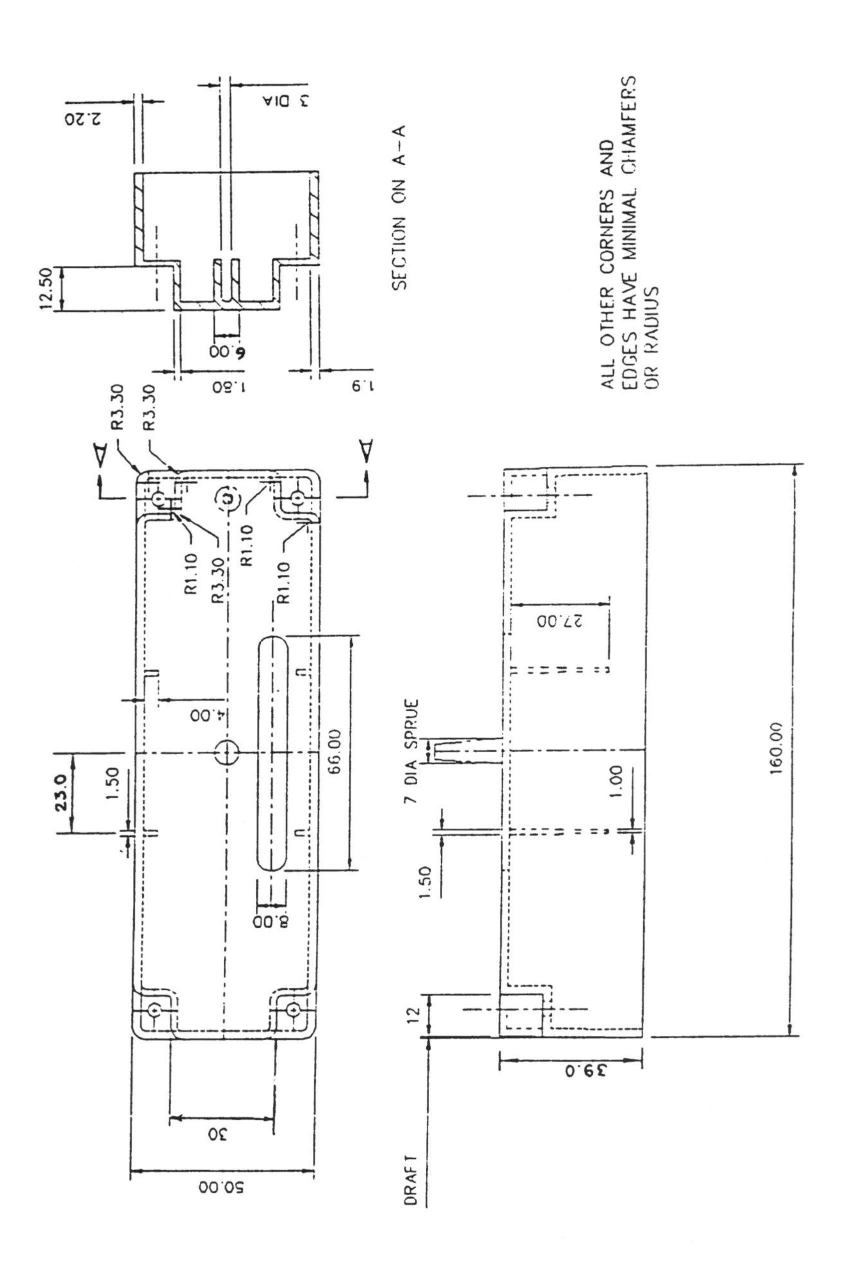

Figure 1 - Component geometry.

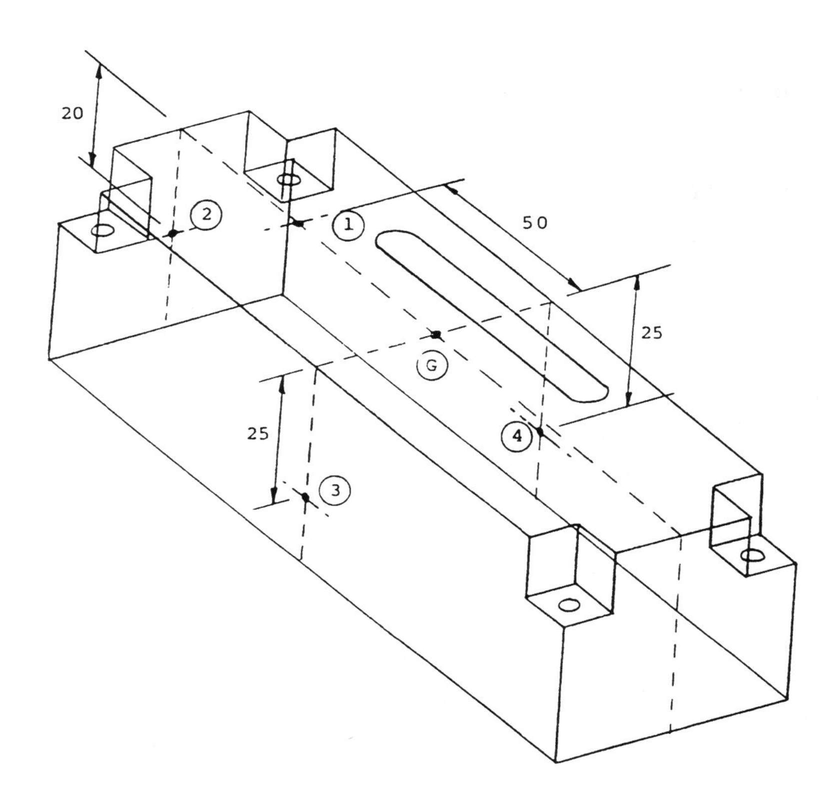

Figure 2 - Location of pressure transducers.
(G=gate, 1- 4 = transducers.)

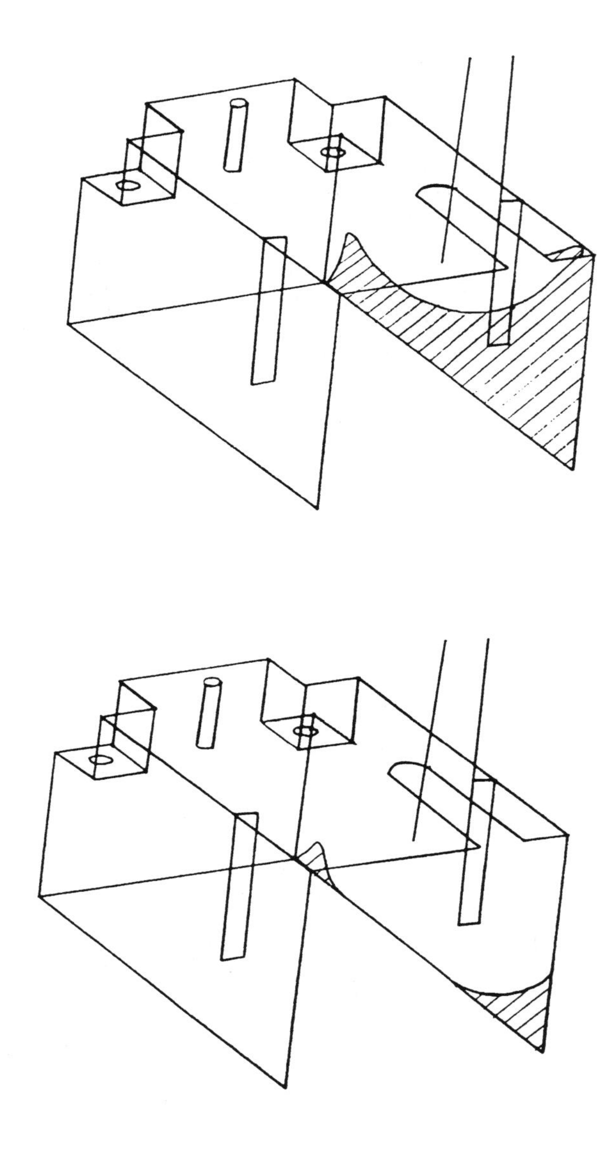

Figure 3 - Short shots at approximately 80% - 90% fill.

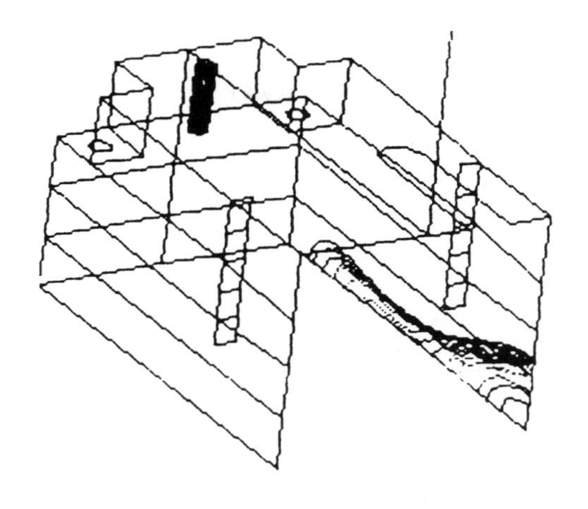

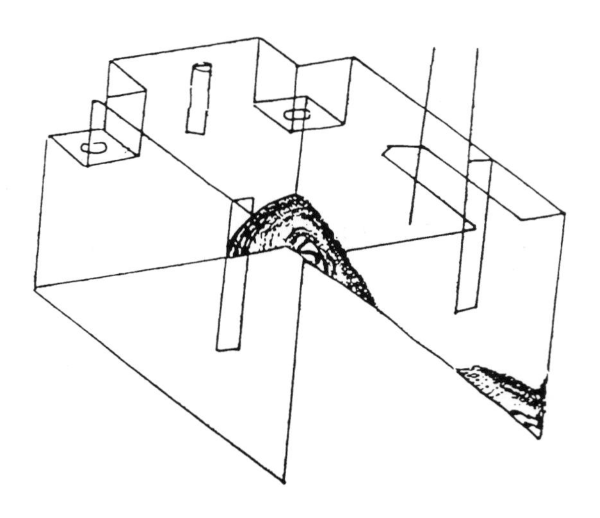

Figure 4 - Simulations for 97% - 100% fill for the glass filled nylon using Moldflow and Mefisto.

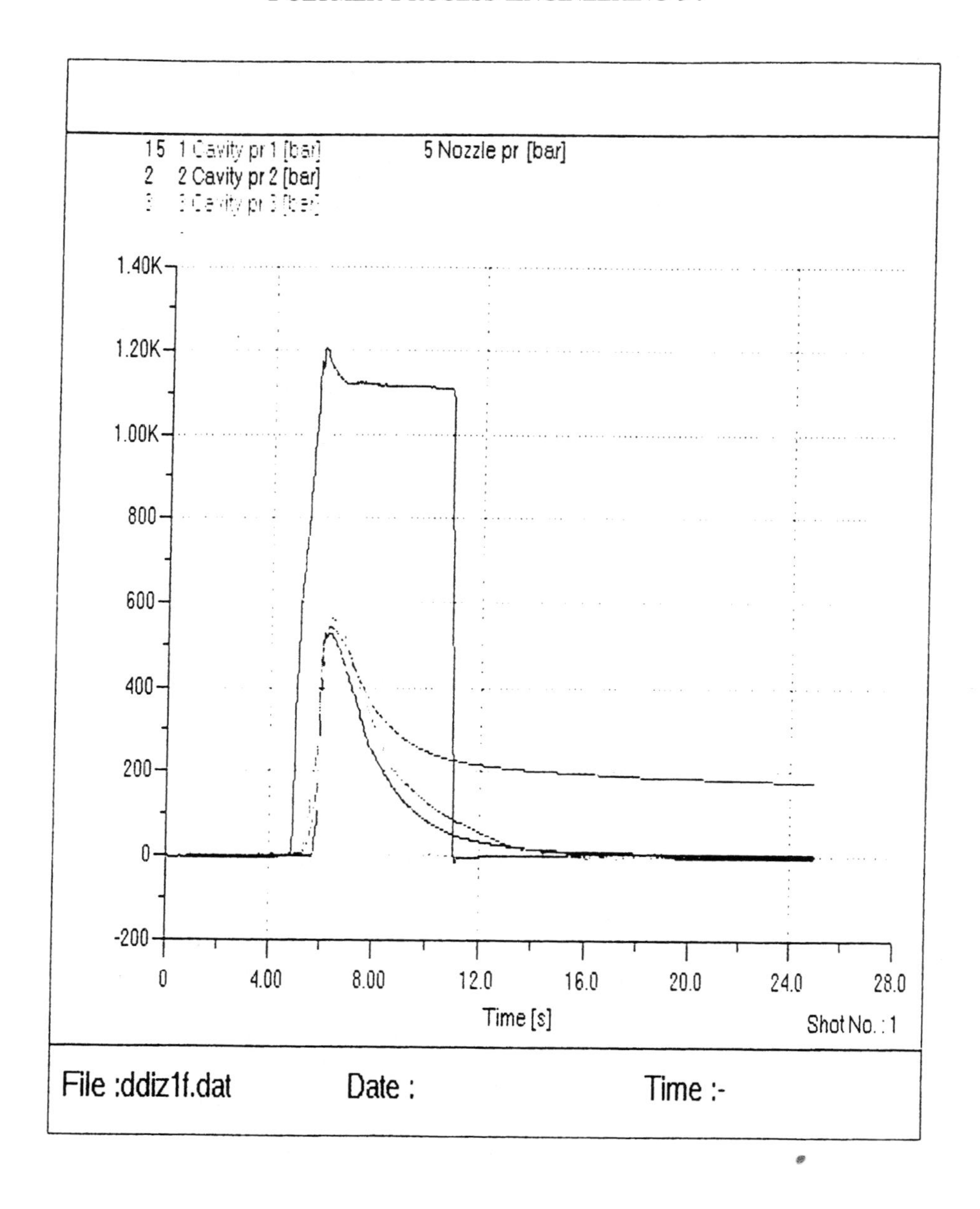

Figure 5 - Pressure vs time profile for moulding cycle.

Gas-assisted Injection Moulding Modelling

PATRICIA GORMAN
Plastic Moulding Consultants Ltd.

ABSTRACT
Gas-assisted injection moulding (GAIM) is a complex but versatile process for manufacturing plastic parts combining both thick and thin sections with less warpage, good rigidity, better surface finish. and lower clamp tonnage than conventional injection moulding. Computer-aided engineering (CAE) analysis of the gas assist process is a powerful tool for mould and part designers and process engineers for investigating potential problems and for establishing engineering know-how in this relatively new field.

OVERVIEW

This paper will give a brief description of the gas-assist process and of the many benefits and challenges associated with it. C-MOLD Gas-Assisted Injection Molding will be described, as an example of computer-aided engineering (CAE) tools available to the part, mould and process designer, including a typical example of its use in modelling the gas-assisted injection moulding process.

WHAT IS GAS-ASSISTED INJECTION MOULDING?

The gas-assisted injection moulding process comprises a partial (short) shot or full shot injection of polymer melt into the mould cavity, followed by an injection of compressed gas into the core of the polymer melt through the nozzle, sprue, runner or directly into the tool at one or several points. The compressed gas is usually nitrogen due to its low cost, ready availability and non-reactivity with the polymer melt. Depending on the machine and gas-injection system used, gas is injected under either gas-pressure control or gas-volume control. With gas-pressure control, gas is injected at a regulated pressure profile (constant or step). With gas-volume control, gas initially is metered into a compression cylinder and injected under compression with movement of the plunger.

During the gas injection stage following a short shot polymer injection, the gas normally takes the path of least resistance to catch up to the melt front where the pressure is lowest. To facilitate the gas penetration in a desirable pattern, the moulded part typically has a network of built-in, thick-sectioned gas channels. In principle, the gas penetrates and hollows out the network of gas channels, displacing molten polymer at the hot core to fill the entire cavity. This is known as the primary gas penetration stage. When the cavity is full, gas will continue to penetrate through thicker, hotter sections compensating as the polymer cools and shrinks. This is the secondary gas penetration, when the gas is used to pack out the part. Figure 1 shows the partial shot method.

In the full shot method, gas is introduced only for packing. In both the primary and secondary gas penetration stages, the relatively inviscid nature of gas allows efficient pressure transmission independent of flow length, hence the uniformity of pressure distribution possible with this process. Figure 2 shows the full shot method.

ADVANTAGES AND DISADVANTAGES OF THIS PROCESS

The potential benefits of using gas assisted injection moulding are enormous. Part cost can be reduced by:-

- Reducing the quantity of material used in the part by coring out thick sections and by allowing thinner wall sections elsewhere than would otherwise be possible;
- Reducing the quantity of cold runner scrap material by coring out the runner system;
- Producing the same part on a smaller machine with lower running costs;
- Decreasing cycle time by reducing wall thickness and hence time for cooling;
- Decreasing cycle time by producing a part with less moulded-in stress which can therefore be ejected safely at a higher temperature;
- Allowing consolidation of several assembled parts into one complex part containing both thin and thick sections.

Part quality can simultaneously be improved by:-

- Allowing flatter parts with less moulded-in stress to be produced;
- Increasing part rigidity;
- Reducing sink marks and improving surface finish.

On the other hand, several challenges are posed by the technology. As the process involves the dynamic interaction of two dramatically dissimilar materials flowing within generally complex cavities, the product, tool and process designs for gas-assisted injection moulding are quite complicated. Furthermore, previous experience with the conventional injection moulding process is no longer sufficient to deal with this process, especially in designing the gas channel network and optimising the processing window. Problems can arise due to:-

- Melt "race-tracking" i.e. flowing preferentially along gas channels creating gas traps;
- Gas blow-through at the melt front;
- Undesired gas "fingering" or gas permeation into thin areas;
- Excessive sink marks in gas channels where the gas has failed to core out;
- Over-sensitivity of the gas penetration pattern to slight variations in the process or material;
- Short shots;
- Undesired gas weld-line locations.

WHAT DOES C-MOLD DO AND HOW THIS CAN HELP?

C-MOLD Gas-Assisted Injection Molding simulates the initial filling of a cavity with polymer, followed by the injection of pressurised gas through the nozzle, sprue, runner, or directly into the tool, at a specified time, under gas-pressure or gas-volume control or with automatic gas-pressure profiling. The analysis can handle a three-dimensional, thin cavity with a built-in network of thick-sectioned gas channels; and a melt-delivery system that may contain cold or hot, circular or non-circular runners, and valve gates.

Depending on the machine and gas-injection system used, C-MOLD can simulate the process where gas is injected under either gas-pressure control or gas-volume control. With gas-pressure control, gas is injected at a regulated pressure profile (constant or step). With gas-volume control, gas initially is metered into a compression cylinder and injected under compression with movement of the plunger.

With the automatic gas-pressure profiling option in C-MOLD Gas-Assisted Injection Molding, the analysis provides an idealised gas-pressure profile that is calculated based on the melt-flow rate the user specifies. This idealised gas-pressure profile can be used as a reference for specifying the actual machine pressure settings, whether gas injection is under gas-pressure control or gas-volume control.

Furthermore, in the model of the part, multiple gas entrance locations may be specified, and the gas entrance(s) may be different from the polymer entrance(s).

These features make it possible to simulate the various gas-assisted injection moulding processes that are commercially available.

With C-MOLD Gas-Assisted Injection Molding, many potential problems mentioned earlier can be detected prior to making the mould. Gate locations, gas inlets, runner sizes, and part thickness can be determined from the predictions. In addition, optimum process conditions, such as fill time, timer for gas injection, gas injection time, gas pressure and volume, and melt temperature, can be identified.

As mentioned above, C-MOLD Gas-Assisted Injection Molding provides three options to simulate the different methods of gas-assisted injection moulding. All three of these options account for primary gas penetration during the filling stage, as well as secondary gas penetration due to polymer shrinkage during the post-filling stage. In more detail, these options are:-

- *Gas-pressure control*
  The gas-pressure control option allows you to specify a variable gas-pressure profile (constant or step) during gas injection, which includes gas injection in both filling and post-filling stages. An optional delay time is incorporated to account for time that may elapse after resin injection ends and before gas injection begins.
- *Automatic gas-pressure profiling*
  The automatic gas-pressure profiling option allows you to determine reasonable gas-pressure range to achieve a melt-flow rate that you specify. No delay time is used in this option, in order to produce an uninterrupted melt-flow pattern.
- *Gas-volume control*
  The gas-volume control option simulates processes that compress a fixed amount of gas within a cylinder, then push it with a plunger to displace polymer melt in the cavity. Again, an optional delay time is incorporated to account for time that may elapse after resin injection ends and before gas injection begins.

C-MOLD Gas Assisted Injection Molding is based on a hybrid finite-element-finite difference/control-volume numerical solution of the generalised Hele-Shaw flow model, and an innovative material-tracing scheme to account for gas penetration. Figure 3 shows the layers through a cross-section of melt and gas flow. The viscous polymeric melt under non-isothermal conditions is assumed incompressible during the filling stage when it interacts with pressurised gas (primary gas penetration). Upon completion of cavity-filling, the analysis continues through the post-filling stage to account for secondary gas penetration due to volumetric shrinkage of the polymer melt. Because of the large temperature and pressure variations during the post-filling stage, a pvT data model is required to describe polymer behaviour during the entire process.

Although the gas properties may change depending on the process conditions and the state of the gas, their variation is negligible compared with the significant difference between the properties of the polymer melt and the gas. Therefore, the representative properties of the gas are incorporated in the analysis, and there is no need to assign the gas properties.

OBJECTIVES OF PERFORMING CAE ANALYSIS

C-MOLD Gas-Assisted Injection Molding allows designers and engineers to evaluate the processability of a design, optimise design and processing parameters without time-consuming mould trials, and reduce development cost and lead time by numerically evaluating and verifying alternatives.

The analysis provides comprehensive numerical predictions of relevant variables for the entire gas-assisted injection moulding cycle. Using this information, designers and engineers can:-

- **Analyse the flow of polymer melt and gas based on the current tool and process design to reveal problems, such as air traps, uneven gas penetration, undesirable polymer skin thickness, gas permeation into thin sections (fingering), or gas blow-through;**
- **Determine the optimum volume and weight of polymer melt and the best switch-over time to gas injection;**
- **Determine the final gas volume (or, equivalently, hollow-section volume) percentage in the moulded part;**
- **Determine the ideal gas-pressure profile based on a specified melt-flow rate, to achieve a reasonable range of gas pressures and reduce the number of mould trials;**
- **Evaluate processing conditions with different gas pressures (or gas volumes) and switch-over times;**
- **Calculate the spatial distribution of polymer and gas throughout a single cavity or a multi-cavity system;**
- **Judge the resulting polymer skin-thickness distributions for mechanical strength considerations or other concerns;**
- **Test alternative product or tool designs by changing the delivery and gating systems or the size, locations and connectivity of the gas channels until the desired result is achieved;**
- **Minimise clamp force with low but efficient gas pressure, and thereby reduce energy consumption and machine cost;**

- **Reduce part weight and material costs while maintaining quality standards and part rigidity;**
- **Prevent sink marks on thick sections and reduce the potential for part warpage.**

EXAMPLE : TELEVISION FRONT BEZEL

This example presents a front bezel for a 27-inch colour television, shown in Figure 4. C-MOLD Gas-Assisted Injection Molding is used to validate the modification of a conventional tool design for gas-assisted injection moulding.

The primary reasons for using gas-assisted injection moulding for this application are:-

- **To reduce the clamp force requirement;**
- **To reduce part warpage by achieving a lower and more uniform pressure distribution in the cavity;**
- **To increase part rigidity by incorporating a built-in gas channel network.**

To arrive at an acceptable design for gas-assisted moulding, several analyses may need to be performed. Results from C-MOLD Filling analysis of the part designed for conventional injection moulding (no gas channels) are useful for comparing with the results from C-MOLD Gas Assisted Injection Molding of the proposed part design which incorporates gas channels.

In addition, it is helpful to run an initial analysis of the design for gas-assisted moulding using C-MOLD Gas Assisted Injection Molding's automatic gas-pressure profiling option to obtain a reasonable range of gas pressures that will achieve a constant melt-flow rate. This pressure range serves as a reference when running subsequent analyses using either gas-pressure control or gas-volume control.

Preparing the Design for Analysis: Modelling

To generate a desirable filling pattern for both polymer and gas, a network of gas channels has been designed into the bezel, half of which has been modelled for the purpose of analysis.

The wall thickness over most of this part is 2.7 mm, compared to a wall thickness of 3.0 mm in the part designed for conventional injection moulding. This 10% thickness reduction is acceptable, since the melt-flow characteristics and part strength are to be enhanced by thicker gas channels.

Material Properties

The material used in this example is a polystyrene (PS) resin. The material property constants used in this analysis are given below, and graphs of the viscosity and pvT (pressure-volume-temperature) data are shown in Figures 5 and 6 respectively.

Polymer Material
Skin: ASAHI CHEMICAL/PS ASAHIPS 404

Constant polymer density	ρ	**= 1.0113E+03 (kg/m^3)**
Constant polymer specific heat	C_p	**= 2.0176E+03 (J/kg-K)**

Constant polymer thermal conductivity k **= 1.5392E-01 (W/m-K)**

2-domain modified Tait polymer density

$$\rho = \frac{1}{\nu_o[1 - C\ln(1+p/B)] + \nu_t}$$

where

C	$= 0.0894$
Tt	$= b_5 + b_6 p$
ν_o	$= b_1 + b_2 T_{bar}$
B	$= b_3 \exp(-b_4 T_{bar})$
ν_t	$= 0.0$ or $b_7 \exp(b_8 T_{bar} - b_9 p)$
T_{bar}	$= T - b_5$
b_5	= 3.6750E+02 (K)
b_6	= 3.6000E-07 (K/Pa)

Liquid phase	Solid phase	
b_1 = 9.5196E-04	b_1 = 9.5196E-04	(m^3/kg)
b_2 = 5.9300E-07	b_2 = 2.3200E-07	(m^3/kg-K)
b_3 = 1.7074E+08	b_3 = 2.5015E+08	(Pa)
b_4 = 3.8806E-03	b_4 = 3.6025E-03	(1/K)

b_7 = 0.0000E+00 (m^3/kg)

b_8 = 0.0000E+00 (1/K)

b_9 = 0.0000E+00 (1/Pa)

Cross-WLF polymer viscosity

$$\eta = \frac{\eta_o}{1 + (\eta_o \dot{\gamma}/\tau^*)^{(1-n)}}$$

where $\eta_o = D1\exp\left(-\frac{A1(T-T^*)}{A2 + (T-T^*)}\right)$

$A2$	$= A2T + D3p$
T^*	$= D2 + D3p$
n	= 3.0759E-01
τ^*	= 2.3365E+04 (Pa)
D1	= 2.2200E+12 (Pa-sec)
D2	= 3.7315E+02 (K)
D3	= 0.0000E+00 (K/Pa)
A1	= 2.8122E+01
A2T	= 5.1600E+01 (K)

Transition temperature

T_{trans} = 3.6750E+02 (K)

Gas - Compressed Nitrogen

Process Conditions

In this example, the process is under gas-pressure control. Rather than employing a constant ram speed, a ram-speed profile that will achieve a constant velocity at the melt front is desirable. This is because variation in melt-front velocity is known to introduce variable stress levels and orientation over the part, contributing to part warpage. This ram-speed profile may be obtained by executing C-MOLD Filling or Filling-EZ analysis. The appropriate percent of stroke and percent of speed values may be entered in the ram-speed profile data set within the process condition input file.

In addition, an ideal gas-pressure profile has been determined by running an initial gas-assisted moulding analysis using the automatic gas-pressure profiling option and the relative ram-speed profile. In this analysis, a constant gas pressure of 30 MPa is used; this value is the average of the ideal gas-pressure profile resulting from the initial gas-assisted moulding analysis.

Important process conditions used in this analysis are listed below:-

- **Fill time: 5.0 seconds**
- **Resin-injection time: 4.375 seconds**
- **Timer for gas injection: 4.375 seconds**
- **Gas-Injection time: 15.0 seconds**
- **Gas-Injection control option: Pressure control**
- **Gas pressure: Constant at 30 MPa**
- **Ram-speed profile (relative): Profiled**

% stroke	% speed
0.0	28.0
4.4	29.7
20.0	66.7
30.0	86.2
40.0	90.2
50.0	100.0
60.0	97.3
70.0	89.2
80.0	85.7
90.0	60.1
100.0	30.9

- **Inlet melt temperature: 220 °C (493 °K)**
- **Mould temperature : 50 °C (323 °K)**

Parameters

The numerical solution of the analysis is controlled by the analysis parameters. These include: the number of layers across the full gap; convergence criteria and maximum numbers of iterations for pressure, flow rate, and temperature calculations; numbers of design and detail outputs written to the results data files; and the heat-transfer coefficient at the mould-melt interface.

Default values of all analysis parameters are used in this example.

Results of the Analysis

The results of the filling analysis performed on the bezel designed for conventional injection moulding predict a required clamp force of 1,563 metric tons to achieve a constant ram speed, and a maximum entrance pressure of 119 MPa at the end of filling (time = 5 seconds). In comparison, the results of the gas-assisted moulding analysis of the modified part design with gas-channels are described below.

At the end of resin injection, immediately before gas injection is triggered; the polymer volume is 92% of the total part volume. The filling pattern is balanced. Unfilled regions remain on both of the side walls and on the bottom plate, which will be filled by the displaced polymer melt from the gas channels.

The predicted gas penetration in terms of skin-polymer fraction at the end of filling is shown in Figure 7. Note that the gas hollows out all of the gas channels evenly. Also, a slight degree of gas permeation into thinner corner areas can be seen which needs to be corrected with future design revisions.

With a constant gas pressure of 30 MPa, the predicted clamp-force requirement has been reduced to 716 metric tons, a 55% reduction compared with the original design.

At the same time, the pressure distribution throughout the cavity is more uniform with gas penetration, as shown in Figure 8, where most of the cavity has a pressure distribution in the 15 to 30 MPa range. In the part designed for gas-assisted moulding, the maximum polymer pressure, which occurred at the polymer entrance immediately before gas injection was triggered, was only 59 MPa. This is a 50% reduction compared to the maximum pressure of 119 MPa that occurred in the part designed for conventional injection moulding.

Conclusions

The following conclusions can be drawn from the analysis results:-

- Gas-assisted injection moulding can be employed to reduce the clamp-force requirement and achieve a lower and more uniform pressure distribution, compared to conventional injection moulding;
- It may require several iterations of analyses to determine the optimal gas-channel layout and combination of process conditions that deliver a desirable gas-penetration pattern.

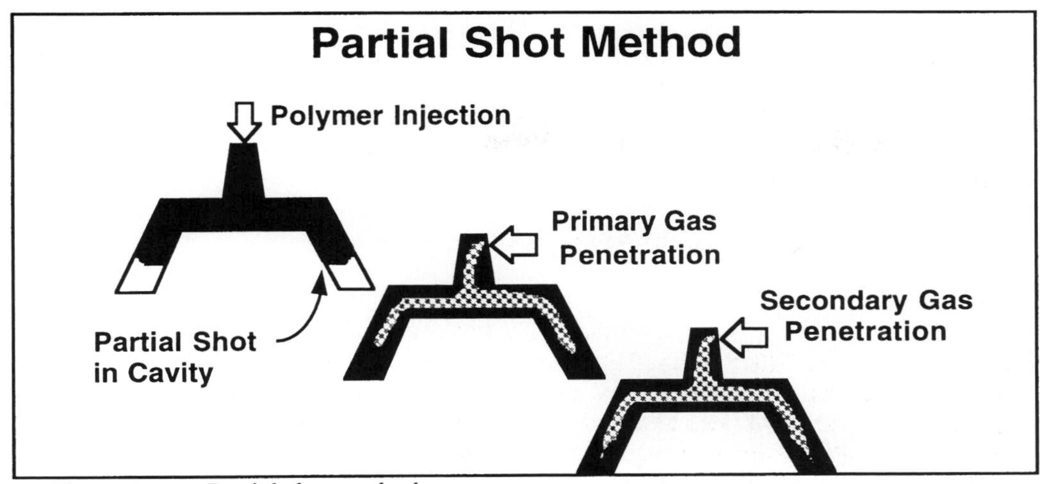

Figure 1. Partial shot method

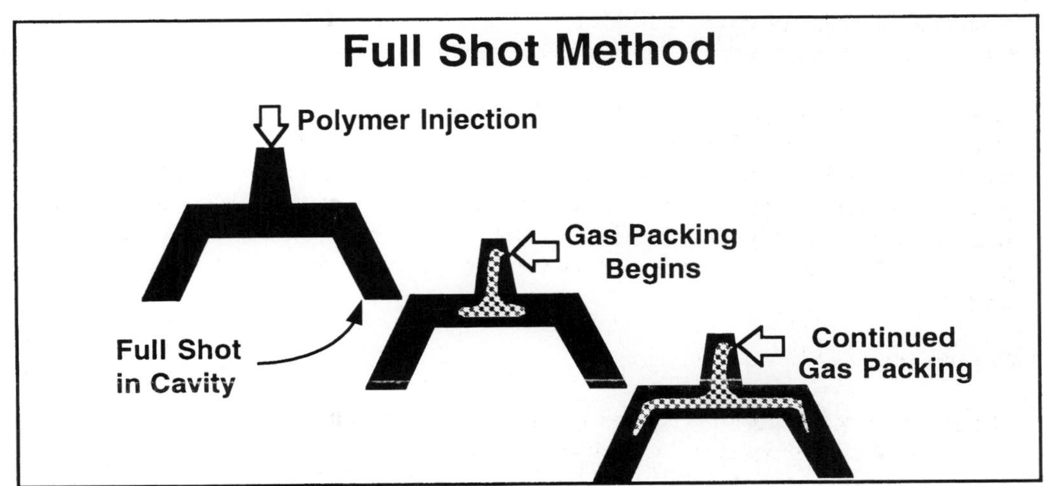

Figure 2. Full shot method

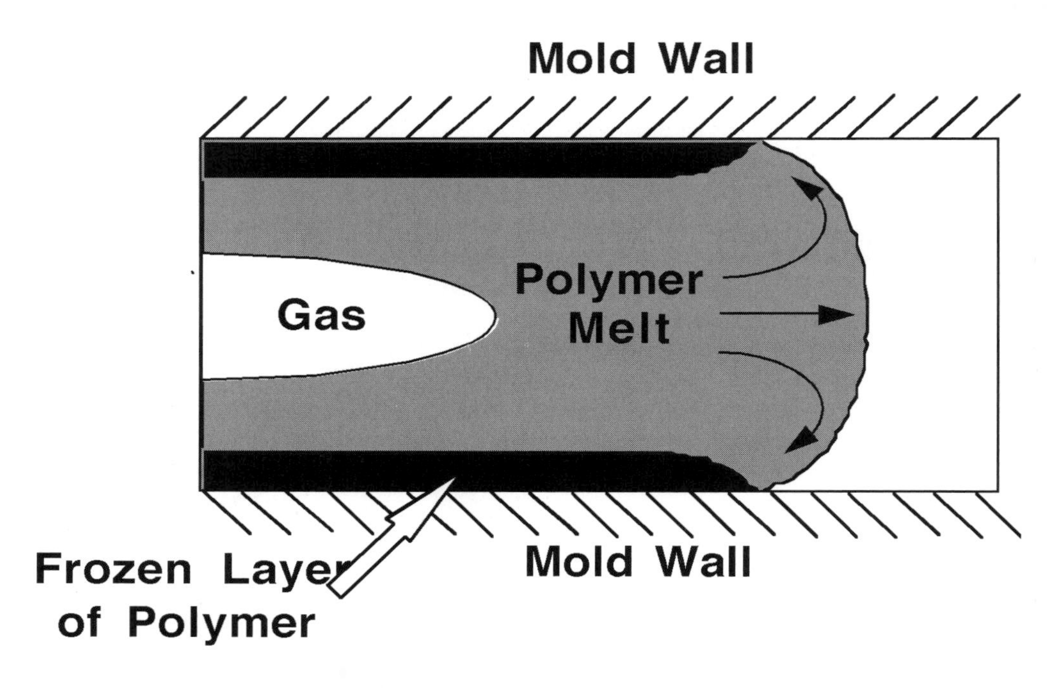

Figure 3. **Polymer and gas flow within the cavity**

Figure 4. **21" TV Bezel**

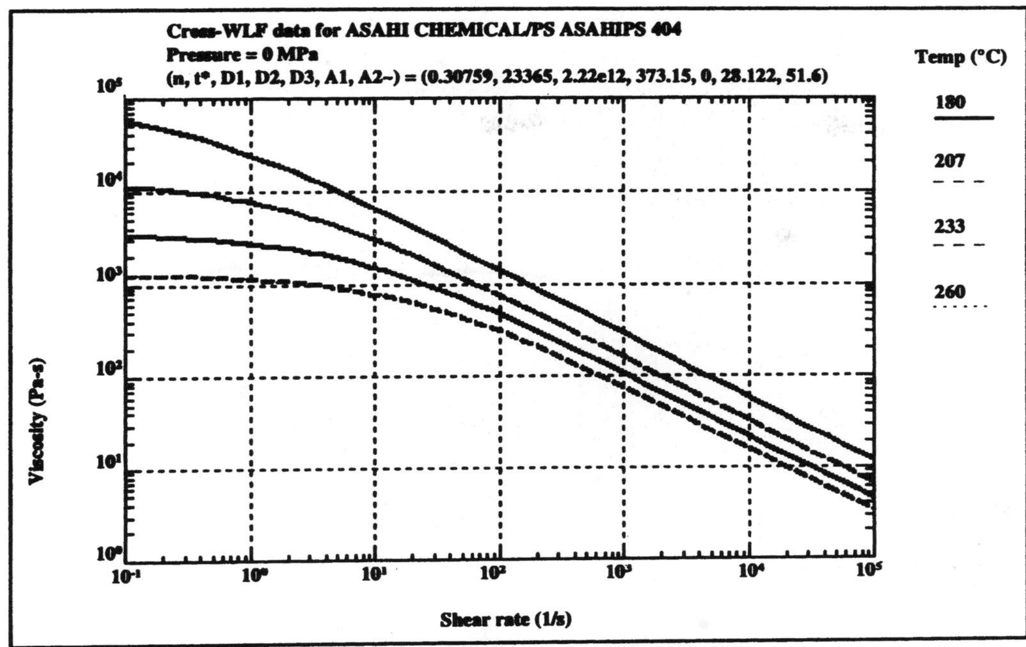

Figure 5. Viscosity data for ASAHIPS 404 POLYSTYRENE

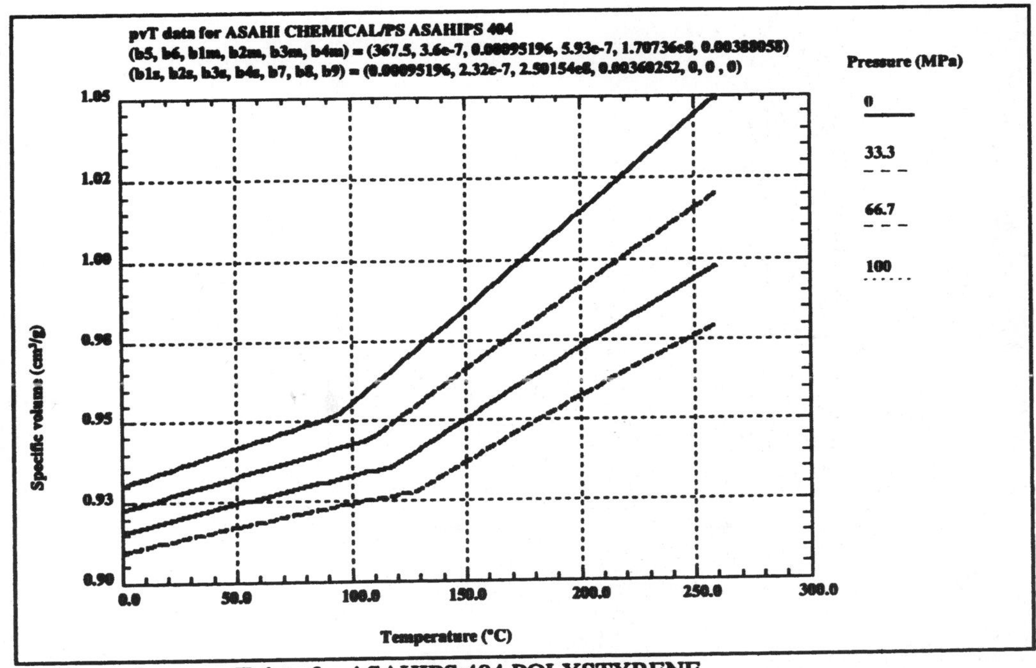

Figure 6. pvT data for ASAHIPS 404 POLYSTYRENE

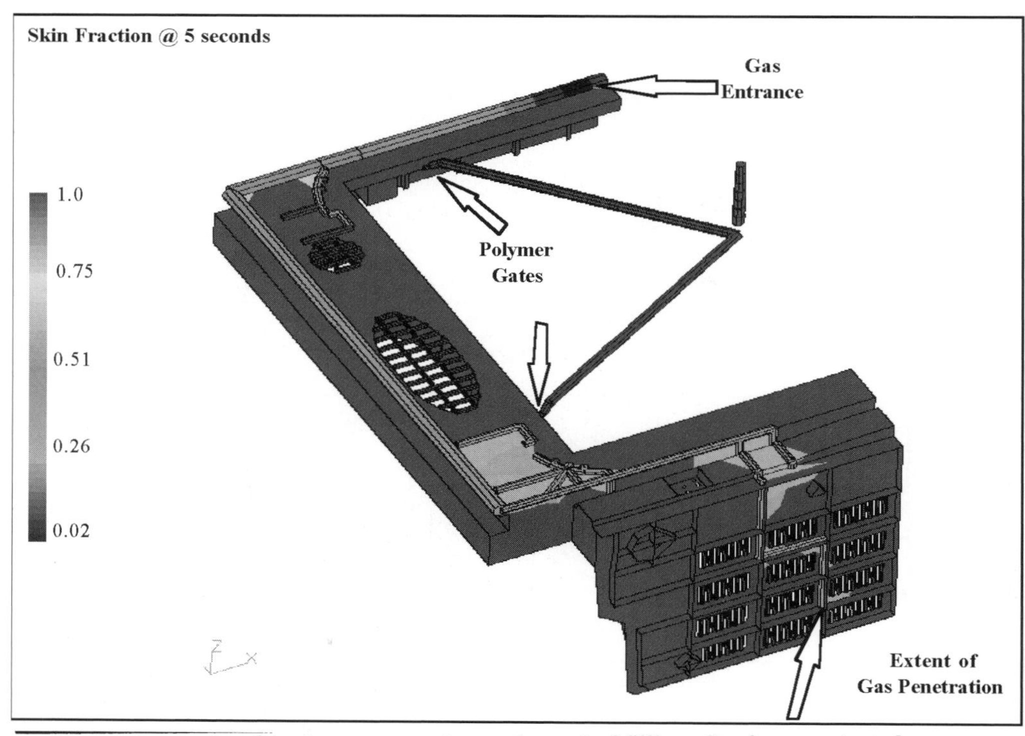

Figure 7. **Predicted gas penetration at the end of filling. Gas has penetrated uniformly into the gas channels as it displaces polymer melt to fill the part.**

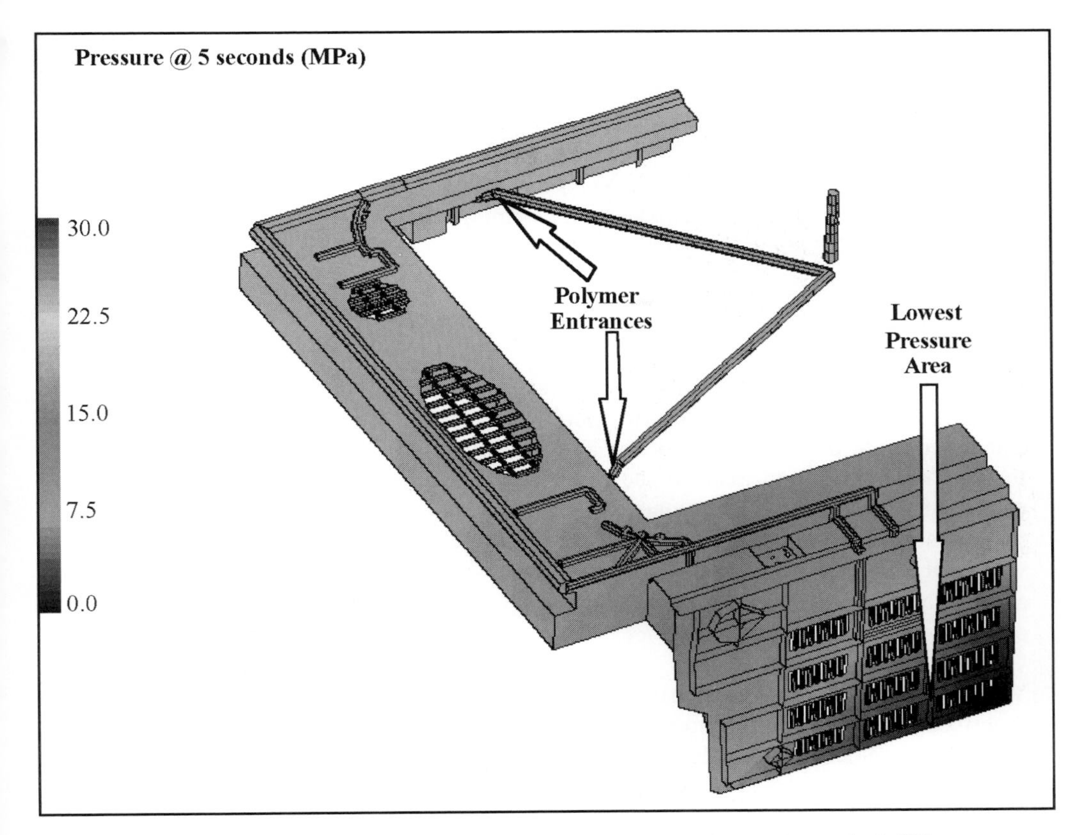

Figure 8. Pressure distribution with gas-assisted moulding at the end of filling

EXTENSIONAL VISCOELASTIC MEASUREMENTS OF POLYMER MELTS

M. RIDES
Centre for Materials Measurement and Technology, National Physical Laboratory, Teddington, Middlesex, United Kingdom, TW11 0LW

ABSTRACT

In many plastics-forming processes the plastic undergoes significant extensional or stretching flow, for example in blow moulding, vacuum forming and film extrusion. The extensional flow behaviour of materials will therefore have a significant affect on their processability and also on the properties of the moulded product. However, established rheological techniques predominantly characterise the behaviour of polymers in shear and it can prove difficult to relate the behaviour measured in shear with that observed in processing. An ability to characterise their behaviour more appropriately will have benefits in materials development and in process and product design.

Various techniques for characterising the extensional flow behaviour of materials are described. The techniques most suitable for use with polymer melts are converging flow and stretching methods. In the former, the extensional flow behaviour is determined from entrance pressure drop-flow rate data typically obtained from capillary extrusion rheometry. In the latter, a specimen is stretched, usually between rotating or translating clamps, and the stresses, strains and strain rates are measured. The advantages and disadvantages of the various approaches are discussed in the context of the needs of the polymer industry, and examples of their application are presented.

Results of converging flow measurements of glass-fibre filled thermosets and thermoplastics indicate the benefits and also highlight some of the difficulties of this approach. In using a stretching technique results obtained for polyethylenes were qualitatively similar to their behaviour measured in shear using a rotational rheometer and were also consistent with extrudate swell measurements made using a capillary extrusion rheometer. The conclusions of a recent international inter-comparison on capillary extrusion rheometry are briefly presented to complement this work on extensional rheometry.

INTRODUCTION

In many plastics-forming processes the plastic undergoes significant extensional or stretching flow, for example in blow moulding, vacuum forming and film extrusion. The extensional flow behaviour of materials will therefore have a significant affect on their processability and also on the properties of the moulded product. However, established rheological techniques predominantly characterise the behaviour of polymers in shear and it can prove difficult to relate the behaviour measured in shear with that observed in processing. An ability to characterise their behaviour more appropriately will have benefits in materials development and in process and product design.

The industrial needs for extensional viscoelasticity testing, specifically in terms of the strains, strain rates and temperatures were assessed by Rides [1]. It was concluded that there was a need for measurements typically up to strains[1] of 2 and strain rates of up to 5 s^{-1} at temperatures of up to 260 °C although higher values may be used. As an extreme case, for example, in wire coating the strain rates were considered to be significantly higher with values up to approximately 100 times greater. Cogswell [2] summarised the industrial needs for measurement techniques in terms of their practicality, performance and cost. He commented that techniques that were then suitable for adoption by industry, specifically the fibre spinning [3, 4] and converging flow [3, 5] methods, were qualitative. There was a requirement for quantitative, accurate techniques that were practicable. Some of the difficulties of extensional flow measurements that limit their accuracy and practicality are now described.

In contrast to shear the measurement of extensional flow properties is difficult and more complex. Whereas reference is normally made only to steady shear flow behaviour, extensional flow is normally described as a transient behaviour. The extensional viscosity may vary as a function of both strain rate and strain (or time), particularly at low values of strain. This transient behaviour is due to the continued development of the molecular orientation caused by the extensional flow field over the range of strains typically encountered in testing and processing. In comparison, in processes that are predominantly shearing flow the shear strain values tend to be significantly higher and the process is dominated by the material's steady state rather than transient shear flow behaviour. In describing the transient behaviour of materials in extension they may exhibit either an *unbounded tensile stress growth*[2] behaviour (for example references 6 and 7), a plateau in the tensile stress value (i.e. *equilibrium extensional viscosity*[3], for example references 8, 9 and 10), or a *maximum* in the tensile stress value [7]. The latter two typically occur at large strains.

Petrie [11] indicated that industrial flows are not normally pure elongational flows and cannot be described using steady state material properties. Industrial processes tend to be

---

[1] Hencky strain and strain rate values are quoted where the Hencky strain is the logarithm of the ratio of the current specimen length to the initial specimen length [12].

[2] The term "stress growth" refers to the increase in stress with time (or strain).

[3] An equilibrium extensional viscosity is dependent on strain rate but not on strain or time.

transient, complex extensional flows. Therefore he questioned the need to be able to determine the equilibrium extensional viscosity rather than, perhaps, the tensile stress growth behaviour. If the determination of equilibrium extensional viscosity values is difficult and expensive and the benefits of so doing are limited then the driving force to do so is also limited. However, in terms of modelling of flow processes there may be benefits in obtaining equilibrium extensional viscosity values to improve the process of selecting the most suitable constitutive equation and for determining the values of the parameters of that equation [13].

In carrying out extensional flow measurements there are four categories of measurement that are normally made: constant strain rate, constant stress, constant force or constant speed. However, industrial processes can rarely, if at all, be described by a single one of these categories. For example although film blowing may be carried out at constant bubble pressure and constant nip (or wind-up) tension, modelling of the process is complex [14] and it may not be possible to approximate it as a constant force process. Extensional techniques can also be described, following Nazem et al [15], as "controllable" or "non-controllable" [16,17]. "Controllable" implies that the instantaneous values of strain rate and stress are uniform throughout the specimen and that the strain rate or stress is held constant with time. The results of non-controllable methods, for example fibre spinning and converging flow analysis [3, 4, 5] tend to be difficult to interpret [18].

In characterising the material's flow behaviour the measurements should preferably be of true material properties (e.g. using constant strain rate or constant stress) rather than those that yield results that are dependent on the measurement method (e.g. constant force or constant speed). However, the former measurements tend to be more difficult and expensive to make and measurements that more closely mimic the process may be considered to be more appropriate.

Comparison of results for the polymer solutions A1 [19] and M1 [20] clearly indicated substantial differences in extensional properties obtained using the different methods: the extensional viscosity varied by three decades over the two decades of strain rate with results presented over virtually the entire region. Petrie [11] commented that the discrepancy was because "the experimental methods were not reproducing the conditions equivalent to the formal definition of extensional viscosity" - i.e. the tests generated flow fields that were not well approximated by the mathematical descriptions used to interpret the experimental data. James et al [20] commented that the rheometers used in the study of the fluid M1 should be viewed as yielding "*an* extensional viscosity rather than *the* extensional viscosity".

Despite the complexity of and difficulties in making extensional measurements, as indicated above, there are substantial benefits in so doing. This paper principally addresses the use of two extensional measurement techniques for polymer melts: "converging flow" or "contraction flow" methods in which the extensional viscosity is determined from the pressure drop in the melt flowing through a converging flow region, and stretching methods in which the specimen is stretched between clamps.

CONVERGING FLOW METHODS FOR CHARACTERISING EXTENSIONAL FLOW BEHAVIOUR

The method and comparison of selected models

A converging flow is a type of extensional flow. It would be expected therefore that extensional viscosity might be obtained from the measurement of pressure drop and flow rate in converging flows. The approach of using converging flow for determining extensional viscosity values is appealing, as experimental data are either readily available or are easily obtained from standard capillary rheometry testing (Figure 1) provided that either Bagley corrections are made (Figure 2) or a zero length die is used to obtain entrance pressure drop values [21]. Thus both shear and extensional viscosity values can be obtained from the same capillary rheometry tests.

Numerous models for determining extensional viscosity from entrance pressure drop measurements have been given in the literature [22-27] and a review of these is presented elsewhere [28].The models can be sub-divided into the categories of constrained flow or free convergence flow depending on whether the entry to the die is profiled thereby preventing the polymer from forming its own flow pattern into the die (Figure 1). Although models in the first category cannot be used to predict the position of the converging flow boundary for comparison with experimental or numerical modelling data, they can be used to predict extensional viscosity values given appropriate (i.e. constrained entry flow) entrance pressure drop data. Furthermore the modelling of a freely converging flow is potentially more complex than that of constrained converging flow due to the additional degree of freedom involved, i.e. the position of the entry flow boundary. The models due to Cogswell [24], Gibson [25] and Binding [26] were selected for analysis of entrance pressure drop data obtained from capillary die rheometry measurements of unsaturated polyester dough moulding compounds (DMCs) and their predictions were compared.

In comparing the formulation of the selected models the following observations were made. Each assumed a fully developed shear flow velocity profile for their analyses. However, Gibson [25] used a spherical coordinate system (as opposed to cylindrical coordinates) in order to describe the problem with the consequence that a different "fully developed" velocity profile to that of the other models was obtained (the centre line velocity was greater in the Gibson model). With the exception of the Binding model, average strain rates were adopted in the formulation of these converging flow models. Binding however calculated the local value of the strain rate from the assumed velocity profile - potentially an improvement. Both Gibson [25] and Binding [26] assumed fully independent power-law equations to describe the shear and extensional viscosities, whereas Cogswell [24] assumed that the extensional viscosity was constant (but power-law for shear viscosity). To determine the relationship between the entrance pressure drop data and extensional viscosity, and also the position of the vortex boundary for freely converging flow, Cogswell and Binding employed the principle of minimum work. The model proposed by Gibson is for application to constrained entry flow only and therefore cannot be used to predict a free convergence entrance angle.

Application of converging flow models

The Cogswell [24], Gibson [25] and Binding [26] converging flow models were applied to the analysis of entrance pressure drop data for an unsaturated polyester dough moulding compound (DMC) that contained approximately 12% by volume of long glass fibres, Figure 3. The three models predicted extensional viscosities of similar magnitude, with those of the Cogswell model being the higher, and those of the Gibson model the lower values. The ratio of extensional to shear viscosity increased with increasing shear or strain rate. Using the Cogswell, Gibson and Binding models respectively, ratios in the ranges ≈ 70-80, ≈ 40-60 and ≈ 35-40 at shear/strain rates of 4-300 s^{-1} were predicted for the DMC.

Entrance pressure drop values, and thus extensional viscosities, were very dependent on the fibre length and duration of mixing for the DMC, Figure 4. Compared with the values for the standard 6mm fibre batch the extensional viscosities of the 3 mm fibre DMC batch were ≈ 60% lower, and those of the 6mm fibre DMC batch with longer mixing duration were ≈ 30% higher. However, the shear viscosities were independent, within experimental scatter, of these variations in fibre reinforcement, Figure 5. The predominant effect of the extra mixing was to cause further break-up of the fibre bundles thereby increasing the resistance of the material to extensional flow (fibres were present in the DMC in the form of bundles).

The measurement of entrance pressure drop values that are required for determining extensional viscosities can be prone to large errors, particularly for unfilled materials where the magnitude of the pressure drop can be small. Results for a polyethylene obtained from an international intercomparison (reference 29 and also reported briefly below) indicated that the scatter in values could be significant, Figure 6. Recommendations for improved testing were made [29].

STRETCHING METHODS FOR DETERMINING EXTENSIONAL FLOW BEHAVIOUR

Review of techniques

Stretching methods, in which the specimen is stretched typically between translating or rotating clamps, is a principal methods for measuring extensional flow properties of polymer melts. A detailed critical review of such techniques is presented elsewhere [12].

Stretching is normally carried out at either constant strain rate, constant stress, constant force or constant speed (constant change of length of the specimen per unit time). The first two are "controllable" methods whereas the latter two are "non-controllable". Constant force and constant speed measurements are more difficult to interpret [16]. This was emphasised by Gupta et al [3] and also by Cogswell [2] who concluded that constant speed or constant force methods provide no information that cannot be provided better by constant stress or constant strain methods. Testing should preferably be carried out using either constant strain rate or constant stress deformations in order to generate quantitative data that are independent of the test method.

These measurement techniques can be further categorised using the types defined below to describe the various methods reported in the literature.

Type 1: Two rotating clamps. (Each clamp consisting of either a single or pair of rotating elements.)

Type 2: Single rotating clamp and a fixed clamp, for example Figure 7. (The rotating clamp consists of either a single or a pair of rotating elements).

Type 3: Translating (i.e. non-rotating) clamp or clamps. (Where a single translating clamp is used the other clamp is fixed in position.)

Type 4: Haul-off or fibre spinning from an extruder.

In reviewing the literature to assess the important aspects of the design of instruments for extensional flow behaviour measurements [12] the following observations were made.

The use of at least one rotating clamp (types 1, 2, and 4) overcomes the restriction that the maximum strain obtainable is limited by the length of the test equipment rather than by the material's response - i.e. its failure. This artificial limitation on strain would also most likely limit the maximum strain rate that could be achieved using the instrument.

On the basis of a comparison of results obtained using two different instruments (type 1 and type 3) the magnitude of the effect of end-errors introduced due to using a non-rotating clamp was found to be negligible [30]. However, the effect of end-errors on the uniformity of the deformation was also observed by others [36] to be very significant. Measurements made using different length to diameter ratios for the test specimen indicated that the length of the specimen should be at least 10 times its diameter [32]. The attachment of etched specimens to clamps using adhesives (typically epoxy) has been found to be generally suitable.

Silicone oil was the most commonly used test medium. It provides both buoyancy for the specimen, thus reducing any sag due to gravity, and also acts as a heat transfer medium. The effect of silicone oil on the specimens was reported in several papers and although absorption was identified there was no effect on the measured properties. However the effect of the oil on the properties at failure was not reported, except for polyisobutylene in alcohols for which there was a significant effect [33]. The effect of silicone oil absorption on ultimate tensile properties of polymer melts requires further investigation.

The uniformity of deformation of the specimen is critical to successful, accurate measurements of extensional flow properties. In this respect the uniformity of temperature is an important factor, particularly for obtaining results at high strains. Considerable effort by researchers has been made to minimise temperature variations, for example the use of a copper trough [34] or double-skinned silicone oil baths [32, 35]. The uniformity of specimen deformation is also affected by the uniformity of the specimen specifically in terms of its dimensions, homogeneity and purity, e.g. no voids.

Having carried out a test it is desirable to check the uniformity of deformation of specimens to assess the quality of the test. Two methods have been reported: either the weight of several samples cut from the deformed specimen were compared, or the specimen's dimensions were measured optically during the test. The latter obviously requires the use of a transparent window that in itself might affect the temperature homogeneity of the specimen. Furthermore the difficulties associated with measuring the diameter of an often transparent filament of molten polymer in an oil bath may also result in unacceptable measurement uncertainty.

The selection or design of the force measuring device is clearly of considerable importance, particularly for measurements at high strains at which the cross sectional area of the specimen will be considerably smaller than it was at the start of the test. Good resolution and high accuracy at low values of force are therefore important for accurate measurements at high strains.

The magnitude of the effect of various errors, for example due to frictional drag of the supporting medium were discussed in the review [12]. Methods for correcting for the effect of these errors were also referenced.

In summarising the current state of measurement capability, extensional flow measurements at strains up to 7, strain rates up to 30 s^{-1} and temperatures up to 350 °C have been reported although these conditions were not attained simultaneously. The availability of equipment to operate at these conditions is very limited. More typically, the upper limits of testing were of strains up to ≈ 4, strain rates up to ≈ 1 s^{-1} and temperatures up to ≈ 200 °C but such equipment is still of limited availability.

Polymer melt behaviour in extension - a case study

In earlier work at the NPL a technique for characterising the viscoelastic behaviour of polymer melts in extension was developed. Measurements of stress retardation and relaxation in extension were carried out using equipment based on a controlled strain Haake Rotovisco RV2 rotational rheometer that had an integral torque measuring device, Figure 7. The rheometer, acting as a winding device, was fitted with a drum to which the specimen was attached. The other end of the specimen was attached to a fixed clamp. Test specimens were prepared by extruding polymer using a capillary rheometer. After loading the specimen the equipment was immersed in a heated silicone oil bath.

The stress retardation and relaxation tests involved stretching the specimen at a high rate for a short period of time by rotating the rheometer drum at constant speed, typically producing a sample strain of approximately 100% in 1 second. The drum was then stopped, thus holding the specimen at constant length and hence strain. The rotational rheometer was used to monitor the increase in torque on loading (retardation), followed by the fall in torque with time due to the relaxation of the specimen.

Two blow moulding grades of high density polyethylene materials were tested: NCPE 3415 and NCPE 3416. These materials were chosen as they exhibited quite different extrudate swell behaviours yet similar shear viscosity behaviours as measured using a capillary extrusion rheometer. NCPE 3415 [36] has melt index values approximately twice those of NCPE 3416 [37].

Results and discussion

Initial testing indicated that the repeatability of measurements was good, Figure 8. The slightly lower values at the start of the relaxation phase for test NES6003 was due to a slightly shorter loading time and consequently a lower peak stress value for that test. Shear and extensional relaxation modulus values (that is the ratio of the stress to the initial strain, where initial refers to the start of the relaxation process) are plotted for both materials in Figure 9. The shear relaxation modulus values were measured using a Bohlin VORM rotational rheometer. The NCPE 3415 grade had higher values of relaxation modulus and relaxed at a marginally slower rate than NCPE 3416. The modulus values were higher in extension than in shear for both materials. The effect of initial strain on the relaxation modulus was also investigated, Figure 10, and clearly showed strain hardening behaviour in extension in contrast with the shear thinning behaviour observed for this material.

Extrudate swell measurements were also made on these two materials. The NCPE 3415 had a percentage swell factor[4] that was in the range 1.5 - 2 times greater than that for NCPE 3416, Figure 11. However the measurement of the shear viscosity at comparable shear rates indicated only a small difference in behaviour, Figure 12. More significant differences in the rheological behaviour of the materials were identified using a rotational rheometer in steady shear mode at lower shear rates, Figure 12, and also by oscillatory rheometry [38]. These techniques might prove to be more reliable indicators of extrudate swell behaviour than capillary rheometry. However they are shearing flows and thus it is unreasonable to assume that they will always provide a good indicator of extrudate swell which is a complex flow involving both shear and extensional flow components. This is supported by Ahmed et al [39] who concluded, based on their work on modelling extrusion flows of polyethylenes, that differences in the extensional strain hardening behaviour of materials can have a profound effect on their processability and die swell behaviour and that these differences are not necessarily realised by simple shearing characterisation methods.

RESULTS OF AN INTERNATIONAL INTERCOMPARISONS ON CAPILLARY EXTRUSION

The results of an extensive international intercomparison on the measurement of shear viscosity and entrance pressure drop using capillary extrusion rheometers for a high density polyethylene (HDPE) and a glass-fibre filled polypropylene (GFPP) melt are summarised in the tables below and reported in detail elsewhere [29]. The intercomparison was based on the use of the standard `ISO 11443 Plastics - Determination of the fluidity of plastics using capillary and slit-die rheometers'. A precision statement for inclusion in the standard was proposed [29].

[4] The percentage extrudate swell is defined as the difference in the extrudate and die diameters expressed as a percentage of the die diameter.

Extrusion pressure measurement

Material, test temperature		PE, 190 °C	GFPP, 230 °C
Repeatability	standard deviation	7.0%	13.4%
	95% confidence	19.6%	37.5%

Shear viscosity measurement (corrected for entrance pressure drop and non-Newtonian velocity profile (Rabinowitsch))

Material, test temperature		PE, 190 °C	GFPP, 230 °C
Repeatability	standard deviation	7.2%	8.6%
	95% confidence	20.3%	24.0%
Reproducibility	standard deviation	9.9%	12.2%
	95% confidence	27.7%	34.1%

Entrance pressure drop measurement

Material, test temperature		HDPE, 190 °C contraction ratio of 15 only	HDPE, 190 °C contraction ratios 9.55 -15.5	GFPP, 230 °C contraction ratio of 15 only
Reproducibility	standard deviation	15.1%	17.8%	19.8%
	95% confidence	42.3%	49.8%	55.5%

On the basis of this intercomparison the following principal conclusions were drawn (all repeatability and reproducibility values quoted below are for 95% confidence levels):

- Repeatability of shear viscosity measurements were approximately 20% for HDPE and 24% for GFPP.
- Reproducibility of shear viscosity measurements were approximately 28% for HDPE and 34% for GFPP.
- Reproducibility of entrance pressure drop measurements were approximately 42% for HDPE and 56% for GFPP.
- The repeatability of measurements of extrusion pressure was in the range 20 - 38%.

The single, most significant source of error was considered to be due to the accuracy of the measurement of pressure, in particular the utilisation of the pressure transducer over the appropriate range.

In testing in accordance with the specification of the standard ISO 11443 the uncertainties in the measurement of shear viscosity and entrance pressure drop were estimated to be in the range 2.5 - 19% for the polyethylene and polypropylene materials, and depended most significantly on the range over which the pressure transducer was being used. If used in the lower part of its range it could contribute significantly to the overall level of uncertainty.

DISCUSSION AND CONCLUSIONS

It is clearly apparent that the measurement of extensional flow properties of polymer melts is difficult. The work both presented and referenced to above indicates the principal difficulties of such measurements. Both measurement approaches discussed above, i.e. converging flow and stretching methods, have particular problems associated with their use for polymer melts. Nevertheless extensional measurements can provide valuable information on the flow behaviour of polymer melts in processing that is not obtainable from shear flow methods. In general it is preferable to use the technique that most closely simulates the process for which the testing is required.

The use of converging flow and stretching flow methods has been demonstrated. Differences in the extensional flow behaviour of unsaturated polyester DMCs having differences in fibre reinforcement were measured using the converging flow method, although differences were not observed in their shear flow behaviour. However, for a given material significant differences in the predicted extensional viscosity values were obtained in using different converging flow models.

The use of stretching flow methods was illustrated by the measurement of two blow moulding grades of HDPE. Significant differences in the extensional flow behaviour were identified and their strain hardening behaviour was clearly demonstrated.

The results of the intercomparison on capillary extrusion rheometry were used to quantify the repeatability and reproducibility of this established shear flow measurement method [29]. The reproducibility of entrance pressure drop measurements was also quantified. Conclusions specific to this intercomparison are presented in the preceding section. On the basis of the intercomparison a precision statement for inclusion in the rheometry standard ISO 11443 [21] was proposed. The use of pressure transducers in the lower part of their pressure range was identified as the most likely major source of error in measurements. Results may be improved by more selective and careful use of pressure transducers. The findings of this intercomparison are relevant to the use of converging flow models for determining extensional viscosity data as both shear viscosity and entrance pressure drop data are required. The precision of these data will limit the precision of the determined extensional viscosity values. An assessment of this will be carried out in future work.

ACKNOWLEDGEMENT

The paper was prepared as part of a project on the measurement of the extensional viscoelastic properties of polymers. This project (MMP11) and previous projects under which some of the reported work was carried out were parts of programmes of underpinning research financed by the Engineering, Automotive and Metals Division of the Department of Trade and Industry.

REFERENCES

1 M. Rides, Industrial needs for polymer melt elasticity measurements, National Physical Laboratory Report DMM(A)131, October 1993.

2 F.N. Cogswell, Trans. Soc. Rheol., 16 (1972) pp.383-403.

3 R.K. Gupta and T. Shridar, Elongational Rheometers, in Rheological Measurement, Ed. A.A. Collyer and D.W. Collyer, Elsevier Applied Science, London, 1988.

4 Elongational Flows, C.J.S. Petrie, Pitman, London, 1979.

5 D.M. Binding, Contraction Flows and New Theories for Estimating Extensional Viscosity, in Techniques in Rheological Measurements, Ed. A.A. Collyer, Chapman and Hall, London, 1993.

6 J.M. Meissner, Trans. Soc. Rheol., 16 (1972) pp.405-420.

7 J. Meissner, Chem Eng. Commun., 33 (1985) pp.159-180.

8 H. Münstedt, J. Rheology, 23 (1979) pp.421-436.

9 H.M. Laun and H. Münstedt, Rheol. Acta, 15 (1976) pp.517-524.

10 H.M. Laun and H. Schuch, J. Rheology, 33 (1989) pp.119-175.

11 C.S. Petrie, Rheol. Acta, 34 (1995) pp.12-26.

12 M. Rides, C.R.G. Allen and S. Chakravorty, Review of extensional viscoelasticity measurement techniques for polymer melts, National Physical Laboratory Report CMMT(A)44, October 1996.

13 M. Rides, Numerical simulation of the extrudate swell behaviour of polymer melts, National Physical Laboratory Report CMMT(A)31, June 1996.

14 J.R.A. Pearson, Mechanics of Polymer Processing, Elsevier Applied Science, London, 1985.

15 F. Nazem and C.T. Hill, Trans. Soc. Rheol., 15 (1974) pp.87.

16 Rheological Techniques, R.W. Whorlow, Ellis Horwood, London, 1992.

17 Rheometry, K. Walters, Chapman and Hall, London, 1975.

18 J.M. Dealy, J. Non-Newtonian Fluid Mech., 4 (1978) pp.9-21.

19 N.E. Hudson and T.E.R. Jones, J. Non-Newtonian Fluid Mech., 46 (1993) pp.69-88.

20 D.F. James and K. Walters, A Critical Appraisal of Available Methods for the Measurement of Extensional Properties of Mobile Systems, in Techniques in Rheological Measurements, Ed. A.A. Collyer, Chapman and Hall, London, 1993.

21 ISO 11443: 1995 Plastics Determination of the fluidity of plastics using capillary and slit-die rheometers.

22 Williamson, G.A. and Gibson, A.G., Plastics and Rubber Processing and Applications, 4 No.3 (1984) pp.203-213.

23 Gibson, A.G. and Williamson, G.A., Polym. Eng. Sci., 25 No.15 (1985) pp.968-979.

24 Cogswell, F.N., Polym. Eng. Sci., 12 No.1 (1972) pp.64-73.

25 Gibson, A.G., Composites, 20 No.1 (1989) pp.57-64.

26 Binding, D.M., J. Non-Newtonian Fluid Mechanics, 27 (1988) pp.173-189.
27 Humphries, C.A.M. and Parnaby, J., Proc. Instn. Mech. Engrs., 200 No.C5 (1986) pp.325-334.
28 M. Rides, Review of converging flow models for determination of extensional flow behaviour of polymer melts, National Physical Laboratory Report, in preparation.
29 M. Rides and C.R.G. Allen, Capillary extrusion rheometry intercomparison using polyethylene and glass-fibre filled polypropylene melts: measurement of shear viscosity and entrance pressure drop, National Physical Laboratory Report CMMT(A)25, May 1996.
30 H. Münstedt and H.M. Laun, Rheol. Acta, 20 (1981) pp.211-221.
31 A.E. Everage and R.L. Ballman, J. Applied Polym. Sci., 20 (1976) pp.1137-1141.
32 G.V. Vinogradov, V.D. Fikhman, B.V. Radushkevich and A.Ya. Malkin, J. Polym. Sci., A-2, 8 (1970) pp.657-678.
33 M.K. Kurbanaliev, G.V. Vinogradov, V.E. Dreval, A.Ya. Malkin, Polymer, 23 (1982) pp.100-104.
34 J. Meissner, T. Raible and S.E Stephenson, J. Rheology, 25 (1981) pp.1-28.
35 D. Froelich, B. Muller and Y.H. Zang, ACS, Rubber Div., 128th Meeting - Fall Cleveland, Ohio, October 1985, paper 66, pp.23.
36 NCPE 3415 High density polyethylene for blow moulding technical information sheet, BM 0177 1989 10/11, Neste Chemicals.
37 NCPE 3416 High density polyethylene for blow moulding technical information sheet, BM 0494 1990 11/4, Neste Chemicals.
38 M. Rides, C.S. Brown, C.R.G. Allen, D.H. Ferriss and P.A.J. Gibbs, Extensional viscoelastic behaviour of high density polyethylene melts: a new rheometer for stress relaxation measurements in extension, National Physical Laboratory Report CMMT(B)57, April 1996.
39 R. Ahmed, R.F. Liang and M.R. Mackley, The experimental observation and numerical prediction of planar flow and die swell for molten polyethylenes, J. Non-Newtonian Fluid Mechanics, 59 (1995) pp. 129-153.

Filename: BRAD97#13_EXC_OSCILL_RPT.DOC, 10 April 1997, MR1/98

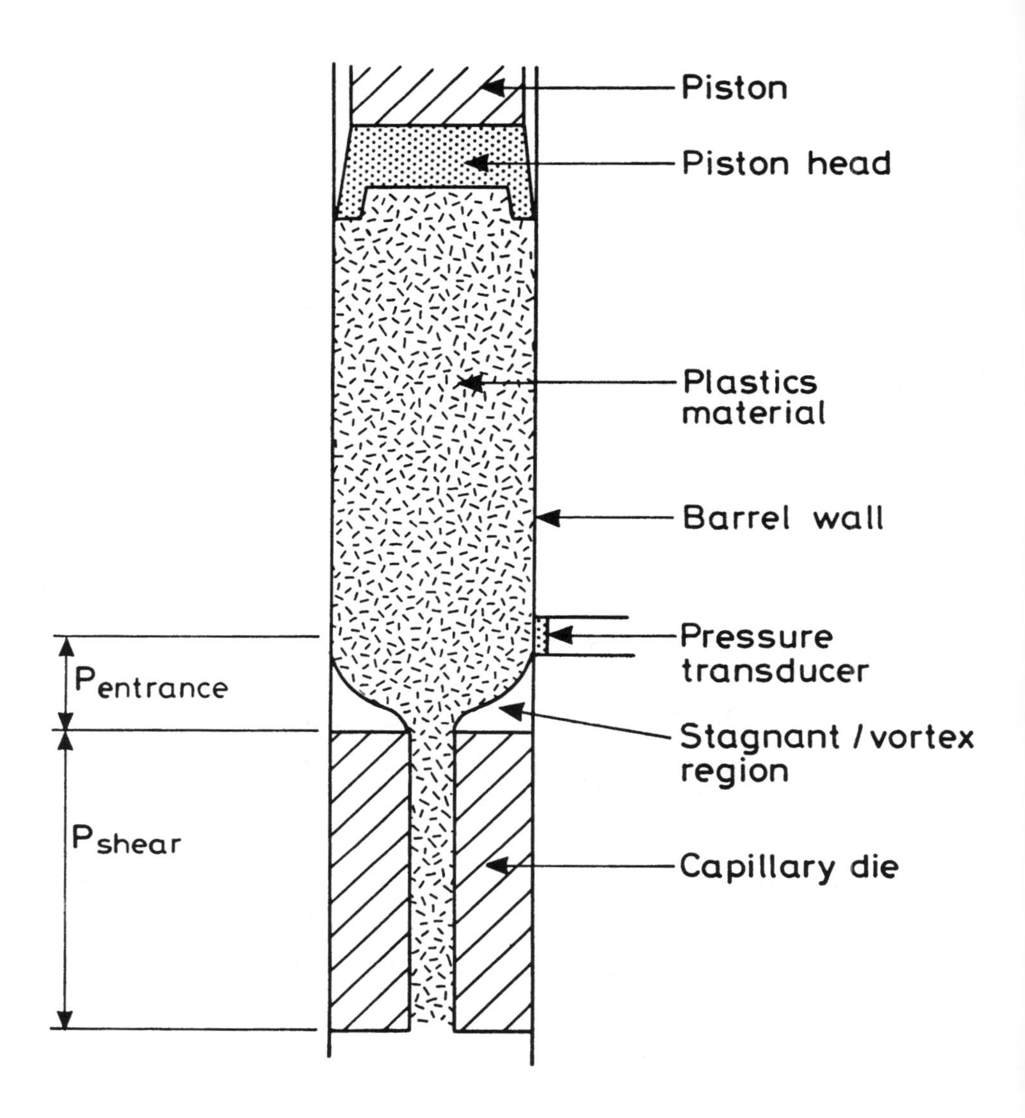

Figure 1. Contraction flow in a capillary extrusion rheometer.

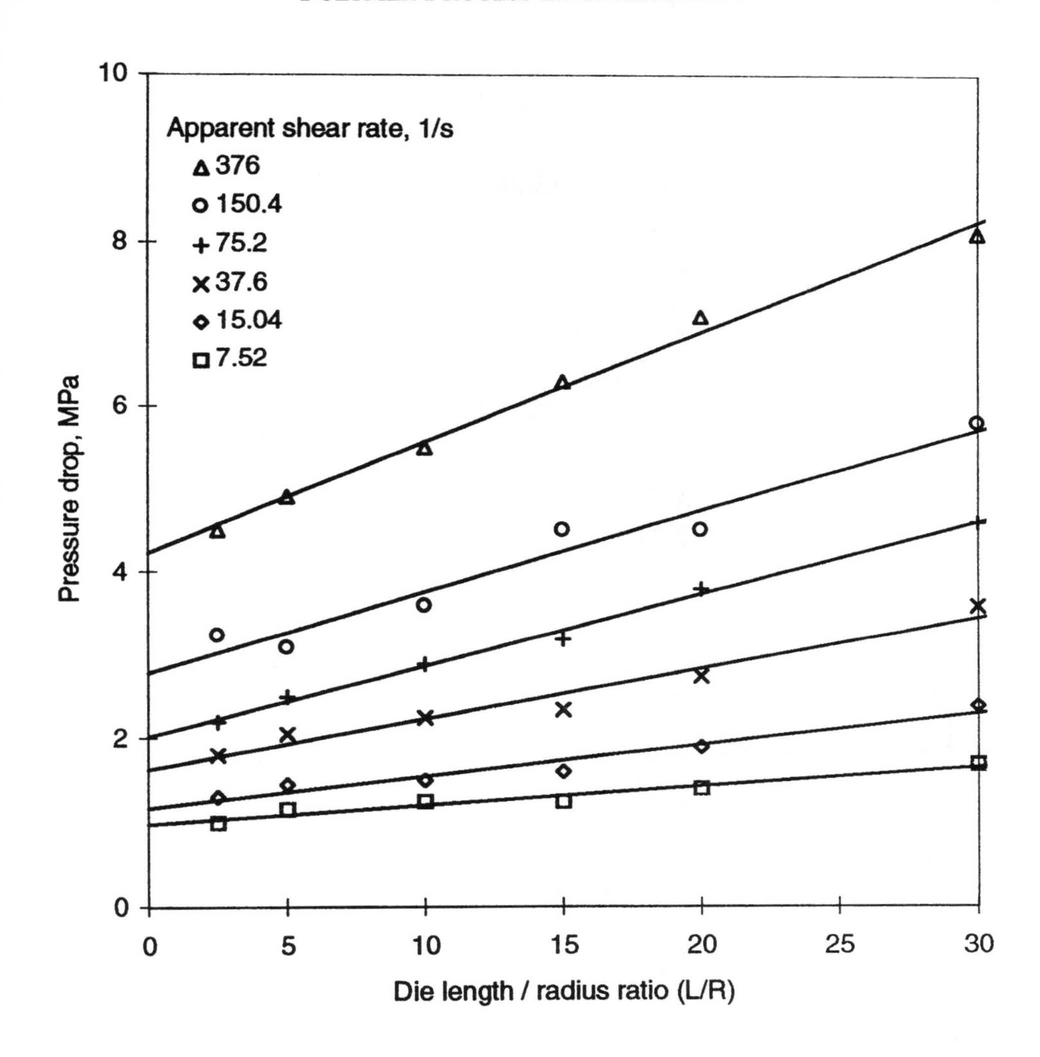

Figure 2. Bagley plot for an unsaturated polyester dough moulding compound (DMC L7049III) at 50 °C.

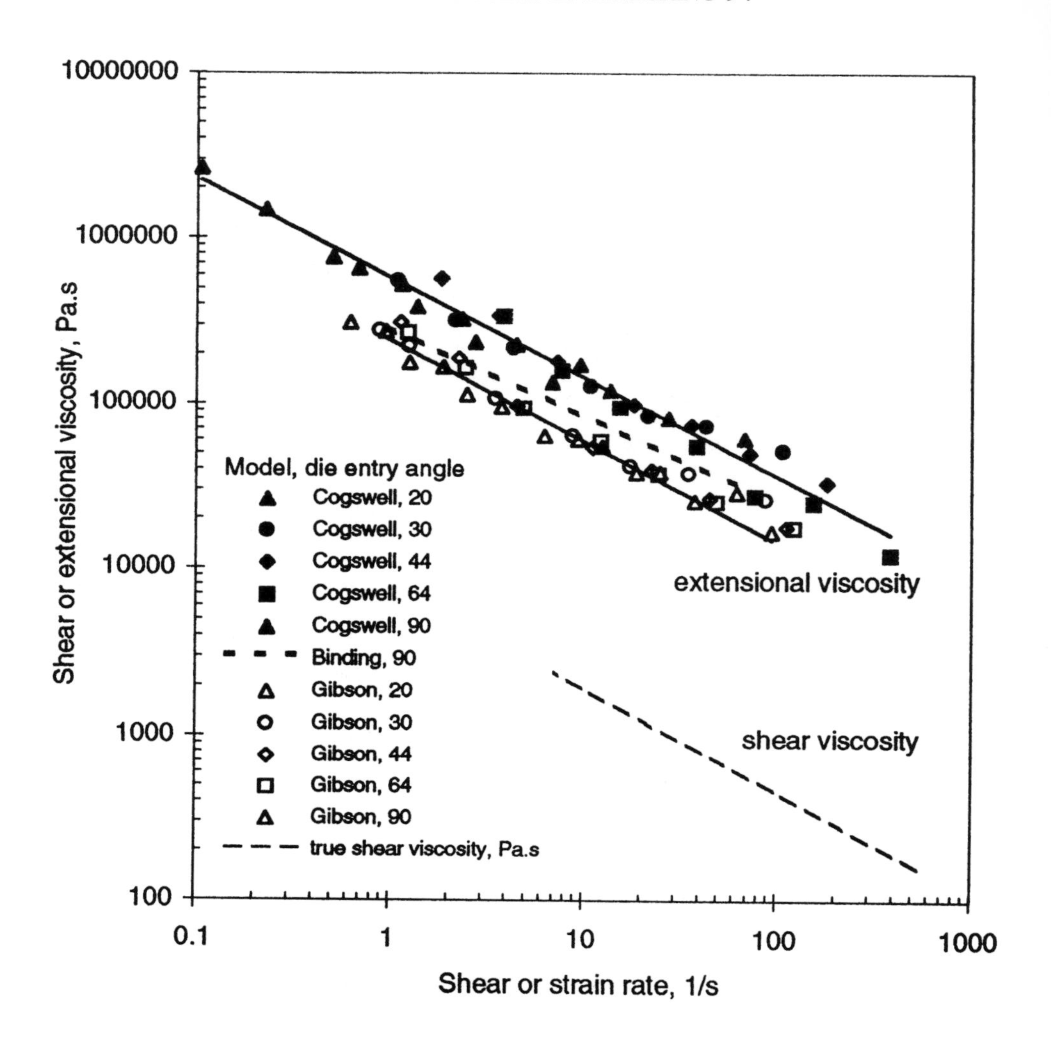

Figure 3. Comparison of extensional viscosity values obtained from entrance pressure drop measurements using the Binding, Cogswell and Gibson converging flow models for an unsaturated polyester DMC (L7049 VII) at 50 °C using various die entry cone angles (legend gives die entry angle in degrees where 90° corresponds to a right-angled entry profile).

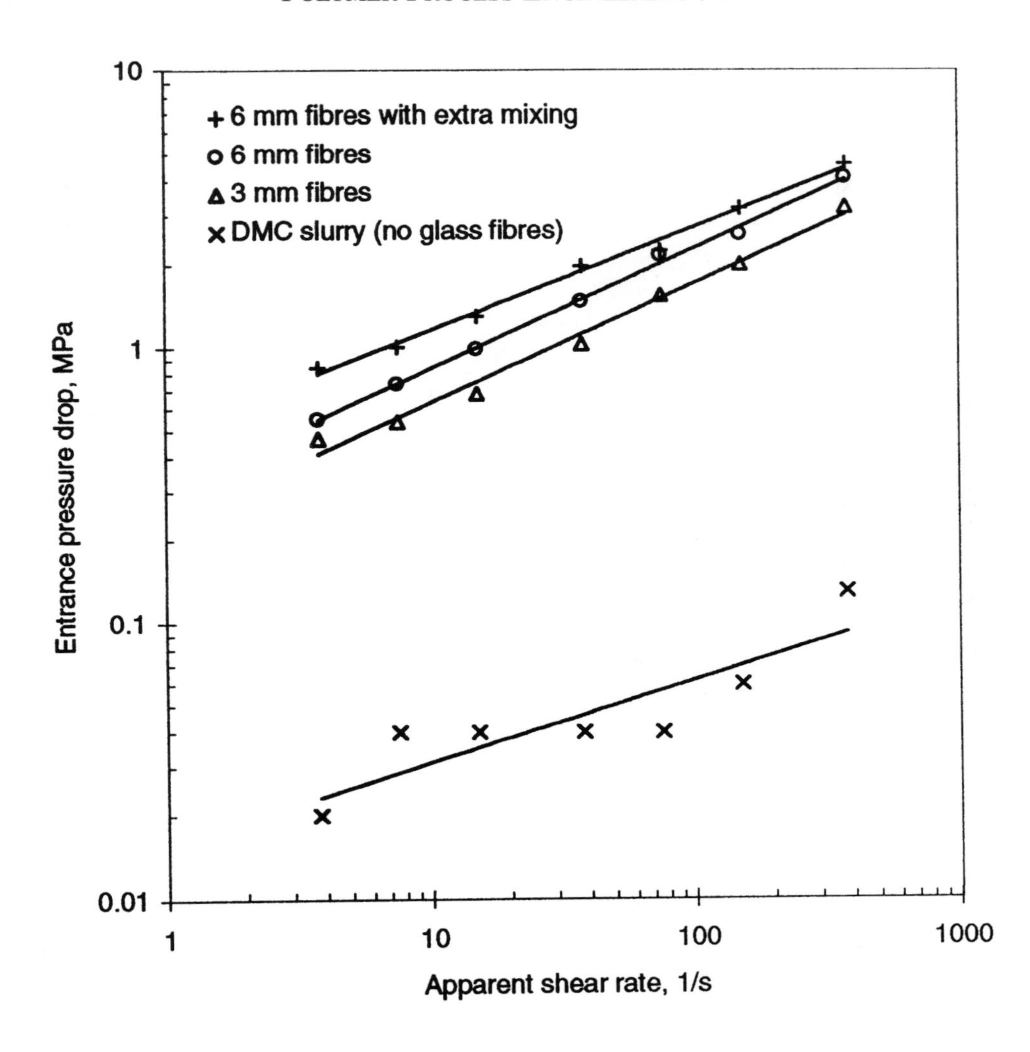

Figure 4. Effect of differences in glass fibre content on the entrance pressure drop values of an unsaturated polyester dough moulding compound (L7049) at 50 °C.

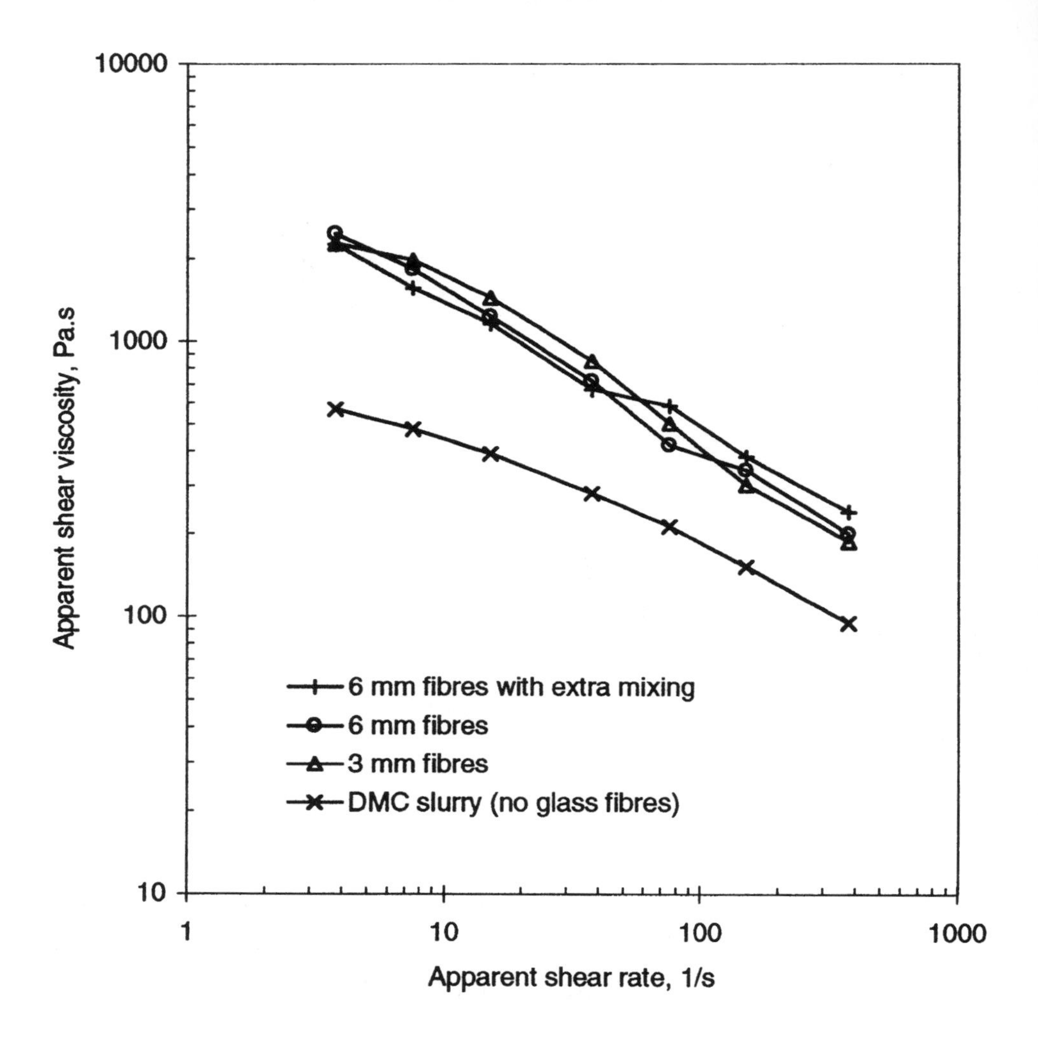

Figure 5. Effect of differences in glass fibre content on shear viscosity of an unsaturated polyester dough moulding compound (L7049) at 50 °C.

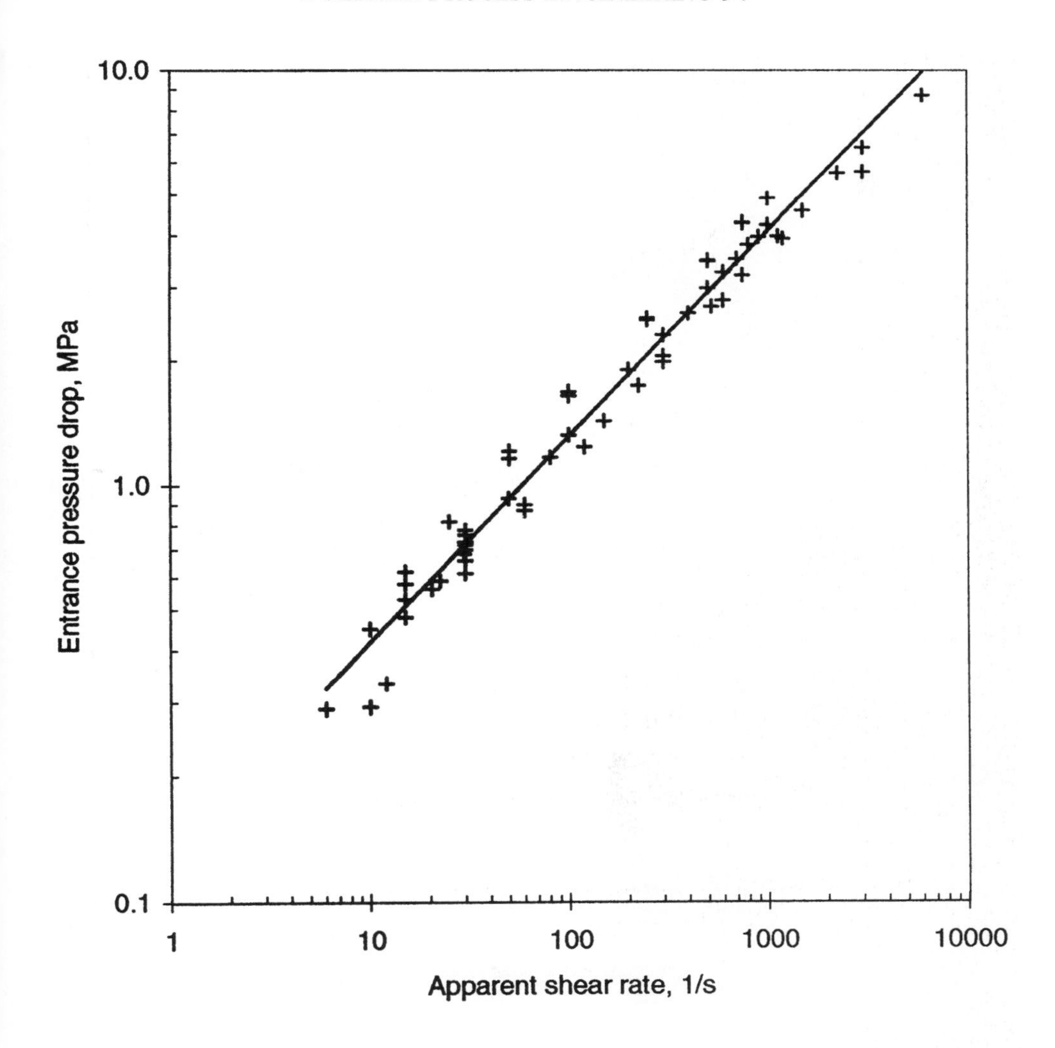

Figure 6. Scatter in entrance pressure drop values for HDPE at 190 °C with apparent shear rate using a barrel to die diameter contraction ratio of 15, with best-fit straight line to the data

Figure 7. The extensional rheometer showing the sample clamped between the rotating head and the air-cooled fixed clamp and suspended above the silicone oil bath.

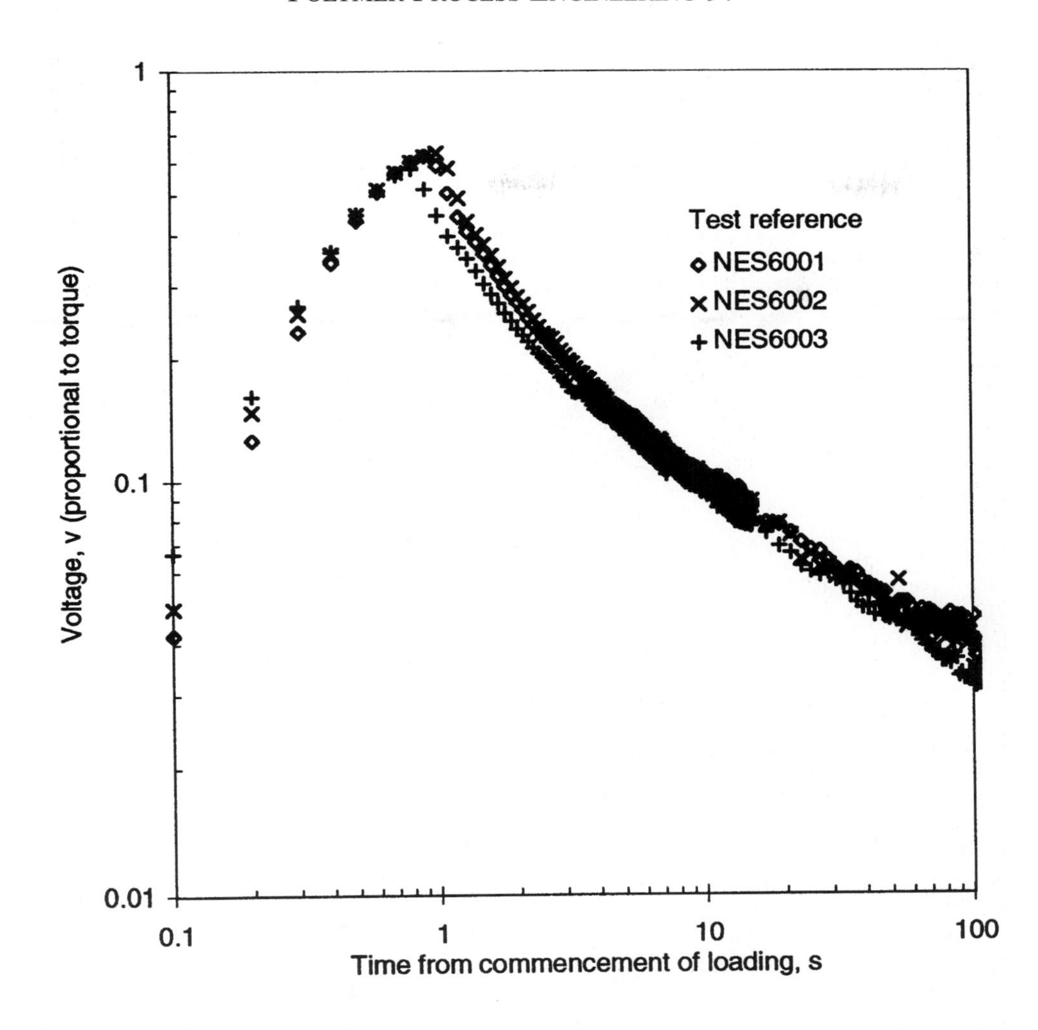

Figure 8. Repeat extensional stress retardation and relaxation measurements of NCPE 3416 at 190 °C.

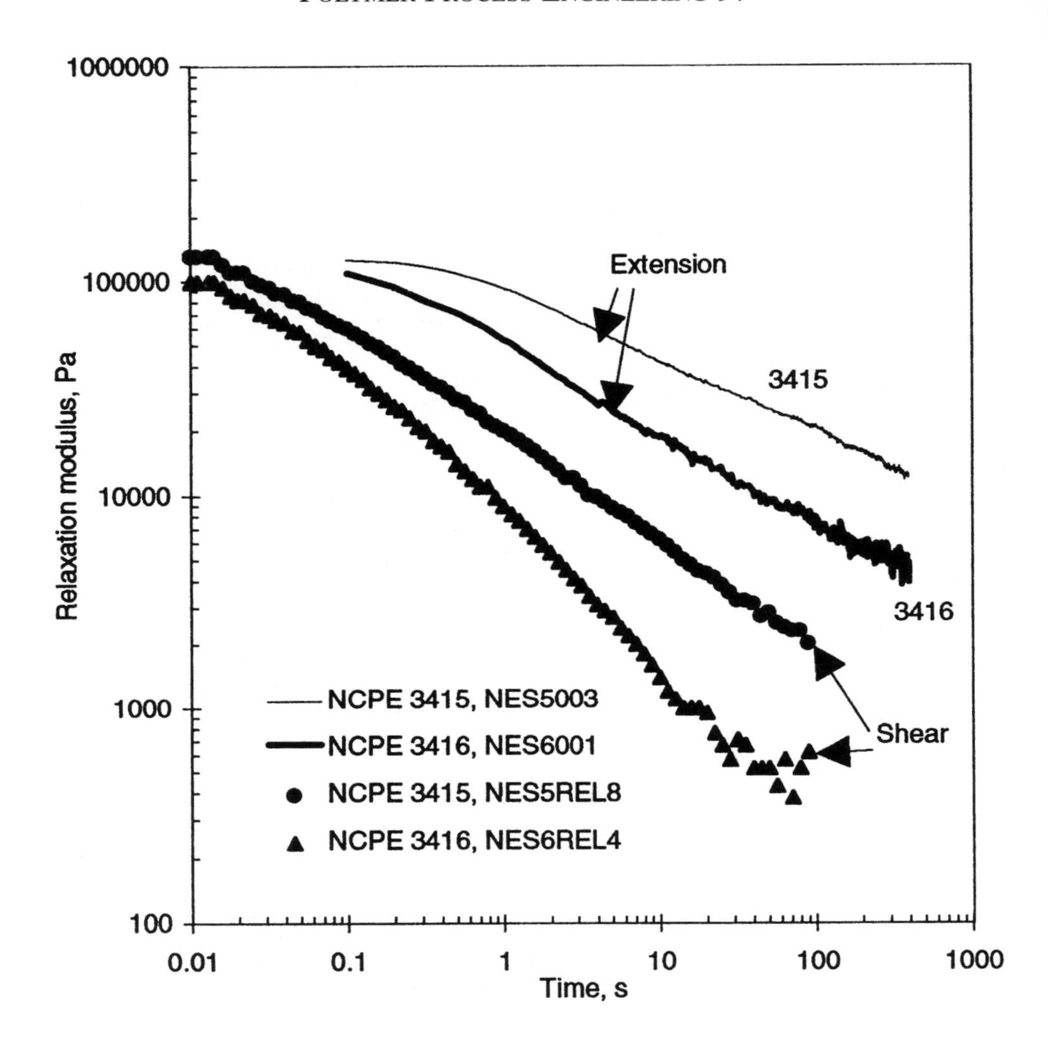

Figure 9. Comparison of extensional and shear stress relaxation behaviours of NCPE 3415 and 3416 polyethylene at 190 °C.

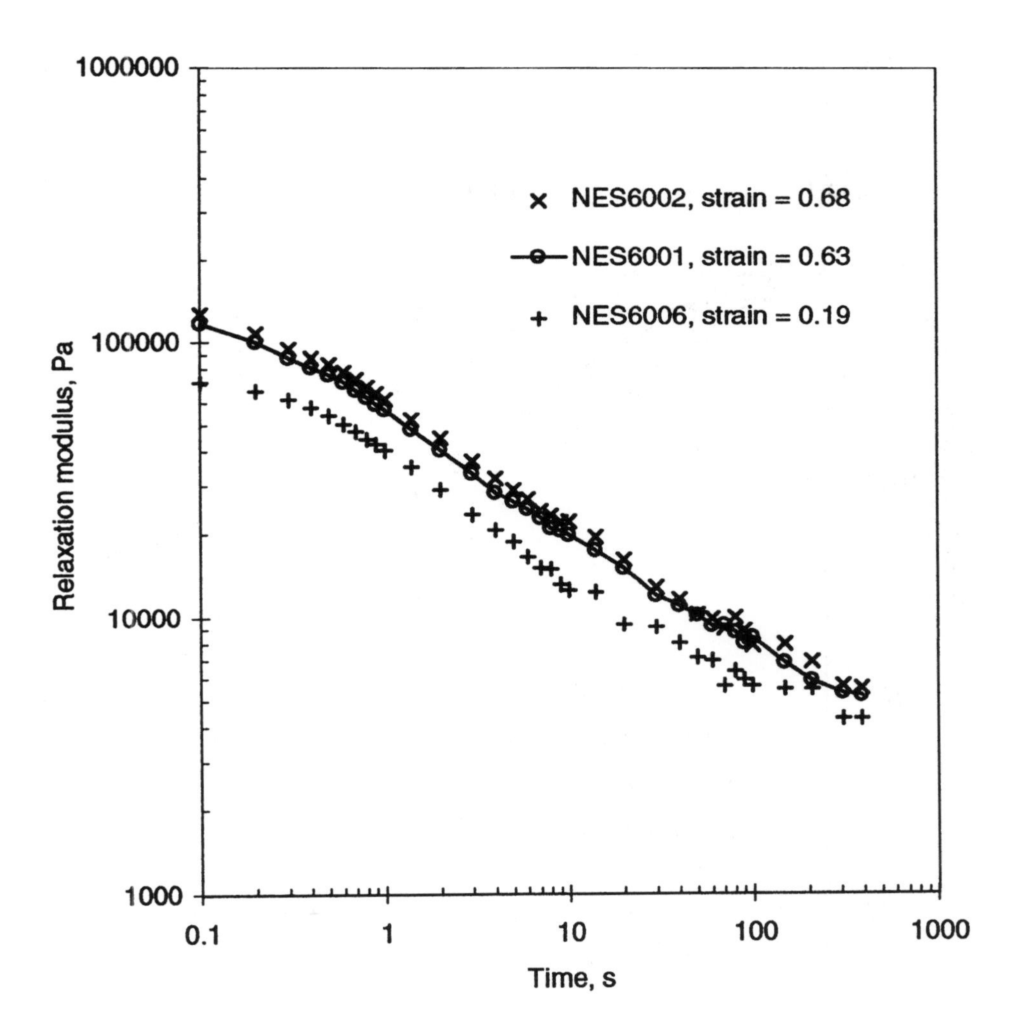

Figure 10. Effect of the initial strain on extensional relaxation modulus of NCPE 3416 polyethylene at 190 °C.

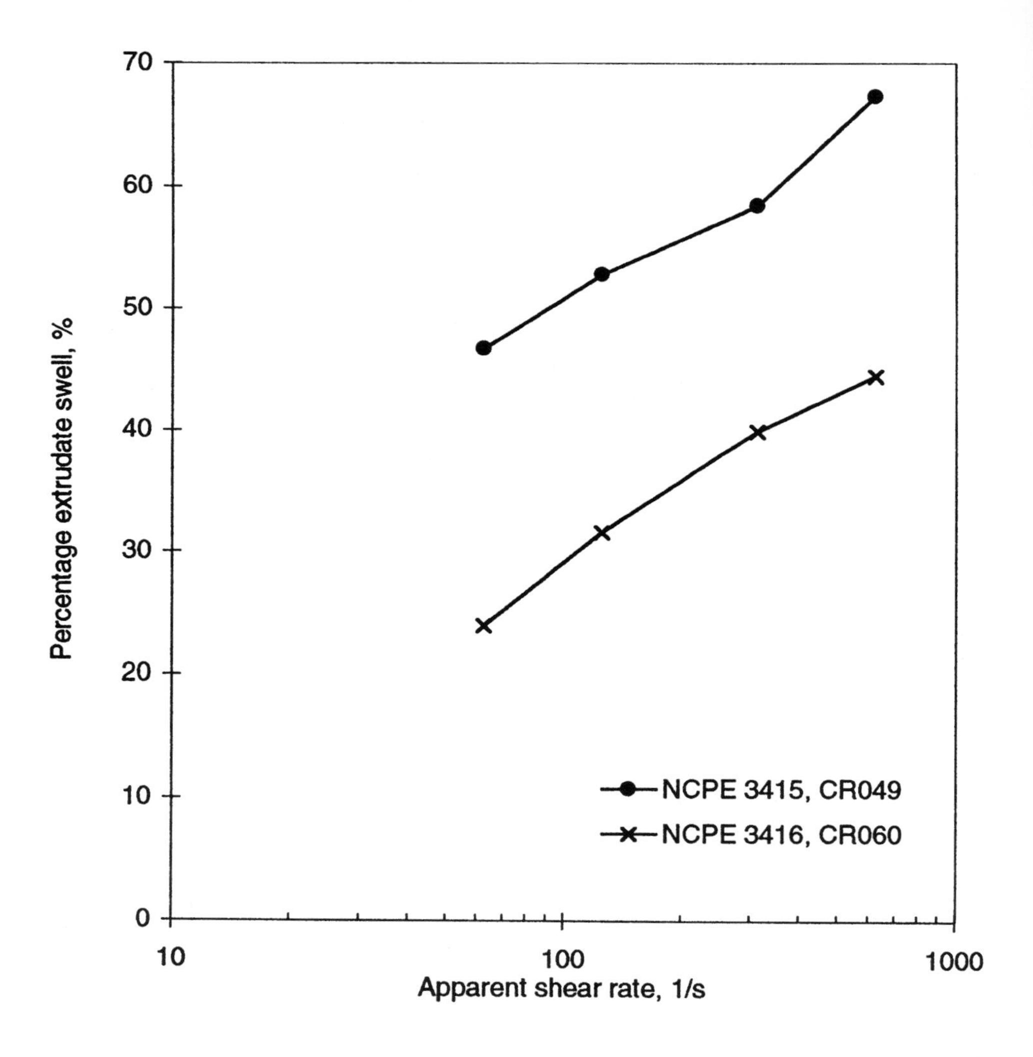

Figure 11. Comparison of the extrudate swell behaviours of NCPE 3415 and NCPE 3416 polyethylenes at 190 °C obtained using a die of length 30 mm and diameter 2 mm.

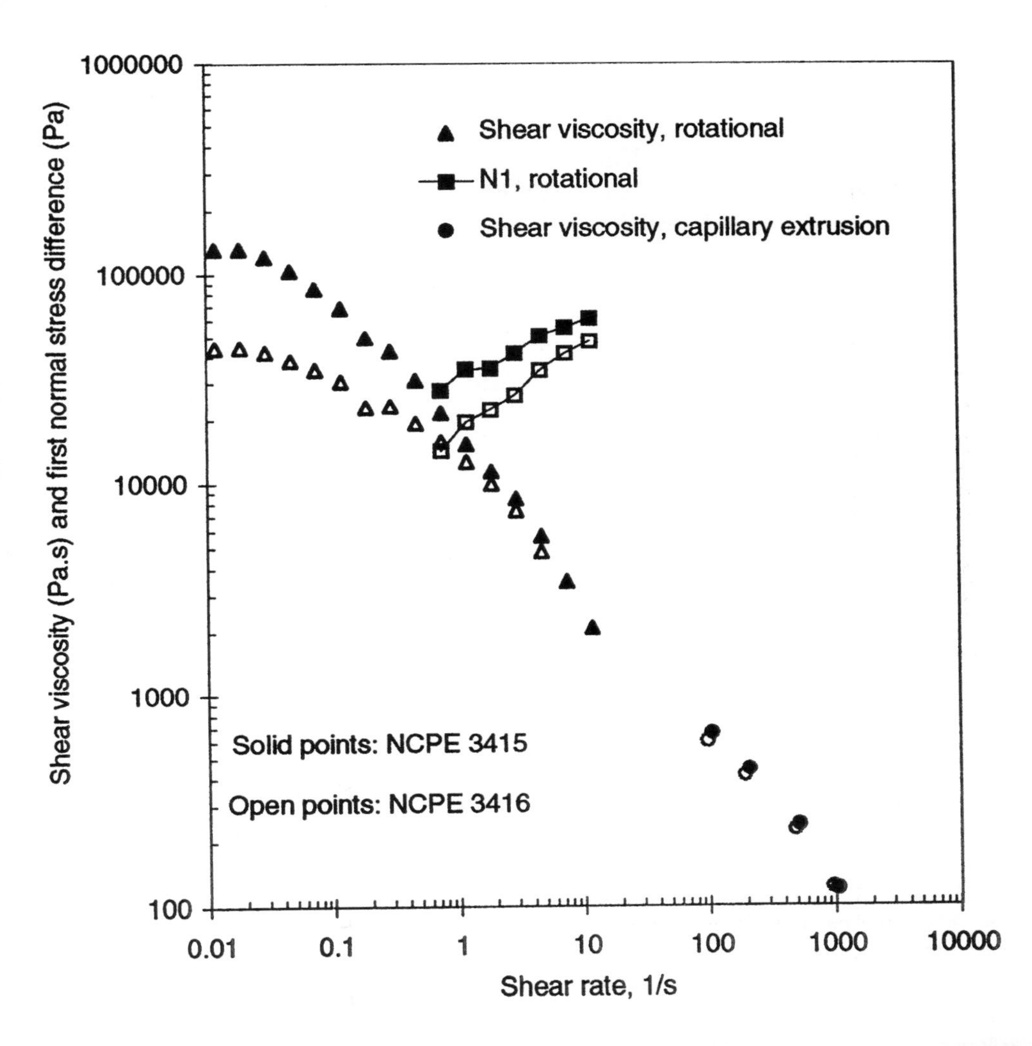

Figure 12. Comparison of the steady shear behaviour of NCPE 3415 and NCPE 3416 at 190 C.

Flow visualisation for extensional viscosity assessment

C Nakason, M Kamala, M Martyn and P D Coates

IRC in Polymer Science & Technology, Mechanical & Manufacturing Engineering, University of Bradford, Bradford UK

ABSTRACT

A flow cell has been used together with a video imaging system and transputer-based particle tracking to observe the convergent flows of polymer melts in steady state single screw extrusion through a slit die. Stress birefringence provides principal stress difference measurements, which can be obtained from the digital image analysis of isochromatic fringes. These can be combined with flow velocity measurements (hence strain rates) for the full entry flow field. This technique provides useful information for assessment of any vortex and entry flow profiles for different flow conditions and polymer types, and is a route to assessment of extensional viscosity. It may also be used in validation of computational fluid dynamics modelling of such flows.

1. INTRODUCTION

Quantification and understanding of polymer melt rheology in complex flow fields is a fundamental prerequisite for accurate flow modelling, processing, and future polymer development. Of particular interest is the identification of polymer properties responsible for differences in flow behaviour observed in geometric contractions (and divergences) such as those encountered in injection nozzles, gates, extrusion dies and capillaries. A key experimental technique for such studies is flow visualisation.. This technique is capable of providing rapid, accurate, quantitative stress and velocity field data, together with other insights into the behaviour of polymer melts in complex flow fields. The stress field, and evidence of vortices have traditionally been the main interest in previously published work, often for comparison with computational fluid dynamics modelling of the melt flow, typically in 180 degree entry angle contraction flows [1,2].

By combining the measurement of stress and velocity (hence strain rate) it is also possible to obtain an assessment of extensional viscosity of the melt, under process conditions. Extensional viscosity is of considerable importance in processing but it is a difficult parameter to measure. Few techniques are available which provide pure extensional flows of melts; all available techniques are constrained, for example by non-constant strain rates or a very limited range of extensional strain rate which does not relate to processing conditions. The flow visualisation technique discussed here offers some advantages, in particular the fact that measurements are made under actual process histories, at process strain rates, but it also suffers significant disadvantages, including a non-constant strain rate and non-isothermal conditions. Consequently, this approach can only provide apparent extensional viscosities.

2. EXPERIMENTAL DETAILS

The flow behaviour of a range of polymer melts passing through an abrupt contraction geometry have been studied using a flow visualisation cell. The cell was fitted to a computer-monitored 38mm single screw extruder (Betol BK38).

The results presented here are for a low density polyethylene (LDPE) a polymer typically exhibiting long chain branching. The work reported relates to a specific LDPE, Novex EXP2184 (MFI= 14.4g/10min at 190C), originally supplied by BP Chemicals.

A schematic diagram of the flow visualisation system is shown in Figure 1. The cell has replaceable inserts allowing different convergent geometries to be studied. The studies reported here involved a 180° entry slit die geometry of 10 mm depth, 1 mm width and 28 mm length, with an entry contraction ratio of 15:1. The dimensions were chosen in an attempt to provide planar and elastic invariant flow conditions, as suggested by Otter and Han [3-5]. The flow cell and imaging system were configured so as to study the whole converging flow region in the upstream reservoir immediately prior to the slit entrance, and the slit entry region.

Birefringence is the difference in refractive indices in a material with direction of a light beam passing through the material. In the case of polymer melts this effect can be caused by polymer chain orientation in flow fields. The equipment used to study such 'flow birefringence' comprises an optical bench with defined mercury vapour light source, polariser, analyser, quarter wave plates and 546.1 nm interference filter.

When a beam of polarised light passes through stressed flowing polymers it is split by the birefringence effect. The direction of one path is coincident with one of the principal stress directions. The split light paths are retarded, which results in interference fringe patterns. The relative retardation, R, (or fringe order) can be related through the stress optical law [5-8] to the principal stress difference, $\Delta\sigma$, by:

$$\Delta n = C\Delta\sigma = \lambda R / d \qquad (1)$$

where Δn is the birefringence, d is the propagation length of light through the polymer melt, λ is the wavelength of the transmitted light (546.1nm) and C is the stress optical coefficient.

Assuming that the orientation of the optical axes are coincident with the principal stress axes, shear stress τ_{xy} and normal stress difference τ_{xx} -τ_{yy} are derived as follows:

$$\tau_{xy} = (\Delta\sigma / 2)\sin 2\theta \qquad (2)$$

$$\tau_{xx} - \tau_{yy} = (\lambda N / Cd)\cos 2\theta \qquad (3)$$

where the principal stress direction, θ, and fringe order, N, are obtained from the isoclinic and isochromatic fringe patterns respectively [5,8,9].

To aid further the visualisation of flow patterns, a trace amount of coloured masterbatch material was introduced into the extruder, to highlight the molten polymer flow. Flow patterns of the polymer melt were then recorded using the image capture system. The digitised images obtained then allowed accurate measurement of fringe positions. In addition, the melt flow convergence profile, also known as the 'natural entry profile' could be determined.

A particular advantageous feature of the experimental technique developed is that real-time particle streak velocimetry [10,11] could be performed, almost simultaneously with the image capture. A trace amount of silicon carbide (SiC) particles of nominal size around 150μm, were added to the polymers. A CCD video camera was used to capture images of the moving particles at 40 msec intervals (i.e. normal video frame rate). A PC-based frame grabber (Data Translation) was used to digitise the captured images, threshold them and display the result on a second monitor, in real time. Moving particles create streak patterns in the captured images. Such streaks were analysed automatically using a transputer system using in-house software, to obtain the flow velocity vector field for the full entry region. Analysis of particle movement could also be performed on the digital images on the PC, using a digitising cursor positioned manually. This technique was found to be useful for velocity measurements on the flow centre line, where large accelerations occur.

3. RESULTS & DISCUSSION

Representative isochromatic fringe patterns for LDPE are shown in Figure 2. It can be seen that the number of fringes strongly depends on the melt flow rate depicted here by slit wall shear rate, that is, the number of fringes increase as the slit wall shear rate is increased. The shape and position of isochromatic fringes represent the loci of points of constant principal stress difference (PSD) and maximum shear stress. At start up of the extrusion process, the fringes appear just at the entry corners of the slit and develop from the corners toward the axial centre line. As the flow rate (extruder screw speed) and hence slit wall shear rate were increased, the combined fringe splits into two parts which simultaneously move in different directions: one part moves into the slit and the other move into the upstream reservoir of the flow cell. The characteristics of isochromatic fringes are dependent on the rheology of the melts.

Along the axial centre line of the flow cell, polymer molecules travelling from the reservoir are subjected to pure extensional flow. The PSD reaches its maximum just before the entrance of the slit is reached corresponding to a region of maximum extensional strain rate. A plot of the centre line PSD for LDPE is shown in Figure 3. Such data has traditionally been of major interest to the flow modelling community: model predictions typically include fringe patterns and centre line PSD (e.g. ref [1]).

Shear stress and first normal stress difference distributions in the convergent flow field near the slit entrance were quantitatively determined using equations 2 and 3. The procedure needed isochromatic and isoclinic fringe patterns together with the appropriate stress optical coefficient (SOC) of the materials at the processing temperature. The SOC was determined from the fringe pattern in the slit, i.e. in a region of steady shear flow. A typical stress field determined using this enhanced visualisation technique is shown in Figure 4.

Flow patterns of polymer melts shown by trace materbatch were observed under several different conditions. Figure 5 shows the images of natural entry profiles for the LDPE. It was observed that vortices were not created at very low slit wall shear rates. However, vortex growth begins at a low flow rate and initially increases slightly with flow rate from the contraction corners to a position upstream in the reservoir, as the recirculation begins. The vortex then continues to grow with increasing flow rate. Natural entry profiles were readily determined from their digitised images. The entry profiles shown in Figure 6 are typical of a branched polymeric melt. The natural entry profile patterns were observed to be strongly dependent on the slit wall shear rate (melt flow rate).

The particle streak velocimetry technique was successfully used to determine local velocity vectors in the flow field as shown in Figure 7 (where the data is actually for a different LDPE grade, at 190C). Raw data is actually generated as 2-d (x and y) components of the velocity vector. The secondary flow (recirculation) of the melt in the corners of the upstream reservoir is evident. However, it is also apparent that the automated technique has to cope with a wide range of velocities, which can be very high at the centre line, but very low in the vortex regions. Both extremes cause measurement difficulties, which are currently being explored. Work is also proceeding to develop the technique to provide a more accurate quantification of velocity profiles at any section through the flow.

Apparent extensional viscosity values can be determined if local stress and strain rate values are known. In particular, the extensional stress (from PSD) and extensional strain rate can be obtained at points along the centre line of the flow. The strain rate may be calculated by various means; as indicated in section 2, manual digitising of particles has been found useful for centre-line calculations of velocity and hence velocity gradient.

However, a simple model can also be used to obtain an apparent extensional viscosity value. In the model, melt flow inside the natural entry profile is approximated to be plug flow, i.e. plane sections remain plane, and a single axial velocity applies to all points across the flow at a given axial position, x. This is equivalent to ignoring the shear flow occuring inside the natural entry profile. The volumetric flow rate, Q is obtained from measured extruder output, kg/hr, using melt density in the die to convert to volumetric values. Consequently, the mean axial velocity v_{xx} at any axial position x is given by

$$v_{xx} = Q/A \qquad (4)$$

where A is the cross-sectional area.

However,
$$A = x.2y \qquad (5)$$

where y is the co-ordinate of the natural entry profile at axial position x (hence 2y is the 'width' of the natural entry profile at position x); y(x) is the function representing the natural entry profile.

By definition, the axial extensional strain rate $\dot{\varepsilon}_{xx}$ is given by the velocity gradient, i.e.

$$\dot{\varepsilon}_{xx} = \frac{dv_{xx}}{dx} \qquad (6)$$

It is therefore possible to use the entry profile and PSD data for the flow to obtain an assessment of the apparent extensional viscosity versus axial strain rate. Figure 8 shows the calculated axial extensional strain rate for the 255 s^{-1} slit wall shear rate flow conditions but also including data from the manual digitising of particle movements, referred to in section 2. Good agreement is observed between the two methods for calculation of axial strain rate.

Apparent extensional viscosity is simply obtained from the ratio of stress to strain rate. Fig. 9 shows the resulting apparent extensional viscosity as a function of axial extensional strain rate, using experimental data points from the 125s^{-1} , 255s^{-1} and 600s^{-1} tests. Figure 9 also includes values calculated by interpolation of PSD for the 255s^{-1} case. The data from all three tests lie around the same straight line. Apparent extensional viscosity appears to be falling with strain rate. The accuracy of this assessment is limited; it is also possible that temperature rise in the flow towards the slit entry will contribute to errors in the assessment.

Convergent flow modelling, based upon various velocity fields and the normal continuum mechanics assumptions, has recently been discussed in detail and compared with experimental results for polyolefin melts from in-line rheometry in injection moulding [12], hence at higher strain rates than in the work presented here.

4. CONCLUDING COMMENTS

Flow visualisation provides a useful means of measuring local tensile, shear and normal stress difference and velocity distributions in polymer melts flowing through a complex geometry. Furthermore, the system has been used to visualise the natural entry flow patterns including vortex growth and other characteristics of convergent flows. Such data is currently being used to validate finite element modelling of these flows. The data from the visualisation technique is also being used to determine local apparent extensional viscosity of polymer melts, but the measurements are subject to non-constant strain rate and temperature conditions.

Initial particle streak velocimetry studies have been useful in determining velocity distributions inside the flow cell. Some difficulties were experienced in determining velocities along the centre line and in the slit due to the relatively high speed flow close to and in the slit. A high speed camera is being used to address this. Also, a laser light sheet will be used to illuminate spherical particles inside the melts, to help study 2-d and any 3-d effects in the flow fields.

References

1. R. Ahmed, and M.R. Mackley, J. Non-Newtonian Fluid Mech., 56, 127-149, (1995).
2. H.J. Park, D.G. Kiriakidis, E. J Mitsoulis. and K.J. Lee, J. Rheol,. 36 (8), 1563-1583, 1992
3. J.L. Den Otter, Doctoral Dissertation, University of Leiden, 1967.
4. C.D. Han, K.U. Kim; Polym. Eng. Sci., 11,395, 1971.
5. C.D. Han, L.H. Drexler; J.Appl. Polym. Sci; 17, 2329, 1973.
6. A.J Durell, and W.F Riley,. Introduction to Photomechanics. Prentice-Hall, INC. 1965.
7. A.W Hendry, Photo-Elastic Analysis, Pergamon Press, New York, 1966.
8. J.L.S Wales, The Application of Flow Birefringence to Rheological Studies of Polymer

Melts. Delft University Press. Delft, 1976

9. K. Nakamura,., K. Ishizaki, Y Yamamoto,., T. Amazutsumi,., and A. Horikawa,., J. Text. Mach. Soc. Japan, 24(3), 61-89, 1978.
10. P.Olley, M. Kamala M.Martyn, C. Nakason and P.D. Coates, Proc PPS European Meeting, 1-5, Sept. 1995.
11. P.D. Coates, C.Nakason, M. Kamala, M Martyn and P. Olley, Proc. PPS 12th Annual Meeting, 27-31 May 1996.
12. D. J. Groves, M. T. Martyn and P.D. Coates, Plast.Rubb.Comp.Process.Applicns, in press (1997).

Acknowledgements

We are grateful for support of the Royal Thai Governement for CN and the University of Bradford for MK.

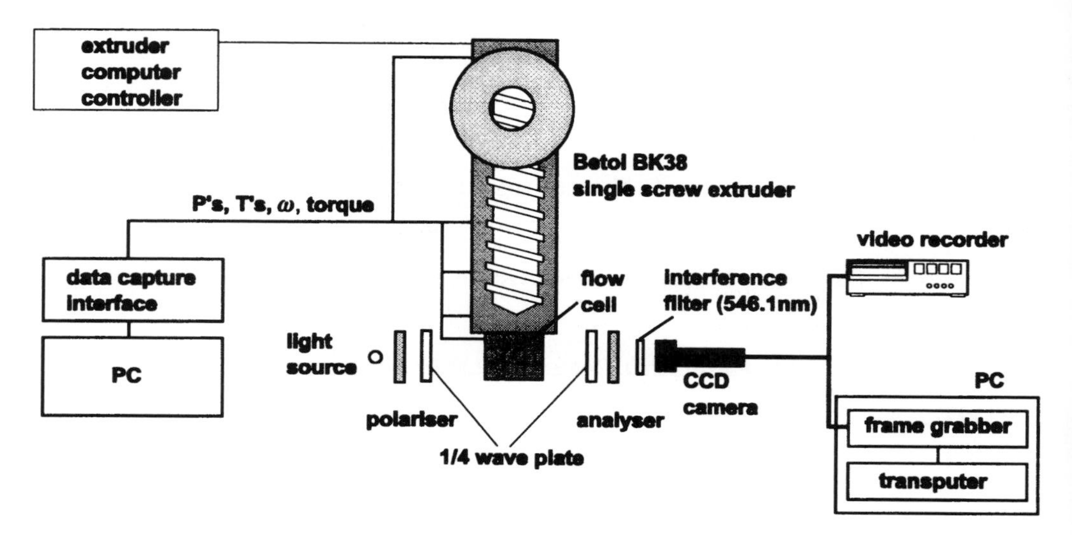

Figure 1 Flow visualisation cell, image capture and image processing facilities.

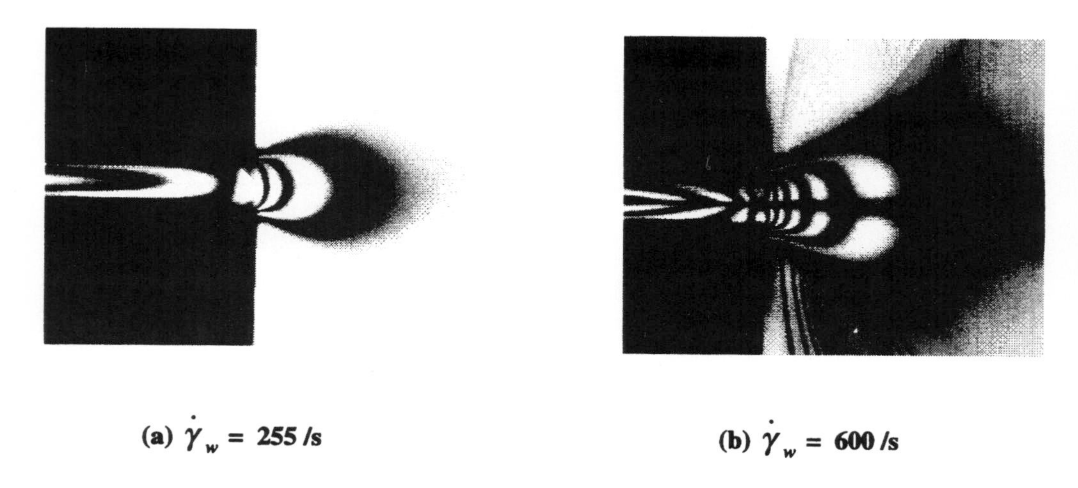

(a) $\dot{\gamma}_w$ = 255 /s **(b) $\dot{\gamma}_w$ = 600 /s**

Figure 2 Isochromatic fringe patterns of LDPE on flow through 180° abrupt geometry at 190°C.

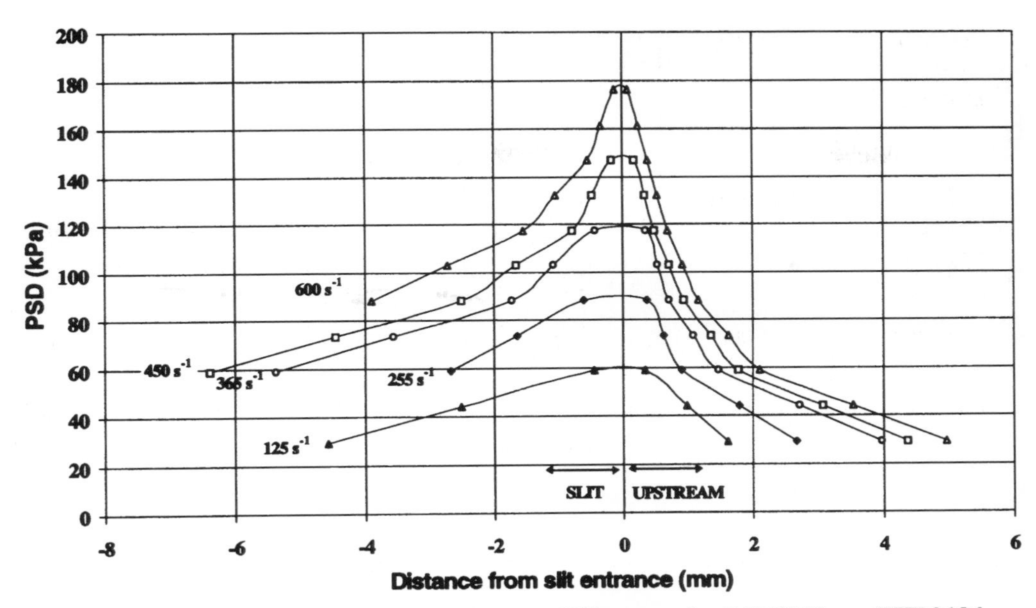

Figure 3 Measured centre line principal stress differences for LDPE NovexEXP2184 melt at 200°C and various slit wall shear rates.

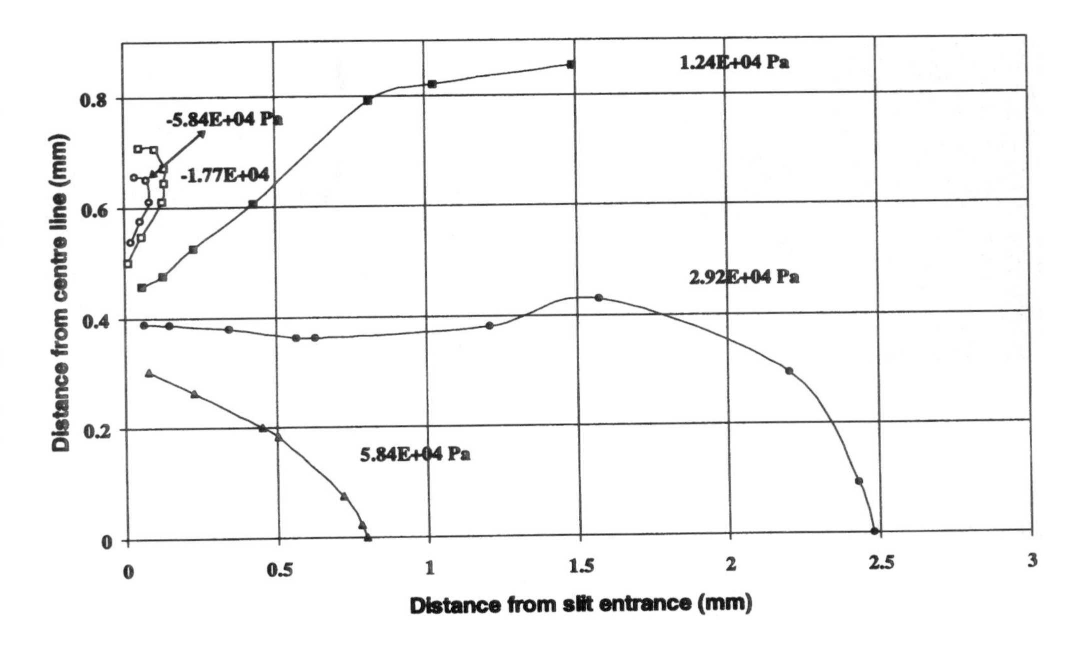

Figure 4 First normal stress difference profiles in the entrance region for LDPE, Novex, EXP2184 at 200°C and slit wall shear rate of $255 s^{-1}$.

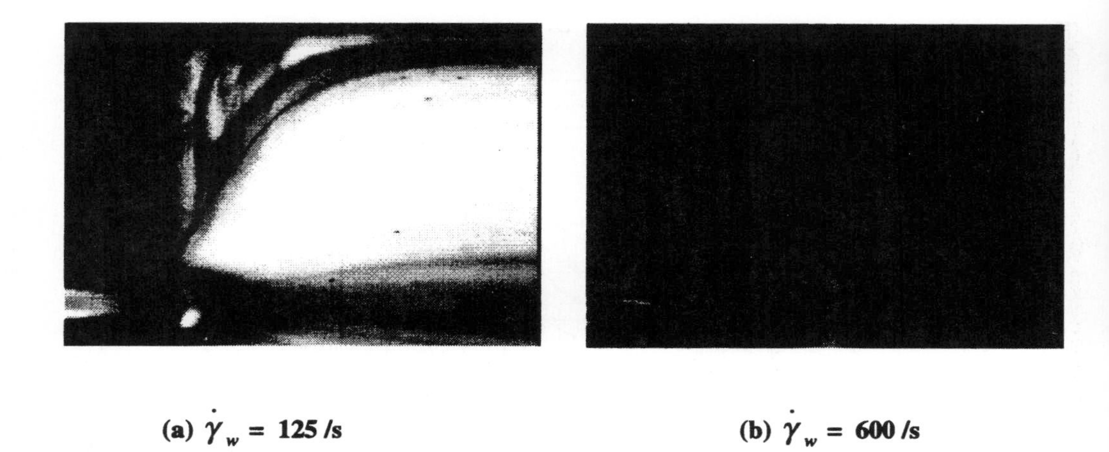

(a) $\dot{\gamma}_w$ = 125 /s (b) $\dot{\gamma}_w$ = 600 /s

Figure 5 LDPE melt on flow through a 180° abrupt geometry at 200°C.

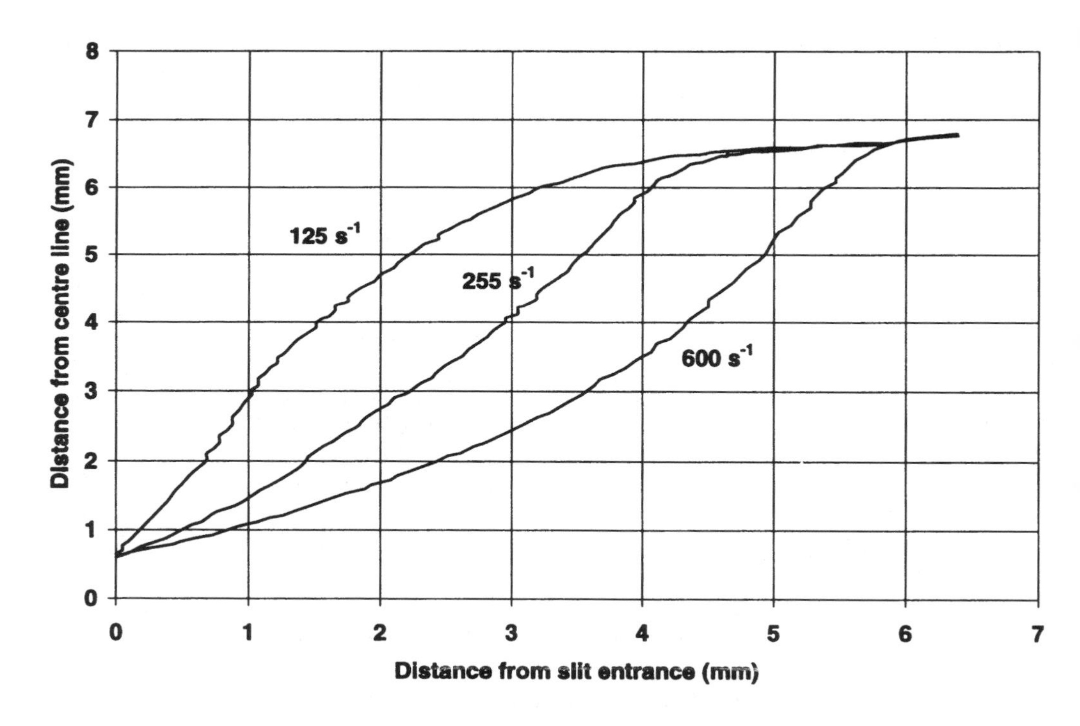

Figure 6 Natural entry profiles for LDPE, Novex, EXP2184 at 200°C and various slit wall shear rates.

Figure 7 Local velocity vectors generated from real time computation of tracer particles in an LDPE melt during flow through a 180° abrupt geometry at 190°C.

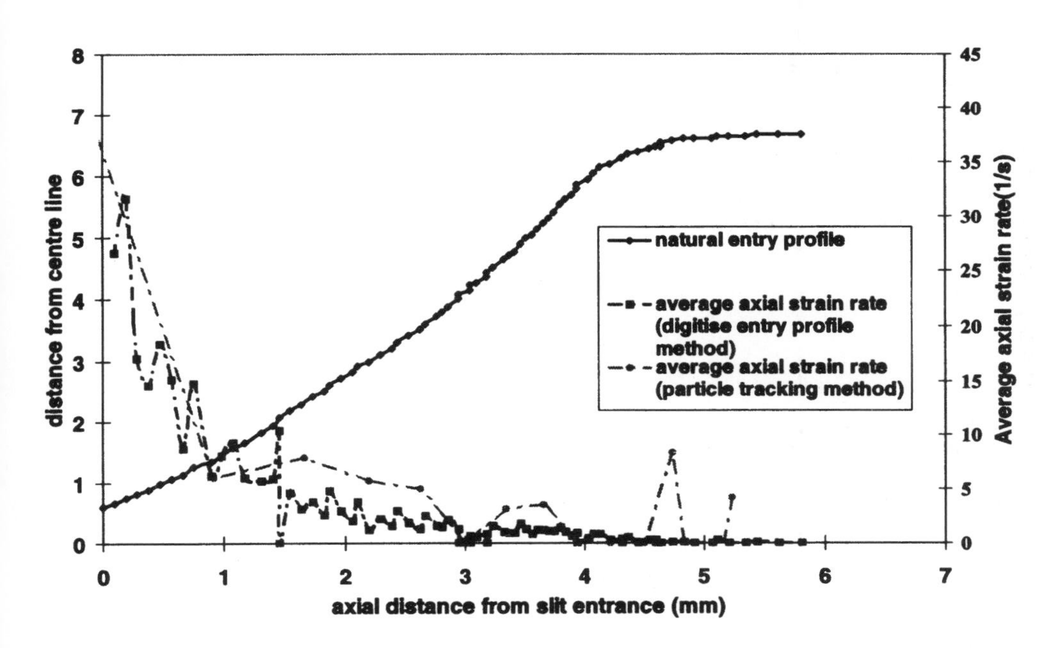

Figure 8 Average axial strain rate calculated by two methods, versus axial position for LDPE NovexEXP2184 at 200°C and a slit wall shear rate of 255s^{-1}; natural entry profile also included.

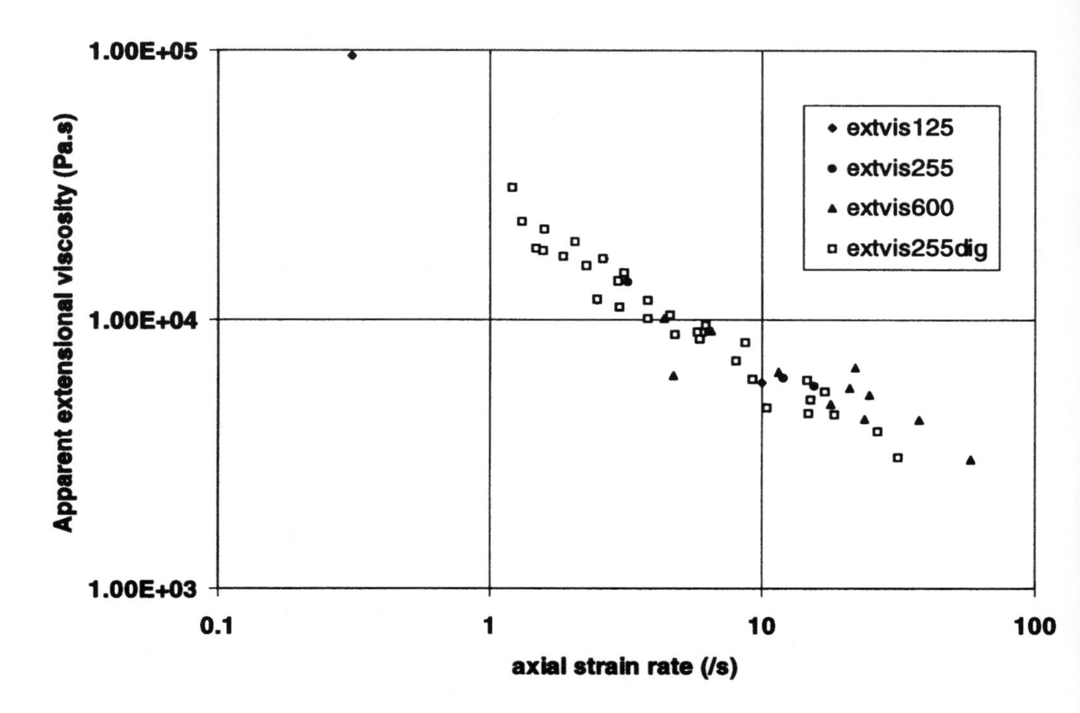

Figure 9 Apparent extensional viscosity of LDPE NovexEXP2184 at 200°C versus axial extensional strain rate, using data from various slit wall shear rate tests.

Biaxial Extensional Rheometry of PET Relevant to Process Conditions

C. Gerlach*, C.P. Buckley*, D.P. Jones†

** Dept. of Engineering Science, University of Oxford, Parks Road, Oxford OX1 3PJ, UK*

† ICI Polyester, PO Box 90, Wilton Centre, Middlesbrough, Cleveland TS90 8JE, UK

ABSTRACT

Improved modelling of stretching flows in polymer processing requires the development of a new generation of constitutive models. These models will be physically-based, able to simulate the intricacies of polymer response to a degree not seen previously, and will need to be validated under conditions truly relevant to industrial processes. This paper reports recent progress towards a methodology for the creation of such models, applied to the case of poly(ethylene terephthalate) (PET) film. The authors propose an integrated approach involving: (a) development of a Flexible Biaxial Film Tester (FBFT) for high-rate biaxial extensional rheometry of polymer films, (b) development of a mathematical model of the constitutive behaviour of films during biaxial stretching, (c) use of the testing machine to validate the model and obtain its parameters for particular polymer films of interest, and (d) embedding of the material model within a continuum (finite element (FE)) model of the process itself. Here the authors discuss the design of the FBFT, and outline how the data it provides may be applied in a FE model for simulating hot drawing PET film.

1 INTRODUCTION

Poly(ethylene terephthalate) (PET) is a commercially very successful polymer with substantial market shares in large-scale industrial applications such as films and bottles. The technological processes for manufacturing PET films and bottles differ in detail but both are characterised by rapid biaxial stretching flows just above the glass transition. Here, in a narrow window of stress/strain/temperature-history, the material experiences significant molecular orientation and deformation-induced crystallisation, allowing the achievement of enhanced room-temperature properties which are required in the end-product. PET is also one of the best studied engineering polymers, and the literature records several investigations of its deformation behaviour. The main features of its constitutive response are therefore well-known.

Surprisingly, however, the database truly relevant to industrial processes is very limited indeed. This is because of the inherent difficulties of reproducing such conditions in a controlled laboratory experiment; especially the large multiaxial deformations under high rates of strain and temperature change. But in view of the polymer's strongly path-dependent thermomechanical properties, it is essential that any constitutive relation is developed and validated under conditions which are relevant to the range of their subsequent usage. It cannot be assumed that results from relatively slow uniaxial tests with conventional testing machines will be related in a simple manner to the behaviour of the polymer in an industrial film drawing process. There are many aspects of PET processing for which accurate modelling with finite element (FE) analysis would be desirable. The urgent need is for a methodology which combines the obtaining of relevant rheometry data with its application to numerical modelling of the polymer in the FE context.

The purpose of the present paper is to propose an integrated methodology for meeting this need. Four steps are required. First, suitable equipment must be produced for relevant biaxial rheometry studies. Second, a mathematical representation (constitutive model) of the material deformation behaviour must be developed. Third, the parameters of the model must be determined from the rheometry data. Fourth, the material model must be incorporated within a continuum (FE) model of the process for solving problems of practical importance. In the present paper, we focus on two aspects: (1) development of a new testing machine - the unique Flexible Biaxial Film Tester (FBFT) - designed specifically to generate experimental data under conditions approaching those experienced in commercial film drawing, bottle blowing and thermoforming, (2) development of an FE model of PET, capturing the large strain, multiaxial, nonlinear viscoelastic behaviour in a system of nonlinear differential equations which are solved numerically during FE analysis. The constitutive model employed is the glass-rubber model described by us elsewhere [1], improved more recently to include entanglement slippage and stress-induced crystallisation [2]. It was implemented in the FE context via the 'user-defined material' subroutine UMAT in the FE tool ABAQUS/Standard.

2 THE FLEXIBLE BIAXIAL FILM TESTER

The study of biaxial stretching flows with rapid changes of strain and temperature is difficult to achieve in the laboratory. Standard thermomechanical testing equipment is generally limited to uniaxial deformations and modest rates of strain and temperature change.

Biaxial testing of films has been possible for some time, however, with the pioneering Long Extension Tester (LET, TM Long Co.), well-known for its innovative scissors-like movement of the pneumatic specimen grips. The film is heated to the test temperature by a combination of heat conduction and forced convection, and the hydraulically driven movement of two perpendicular load bars (with respect to their fixed counterparts) extends the specimen biaxially to a nominal strain of up to 6 at constant nominal strain rates of up to 8 s^{-1}. More recently, some research laboratories have described machines aiming to improve on the LET to meet specific research needs, notably at Oxford [3], Loughborough [4] and Leeds [5]. A particularly novel approach to extensional rheometry of polymer melts has been pioneered at the ETH Zürich, where moving belts are used to grip polymer films and provide relatively rapid stretching flows: the uniaxial testing machine described by Meissner and Hofstettler [6] is understood to have been enhanced recently to provide biaxial stretching.

We believe, however, that the purpose-built biaxial film tester developed in our laboratory provides an unusual degree of flexibility in experimental sequence of temperature and deformation, with the specific aim of getting close to process conditions. It was therefore termed a Flexible Biaxial Film Tester (FBFT). We have already reported briefly on the design and certain novel features of the FBFT [3]. Since then it has been further developed and enhanced, and used successfully for a number of experimental studies [2,7]. This encourages us to give here a fuller account of its individual parts and their assembly into a complete biaxial testing system.

2.1 Design of Biaxial Stretching Frame

The biaxial stretching frame consists of four aluminium load-bars forming a rectangular 'picture frame' which is driven by two orthogonal reverse-thread leadscrews, such that opposing sides form one drawing axis with the load-bars moving in opposite directions for the same distance, leaving the specimen centre stationary. As shown in Figure 1, each load-bar is fitted with specimen grips which are mounted on linear bearings providing low friction sideways movement along the load-bar. In total we use twenty-four grips which are interconnected by a scissors-mechanism via the four corners of the 'picture-frame'. As the corner pieces are mounted on the linear guides of neighbouring load-bars, the combined inwards or outwards movement of opposite load-bars, as their corresponding leadscrew rotates, causes the grips on the perpendicular load-bars to either close-up or open-out while remaining equi-spaced (see Figure 1). This scissors-like action is common in biaxial testing machines such as the LET, but with the combination with low friction sliding of the grips and synchronised movement of opposite load-bars, the FBFT minimises lateral forces and loading asymmetries. The FBFT allows nominal strains of up to 4 and in order to provide nominal strain rates of up to 32 s^{-1}, it employs leadscrews with an unusually high pitch of 50 mm and each leadscrew is driven by a fast-accelerating brushless DC servo-motor which achieves high speeds and high torque retention through a low rotor inertia. The two motors can be controlled independently via an indexer board which is integrated in hard- and software in the control system outlined below.

Each grip is pneumatically operated, providing a constant clamping pressure even when the sample undergoes substantial thinning during deformation. At the end of the test a vacuum returns the miniature pistons allowing release of the sample. The pistons are manufactured from glass fibre reinforced PEEK which thermally insulates the sample from the grip housing. Furthermore, the contact-surfaces of each piston and anvil are machined with ridges carefully aligned parallel to the sample edges to restrain the specimen from slipping out in the load direction while enabling it to flow laterally out from under the grips.

2.2 Temperature Control

The FBFT is designed to produce not only deformations relevant to manufacturing processes but also the thermal sequence, i.e. film temperatures above the glass transition (ca. 70°C) combined with high rates of heating and cooling. Clearly, the achievement of high rates of temperature change is not possible if thermal stablility is provided by a conventional isothermal heating chamber. Instead, temperature control is achieved by impingement on the top film surface of a variable temperature air jet. The back surface of the sample is effectively insulated by static air enclosed between the sample and a soft rubber sheet which follows the deformation reversibly. Since we first reported on the FBFT [3] the temperature control system has been modified. Currently, air at room temperature is compressed by two high-velocity fans and fed through high-performance hot air-guns where it is heated rapidly to the test temperature. Before the air jets impinge on the specimen, they are directed through a thermally insulated aluminium cone structure which leads into an expandable silicone rubber jacket. The silicone rubber jacket prohibits mixing of the controlled air stream with cold air but minimises the amount of exchangeable air volume by following closely the biaxial movement of the stretching frame. This was achieved by clamping it to the four corners of the stretching frame and additionally guiding it by miniature 'runners' sliding along metal rods which are attached also to the corners.

The air temperature is regulated by close-loop control adjusting the power supply to the air-guns in response to the temperature measured downstream from the air guns using thermocouples placed in the high velocity air streams. The thermocouples are connected to a PDI-temperature controller which enables a temperature change of $45°Cs^{-1}$ to a given set point with minimum overshoot. During an experiment this set-point is under computer control; the temperature controller being linked to the control computer via a RS-232 serial interface. The low thermal inertia of the hot air-guns allows cooling of the specimen back to ambient temperature by simply switching off the power with the fans left running, the cooling rate of $20°Cs^{-1}$ then falls short of the heating rate.

2.3 Stress/Strain/Temperature-Measuring Instrumentation

The drawing forces are measured on two centre grips of the perpendicular axes using subminiature diaphragm-type load cells. Each load cell is mounted in a housing, which measures the axial movement of the specimen grip but remains unaffected by any lateral movement of neighbouring specimen grips. Drawing forces are converted to nominal stresses using the initial specimen mean thickness and the initial width associated with the specimen grips attached to the load cells. The true stress is calculated from the nominal stress by assuming that the specimen deforms at constant volume.

Prior to each experiment the specimen is marked with a square grid, using a computer and ink plotter. The strain and strain rate can be obtained directly from deformation of the grid, recorded at 25 frames per second using an image capturing system comprising a colour CCD camera and a Betacam video-recorder. The camera is positioned in a cooled mount directly above the centre of the specimen. The images are post-processed on a personal computer equipped with a high-speed DMA frame grabber and commercial image analysis software.

The specimen temperature is determined from a measurement of temperature near the specimen in the free air stream. A calibration graph obtained using an embedded 50 μm thermocouple exposed to the same experimental setup then allows the air temperature to be related to the film temperature.

2.4 Control System

As can been seen in Figure 2, the FBFT forms a computer-integrated system with a control computer co-ordinating the movements of the two servo-motors and the temperature history while simultaneously recording the analogue signals. The orthogonal drawing axes are driven by two servo-motors which are connected to independent drive units. Those drive units are addressed by the same indexer board which is plugged into the control computer expansion bus. As mentioned above, operating temperature is controlled by an independent PDI-controller, addressed by the control computer via a RS-232 serial interface. The transient recorder compromises a system of individual I/O modulus currently collecting signals of the force transducers and various thermocouples. The recorder is capable of high-speed multiplexing (typically at a rate of 20 data sets per second) and communicates with the control computer by the IEEE-488 interface. Tailored computer routines were written in Microsoft Visual C++ providing an integrated, easy-to-use environment for setting up individual components, and then for data acquisition during experiments. The experimental sequence of temperature and deformation, including data acquisition, can be setup interactively or a certain experimental sequence can be retrieved from the harddisk, allowing quick set-up and repeatable experimental procedures.

The full experimental sequence is recorded on Betacam video tape. The video recording is synchronised to the acquisition of the transducer signals by using a photographic flash which provides a light intensity peak in one frame. The flash is triggered automatically by the movement of the drawing axes, which coincides with a sudden increase in drawing force.

3 CONSTITUTIVE MODEL AND ITS APPLICATION IN FE ANALYSIS

For the predictive modelling of biaxial stretching processes, material properties obtained with the FBFT are imported into a Finite Element representation of the polymer employing the constitutive model for amorphous polymers near the glass transition, proposed earlier [1]. This model fits data for PET well close to the glass transition [7], and recent extensions to include entanglement slip and strain-induced crystallisation extend its applicability to temperatures more than 30°C above the glass transition [2]. Since the model is detailed elsewhere [1,2] an outline will suffice here.

3.1 Outline of the Model

The model rests on the assumption that the reversible work done in deforming the material is stored in two independent molecular mechanisms: firstly, the 'distortion' of intra- and intermolecular atom-atom distances; and secondly a thermally activated orientation of the entangled macromolecules against the conformational entropic resistance of a rubber-like network. The equation of equilibrium states therefore that the element stress Σ is balanced by what we may call the bond-stretching stress Σ^b and the conformational stress Σ^c. Further, both components share the same physical space and hence experience the same deformation gradient $\mathbf{F}$ (stretch and rotation) and deformation rate $\mathbf{D}$.

$$\Sigma = \Sigma^b + \Sigma^c \tag{1}$$

$$\mathbf{D} = \mathbf{D}^b = \mathbf{D}^c \tag{2}$$

It is well understood that at sufficient high temperature the elastic stretching of intra- and inter-atomic bonds no longer accounts for the full macroscopic stretch. There are additional diffusional motions of backbone chain segments whose onset gives rise to the glass transition and at higher temperatures occurs faster than the experimental time scale. Thus, the total deformation rate may be written:

$$\mathbf{D} = \mathbf{D}^e + \mathbf{D}^p \tag{3}$$

where

$$\mathbf{D}^e = \frac{1}{2\,G^b}\dot{\mathbf{S}}^b \qquad \mathbf{D}^p = \frac{1}{\mu}\mathbf{S}^b \tag{4}$$

where G^b is the shear modulus from the elastic bond-stretching and $\mathbf{S}^b = \Sigma^b - \sigma_m^b\,\mathbf{I}$ is the deviatoric bond-stretching stress component; σ_m^b is the scalar mean bond-stretching stress which defines the dilation of the material: $\sigma_m^b = K^b\,\mathrm{tr}(\ln\Lambda) + \sigma_{m0}^b$ where Λ is the stretch component of $\mathbf{F}$, K^b represents the bond-stretching constribution to the bulk modulus and σ_{m0}^b is the built-in hydrostatic stress present in rubbery polymers. Whereas many rubber and hybrid glass-rubber constitutive models force incompressiblity the current model naturally predicts the hydrostatic mean stress via its stress balance Eq. 1, leading to zero-strain (or unit stretch) compression stress: $\sigma_{mO}^b = -\frac{1}{3}\mathrm{tr}\left(\Sigma^c(\mathbf{I})\right)$. The stress-dependent viscosity μ is described by a multiaxial generalisation of the generic Eyring flow theory which was amended to include the dependence of the activation step on structure via the fictive temperature T_f. A detailed description of the scalar viscosity function is given in [1] and it will suffice here to indicate the different dependencies symbolically:

$$\mu = \mu\left(T, T_f, I\left(\Sigma^b\right)\right) \tag{5}$$

i.e. the bond-stretching viscosity is dependent on temperature T, the structure which is indicated by T_f and the bond-stretching stress through invariants $I\left(\Sigma^b\right)$.

The large strain deformation behaviour of uncrosslinked amorphous polymers above the glass transition is often described as a hyperelastic stretch of an entangled network of molecules against an entropic resistance. Research in our laboratory showed this description did not fit accurately the behaviour for PET. Instead, we found that PET appears hyperelastic in only a narrow temperature/time-window in the region of 95°C on a time scale of ca. 1s, but there is substantial entanglement slippage at longer times and higher temperatures. This flow process adds to the rubber-elastic (network) stretch, until strain-induced densification seems to intervene to halt the flow. The total rate of deformation $\mathbf{D}$ therefore comprises a deformation rate associated with the change of the conformational entropy of the network $\mathbf{D}^n$ and a deformation rate due to flow by slippage of entanglements $\mathbf{D}^s$:

$$\mathbf{D} = \mathbf{D}^n + \mathbf{D}^s \tag{6}$$

At present there is intense discussion of the exact nature and evolution of this strain-induced densification. No general agreement has been reached and therefore it is expedient to simplify this entanglement flow to a scalar Newtonian viscosity coaxial with the deviatoric conformational stress, which is arrested asymptotically at a critical value of network stretch λ^n_{crit}.

$$\mathbf{D}^s = \frac{1}{\gamma}\mathbf{S}^c \text{ where } \qquad \gamma = \gamma\left(T, \Lambda^n, \lambda^n_{crit}\right) \tag{7}$$

The conformational stress is expressed in terms of the scalar-valued conformational free energy function A^c:

$$\Sigma^c = \frac{1}{\det \Lambda^n} \frac{\partial A^c\left(\Lambda^n\right)}{\partial \Lambda^n} \Lambda^{n\mathbf{T}} \tag{8}$$

where Λ^n representing the principal network stretches: $\Lambda^n = \left\{\lambda^n_1\ \lambda^n_2\ \lambda^n_3\right\}^{\mathbf{T}} \mathbf{I}$. There are many suggestions for this Helmholz free energy function A^c; the present study employs the Edwards-Vilgis free energy function [8].

Procedures have been described elsewhere [2,7] for fitting the key nonlinear functions μ, γ and A^c to data obtained with the FBFT for PET. Other parameters of the model were taken from literature for PET.

3.2 Implementation of Model in FE Tool ABAQUS/Standard

The Eqs. 1-8, together with expressions for μ, γ and A^c may be combined to form a set of non-linear differential equations defining the material stress in terms of strain and temperature. For the purpose of the current study these differential equations were transformed into incremental equations where an explicit time marching procedure was adopted, and furthermore computer routines were written implementing the equations in the non-linear FE analysis tool ABAQUS/Standard via the 'user-material' subroutine UMAT.

ABAQUS/Standard follows the displacement-based FE method approximating first the incremental nodal point displacements and then calculating the internal stress distribution corresponding to those nodal point displacements before it establishes the force balance equations. Only if the equilibrium condition of externally applied forces and internally calculated nodal point forces is satisfied to a sufficient accuracy, the solver will proceed to the next time step but otherwise calculates a new approximation for the incremental displacements. Thus, the basic step in a UMAT subroutine is the calculation of the element stress for a given displacement history. In a large strain formulation the element deformation, or loosely the nodal point displacements, is described by the deformation gradient $\mathbf{F}$ which combines both material straining and rigid body rotation. In terms of a first-order explicit integration algorithm the stress at the current time moment $t + \Delta t$ is calculated from the previously converged stress at time t and the rate of stress at time t. The time-dependent constitutive behaviour is usually expressed in terms of the evolution of internal state variables including the material stress itself. A summary of the numerical integration scheme used in the current study is as follows:

deformation rate: $${}^{t}\mathbf{D} = {}^{t}\mathbf{D}\left({}^{t}\mathbf{F}, {}^{t+\Delta t}\mathbf{F}\right) \tag{9}$$

flow rules: $${}^{t}\mathbf{D}^{p} = {}^{t}\mu^{-1}\left({}^{t}T, {}^{t}T_{f}, I\left({}^{t}\Sigma^{b}\right)\right)\, {}^{t}\mathbf{S}^{b}$$

$${}^{t}\mathbf{D}^{s} = {}^{t}\gamma^{-1}\left({}^{t}T, {}^{t}\Lambda^{n}, \lambda^{n}_{crit}\right)\, {}^{t}\mathbf{S}^{c} \tag{10}$$

viscoelastic comp.: $${}^{t}\mathbf{D}^{e} = {}^{t}\mathbf{D} - {}^{t}\mathbf{D}^{p} \qquad {}^{t}\mathbf{D}^{n} = {}^{t}\mathbf{D} - {}^{t}\mathbf{D}^{s} \tag{11}$$

elasticity: $${}^{t}\overset{\nabla}{\Sigma}{}^{b} = \overset{\nabla}{\Sigma}{}^{b}\left({}^{t}\mathbf{D}^{e}\right) \qquad {}^{t}\overset{\nabla}{\Sigma}{}^{c} = \overset{\nabla}{\Sigma}{}^{c}\left({}^{t}\overset{\nabla}{\Lambda}{}^{n}, {}^{t}\Lambda^{n}\right) \tag{12}$$

The reader will notice that the material rate of Cauchy stress $\overset{\nabla}{\Sigma}$ used in Eqs. 12 is in general different to the element rate of Cauchy stress $\dot{\Sigma}$ required in the FE tool. This is because a rigid body rotation is superimposed on the straining of the material and rotates the material frame with reference to the global frame. Hence, what is required is a mapping algorithm between both stress rates. In the present study, the material rate is interpreted as the objective Jaumann rate of stress but it is worth commenting that this choice currently is arbitrary since no experimental data is available to verify the choice of the physically correct objective rate. Furthermore, there is no general agreement yet as to which objective rate is best for describing finite elastic / finite 'plastic' behaviour in any material. Traditional methods are limited usually to viscoplastic and viscoelastic materials with infinite small elastic strains

But the conformational constituent of the present model does not conform to one of these simplified cases. In particular, the additive decomposition of the elastic-plastic deformation rates is incompatible with the classic multiplicative decomposition of the deformation gradients except for small, non-rotational elastic strains. Although this does not impose complications on the bond-stretching constituent since most of the deformation energy is dissipated in shear relaxation and only small elastic strains accumulate, the conformational constituent will comprise large (rubber) elastic strains since the entanglement slippage is arrested at large macroscopic strains. This leads to the well-known strain-stiffening of rubbery polymers complicating the formulation of the physically correct objective rate of stress. A way out is to confine applications of the material model; if we assume that conformational relaxation has a large relaxation time, the 'rubber' stress is defined directly by the macroscopic material deformation via Eq. 8, i.e. ${}^{t+\Delta t}\Sigma^{c}\left({}^{t+\Delta t}\Lambda\right)$, and hence the need for a mapping algorithm in the integration step vanishes.

3.3 Application to FE Analysis

The three-dimensional FE model was used to simulate the deformation behaviour of PET at conditions relevant to film processing. Prior to these simulations, the material parameters of the consitutive model were fitted systematically to experimental data obtained from the FBFT where high strain rate experiments at constant width and equal biaxial strains were conducted. The result obtained is shown in Figure 3, comparing experimental data with FE simulations for constant-width stretching with a nominal strain rate of 1 s^{-1} over the temperature range 80°C - 100°C. In good agreement with the experimental data the FE simulations predict that at higher temperatures the conformational stress relaxes causing the strain-stiffening to be deferred to higher strains.

As an example of its application to situations of practical interest, the FE model has been used to simulate a drawing process where film is stretched longitudinally between two sets of rollers. In the draw direction the load is constant, while in the transverse direction the width is constant. The FE predictions for constant tensile nominal stresses 2.4-8.8MPa at temperatures 80°C and 85°C are presented as creep curves in Figure 4 and 5 showing the expected highly nonlinear viscoelastic response: upon loading the film elongates initially slowly but then goes through a strain-softening region associated with high strain rates before the creep arrests because of strain-stiffening. This plateau in strain is reached earlier when the temperature is raised and/or the load is increased. In the temperature range considered, the temperature effect is very pronounced with an increase in temperature of 5°C causing the time needed to reach this plateau to be 10 times shorter. Furthermore, this plateau draw ratio (nominal strain + 1) shows a minimum for a given temperature when the load is varied. The load associated with the minimum plateau draw ratio induces the smallest strain rate in the transition region, tension loads higher or lower lead to a more rapid strain-stiffening, and furthermore at smaller tension will cause a rapid increase in the plateau draw ratio which can be explained by entanglement slippage reducing the local orientation (network strain) on the experimental time-scale and therefore deferring strain-stiffening to higher strains.

Comparing those FE predictions to experimental data presented by LeBourvellec et al. [9], it is clear that the present FE model predicts the correct shape of the curves with the drawing time being reduced as the temperature and/or load increases. Furthermore, the FE model predicts approximately the correct final draw ratio but reaches it over a shorter time scale than is seen experimentally. There is no doubt the FE analysis uses a simplified material description for the complex behaviour of stress-crystallising PET, and there is at least one obvious reason why the numerical predictions do not accurately represent the 'true' material behaviour. This is that the current constitutive model uses a single relaxation time for each viscous flow processes, i.e. bond-stretching and entanglement slippage, although it is well-known experimentally that distributions of relaxtion times are required. However, this cannot explain the following two discrepancies which indeed require further work. Firstly, although the experimental work corresponds to the same temperature range and what is understood to be an initially amorphous PET, the loads are slightly smaller than those used in the FE analysis, and yet the creep response is very similar in the time domain. Secondly, experiments suggest that the minimum draw ratio at a specific temperature increases with increase in temperature, whereas the FE analysis predicts that it would decrease. Unfortunately, these discrepancies cannot be explained sufficiently without more information on the actual material state of the PET film specimen used by LeBourvellec.

4 CONCLUSIONS

The work reported in this paper has outlined a fundamentally new thermomechanical testing machine - the Flexible Biaxial Film Tester (FBFT). It was designed to study the properties of polymer film under conditions approaching those expierienced in commercial film manufacture which are characterised by large biaxial strains at very high rates of strain and temperature change. The specific features that enable the FBFT to offer a unique choice of experimental sequences of temperature and deformation are (1) the particular design of the symmetric biaxial stretching frame which is driven by two independent high-speed DC servo-motors, (2) the heating method using air jets impinging an the specimen, and (3) the integration of both in a computer-controlled system which is interfaced via a tailored computer software.

Furthermore, we have described a mathematical model for the non-linear viscoelastic deformation behaviour of stress-crystallising polymers providing glassy response at low temperatures and rubber-like behaviour at high temperatures. The latter was characterised by entropic resistance of a rubber-like network relaxed by the viscous entanglement slippage. The FBFT provided experimental data relevant to industrial manufacture. This was used, firstly, to determine the material parameters required for describing the biaxial stretching flows of PET and secondly to validate a three-dimensonal FE material model.

We also have shown that the FE model matches the observed response of PET for constant-load, constant-width stretching flows characteristic of continious film-drawing between rollers. The current constitutive model, however, deviates in the time-scale of response because of the known simplification of using a single relaxation time instead of a spectrum.

ACKNOWLEDGEMENTS

C. Gerlach is indebted to the UK EPSRC and ICI Polyester for financial support. The authors are also grateful to Dr FPE Dunne for helpful discussions and Mr N. Warland for his skilled technical assistance.

REFERENCES

1 Buckley, C P; Jones, D C. *Polymer*, 36 (1995) 3301-3312
2 Adams, AM. DPhil thesis, University of Oxford, 1997
3 Buckley, CP; Jones, DC; Jones, DP. Proceeding of the 8th Int. Conf. on Deformation, Yield and Fracture of Polymers, Cambridge (UK), 1991
4 Hitt, DJ; Gilbert, M. *Polym. Testing*, 13 (1994) 219-237
5 Sweeney, J; Ward, IM. *Polymer*, 36 (1995) 299-308
6 Meissner, J; Hofstettler, J. *Rheologica Acta*, 33 (1994) 1-21
7 Buckley, CP; Jones, DC; Jones, DP. *Polymer*, 37 (1996) 2403-2414
8 Edwards, SF; Vilgis, Th. *Polymer*, 27 (1986) 483-492
9 LeBourvellec, G; Beautemps, J. *J. Appl. Polym. Sci.*, 39 (1990) 319-339

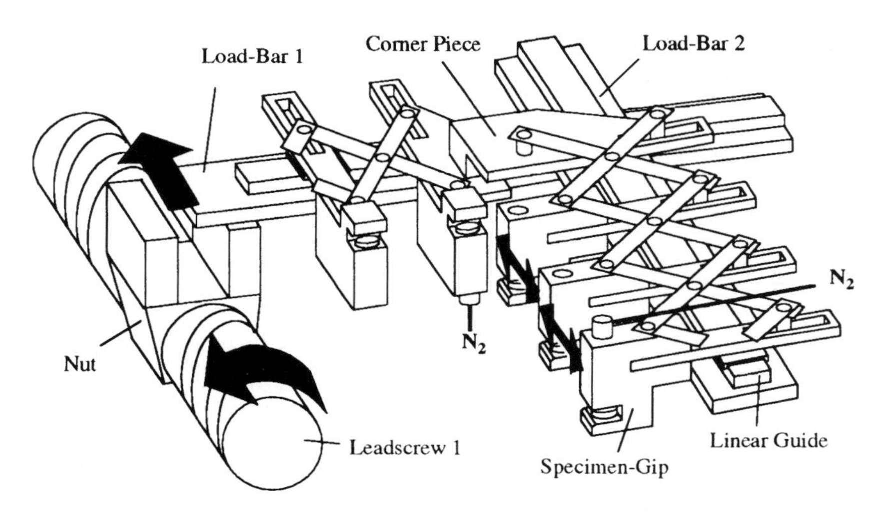

Figure 1 One quarter of the stretching frame of the FBFT showing the interconnected movement of the specimen-grips: the counter-clockwise rotation of the reverse-thread leadscrew causes an outwards movement of load-bar 1 and via the corner-pieces the simultaneous opening of the specimen-grips on load-bar 2.

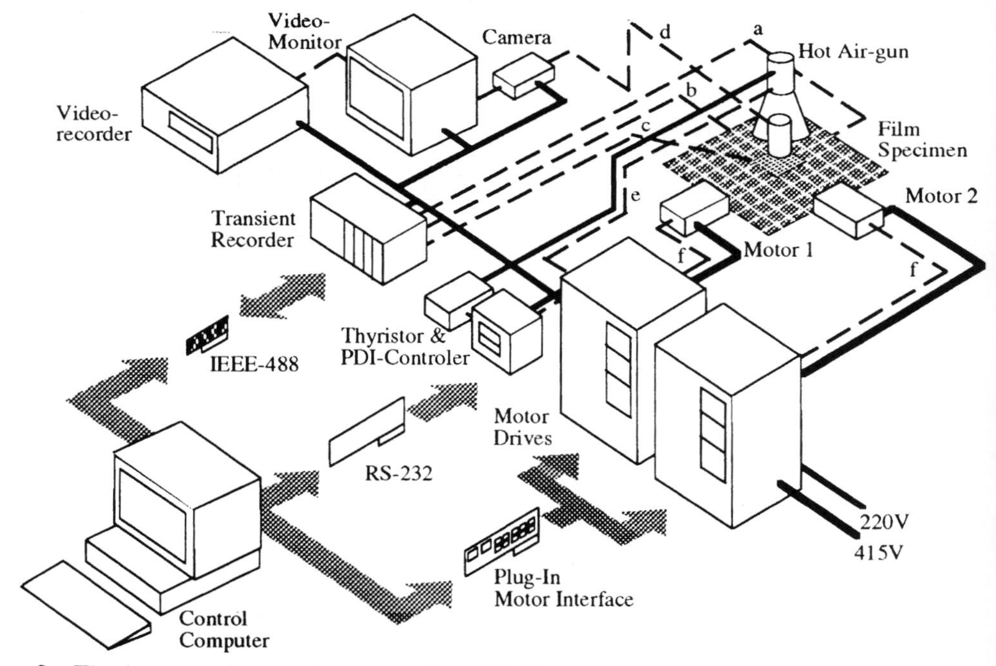

Figure 2 The integrated control system of the FBFT. Data signals are indicated as broken lines: (a) force on axis 1; (b) force on axis 2; (c) free air stream temperature near specimen; (d) video signal; (e) control temperature in high-velocity air stream; (f) encoder control signal of axes 1 and 2.

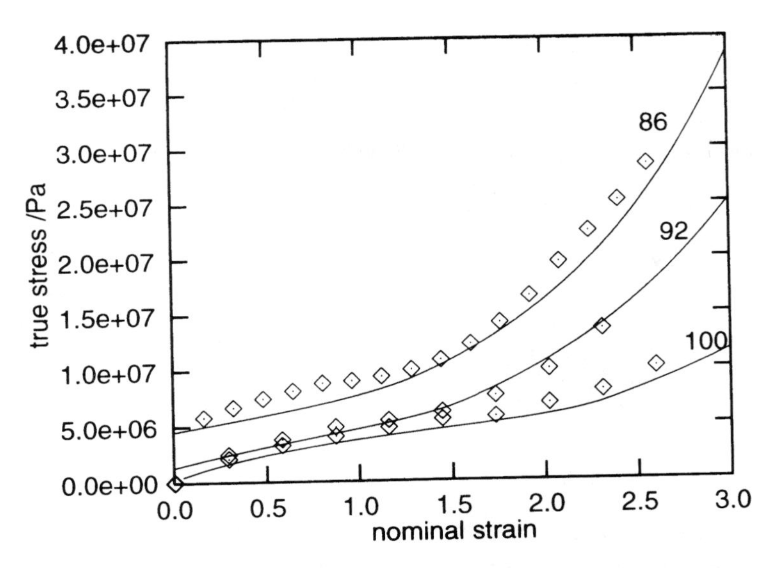

Figure 3 Stress-Strain curves for hot-drawing of PET at constant-width for an extension rate of $1s^{-1}$ and various temperatures (°C). Shown is a comparison of experimentally obtained true stress curves (data points) and numerical predictions (full lines) employing the three-dimensional FE model in ABAQUS/Standard outlined in the main text.

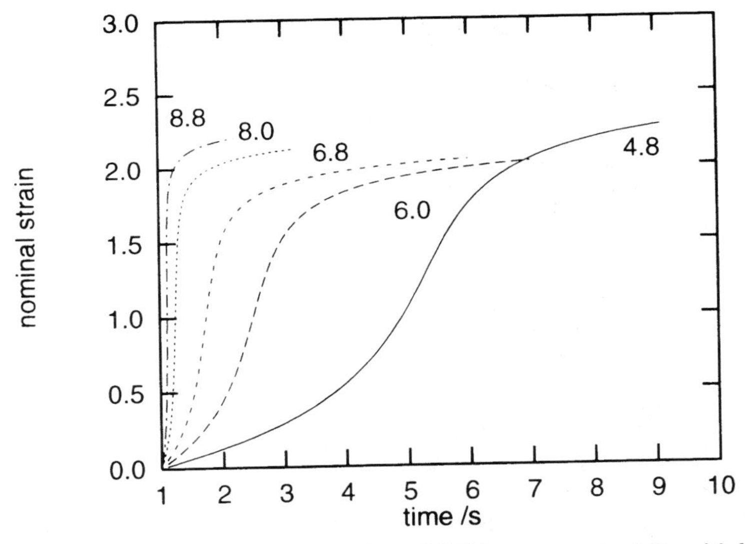

Figure 4 Creep curves for constant load drawing of PET at constant width with load applied from time t =1s, as computed with the FE constitutive model in ABAQUS/Standard for a film temperature of 80°C and various tensile nominal stresses (MPa).

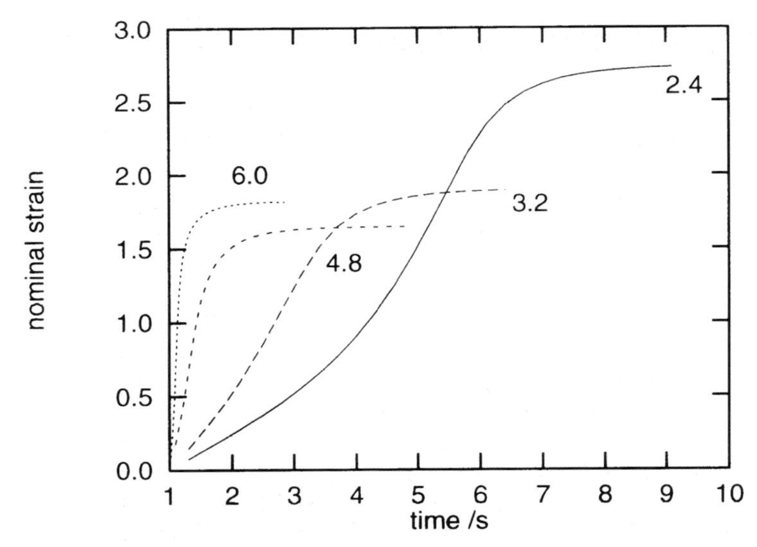

Figure 5 As **Figure 4**, but for a film temperature of 85°C.

In Line and On Line Rheometry of Filled Polyolefins

D Barnwell+, K Martin+, A L Kelly*, M Woodhead*, P D Coates*

+Raychem Ltd, UK,

** IRC in Polymer Science & Technology, Mechanical & Manufacturing Engineering, University of Bradford, Bradford UK*

ABSTRACT

In process (in-line and on-line) rheometry studies are reported for magnesium hydroxide [$Mg(OH)_2$] filled LDPEs and a carbon black filled HDPE. The compounds were processed using a fully instrumented, computer monitored, co-rotating twin screw compounding extruder in the IRC laboratories (University of Bradford). A six sensor in-line rheometer (ILR) slit die and a prototype Rosand on-line rheometer (OLR) were used for the in-process rheometry, and a Rosand (RH7) twin bore capillary rheometer was used for off-line measurement. Each device allowed measurement of shear flow behaviour, the off-line and on-line rheometers also providing die entry pressure loss measurements. In-process rheometry was compared with conventional off-line rheometry for both shear and extensional properties of the LDPEs [0 to 50wt% $Mg(OH)_2$ filler loading] and HDPE [30wt% carbon black]. Good agreement was been observed between the three routes to measurement, for these relatively stable materials.

1. INTRODUCTION

Product quality may be defined in terms of the ability of the product to fulfil a pre-determined role - that is its "fitness for purpose". For synthetic polymer products this is critically determined during two stages, namely initial polymerisation and the post processing of this raw material into a finished product. Polymer melt rheology plays a key role in the assessment of this "fitness for purpose" prior to and during the formation of the completed product. Historically, off-line rheometry in the form of melt index testing has frequently been used to assess the low shear rate flow behaviour of raw polymer feed stocks in the industrial environment [1]. However, an intensifying requirement to manufacture consistent products and also reduce re-worked or scrap material has led to the increasing adoption of alternative forms of rheometry in order to determine more extensively and accurately the processing behaviour of polymers. These alternative forms of rheology include off-line, on-line and in-line rheometry.

Off-line rheometry involves the use of specialised testing equipment to measure polymer melt rheology at a site remote from the processing operation. This has become something of a standard for academic and industrial research due to the accurate and repeatable results which can be achieved using these devices [2,3]. However, this off-line polymer melt characterisation may add a further strain and temperature history to the material and a significant time delay involved in performing the tests, resulting in slower feedback of information to the processing operation.

On-line and in-line rheometers are tools which have the potential to characterise the flow behaviour of polymer melts during processing, thereby reducing time delays involved in performing the tests [4-9]. Also the information provided may be more indicative of the processing behaviour of the polymer and the properties of the finished product since it is derived from melt which has an identical or similar processing history to that constituting the finished product.

The role of in-process rheometry is an expanding one and in view of this the objectives of this work were to compare and contrast the performances of a prototype on-line rheometer and an established laboratory standard off-line capillary rheometer for measurement of the shear and extensional flow behaviour of filled polymer melts. In addition, it was aimed to compare and contrast the performances of these two devices with an in-line rheometer slit die for the measurement of the shear flow behaviour of filled systems. In assessing the measurement capabilities of these rheometers for LDPE compounds, it was proposed to establish their level of sensitivity to $Mg(OH)_2$ filler loading.

2. EXPERIMENTAL DETAILS

A series of in-process rheometry tests were performed at an extruder setpoint temperature of 170C using LDPEs with 0, 5, 15, 30 and 50wt% filler $Mg(OH)_2$ contents, and a 30wt% carbon black filled HDPE was processed at an extruder setpoint temperature of 200C.

A Rosand (RH7) twin bore capillary rheometer (figure 1) was used to perform off-line tests on virgin material at temperatures corresponding to the process setpoints, whilst a prototype gear pump-driven Rosand capillary on-line rheometer (figure 2) was used for the on-line tests. In-line rheometry was performed using a six sensor (pressure and/or temperature) slit die rheometer designed at Bradford University, shown in figure 3. The in-process rheometers were used in conjunction with an instrumented twin screw extrusion compounding line (Betol BTS40, 29:1 L/D ratio). The extruder screws have a trapezoidal flight profile and are of the closely intermeshing co-rotating type.

The shear and extensional flow behaviours of the polymers were investigated through the use of a long capillary die (L/D 16:1) and an orifice die (L/D 0.25:1) fitted to the barrels of the off-line rheometer. The on-line rheometer performs this task using an indexing die block containing different L/D ratio capillary dies, and standard rheometer software then allows a two die test including Bagley [10] correction of rheological data to be performed. This typical OLR melt characterisation took approximately ten minutes, on completion of initial purging. Orifice capillary die pressure drop (P_o) measured using the off-line and in-line rheometers was used to infer and compare the extensional flow behaviours of the polymers [11].

The shear flow behaviour of the polymer melts was determined in-line via the pressure drop along the wide rectangular slit of the extruder die [1.5 x 40 x 120mm slit flow path (aspect ratio of 26.7:1) for LDPE and 3 x 40 x 120mm (aspect ratio 13.3:1) for HDPE] at different feedrates (3 to 15kg/hr). Pressure drop along the melt flow path was measured using a series of five flush mounted pressure transducers (Dynisco PT422A, mercury filled capillary & diaphragm type) which makes the Bagley correction redundant for this rheometer. After an initial settling period, real time data capture from the extruder and ILR was carried out at 1Hz for a period of approximately ten minutes. Bulk melt temperature variation due to changes in the extruder feedrate was estimated using a handheld K type thermocouple (3 mm diameter) immersed in melt emerging from the ILR exit.

3. RESULTS AND DISCUSSION

Results from off-line rheometer tests on the filled LDPEs are shown in figures 4 and 5, which illustrate the extensional and shear flow behaviours of the compounds respectively. Five individual tests were carried out at each filler level and power law curve fits applied to the data. In both figures 4 and 5, the off-line rheometer can be seen to discriminate between different filler levels. However, this is partly due to the sample size of five, and some overlap of the data points from different filler levels can be observed at lower shear rates.

ILR shear flow behaviour of the filled LDPEs which had been processed using the twin screw extruder is indicated in figure 6, which shows shear stress functions for the materials. The closely intermeshing nature of the extruder screws caused significant deformation heating of the polymers, giving rise to bulk melt temperatures in the range 172C to 214C. Shear stress curves in figure 6 are therefore non isothermal, which is particularly evident for the 50wt% filler level flow curve at high shear rates (ie extruder feed rates). With the exception of the flow curve for the 30wt% filler level material, the ILR flow curves show increasing viscosity with increasing filler levels. However, the close proximity of the flow curves for the 0 to 30wt% filler level compounds curves indicates similar flow behaviour.

Results showing the extensional and shear flow behaviours of the filled LDPEs, measured at 190C using the Rosand OLR, are indicated in figures 7 and 8 respectively. OLR tests were carried out at 190C to correspond with the expected mean ILR melt temperature. Individual data points in figures 7 and 8 represent the mean of three OLR tests, and power law curve fits have been applied. The OLR results show similar filler level discrimination in both shear and extensional flow behaviours. Clear discrimination can be seen between flow curves for the unfilled and filled LDPEs, however, only the flow curve for the 50wt% filler level material can be clearly distinguished from those for other filler levels.

P_o trends shown in figure 9 indicate good agreement between the extensional flow properties of the un-filled and 50wt% filled LDPEs measured using the on-line and off-line rheometers. Figures 10 and 11 illustrate the shear flow behaviours of the unfilled and 50wt% filled LDPEs respectively. In both cases good agreement was observed between shear viscosities measured using the off-line and on-line rheometers.

Taking into consideration the significant levels of deformation heating associated with generating a range of shear rates in polymer processed with the ILR, good agreement was also observed between shear flow data measured using this instrument and the other rheometers. Each rheometer tested provided clear discrimination between the shear and extensional flow properties of the compounds with the extreme levels of filler.

Figures 12 and 13 show P_o and shear stress functions measured at 200C for a carbon black filled (30wt%) HDPE using both off-line and on-line rheometers, whilst figure 13 also includes shear stress data from the ILR. Reasonable agreement was observed between shear and extensional flow data from these instruments, the small discrepancies, particularly noticeable in P_o trends, being attributed to differences in OLR and RH7 rheometer melt temperatures.

4. CONCLUDING COMMENTS

- In contrast to the off-line rheometer results, neither the ILR or Rosand OLR showed clear and repeatable discrimination between filler levels in the intermediate range for the LDPE compounds. This may be due to modified flow behaviour of the compounds resulting from their processing on the twin screw extruder, in which case off-line rheometry may give misleading results.

- LDPE and HDPE flow curves generated using the OLR indicate that this instrument is capable of delivering a similar breadth and accuracy of materials processing information to its off-line counterpart, but at closer proximity to the production operation. The gear-pump driven instrument was found to cope well with the heavily filled systems used in this study, over the limited timescales of the experiments.

- Melt characterisation under typical processing conditions using the ILR showed good agreement with other rheometers. However, the melt temperature related changes associated with generation of a range of shear rates limits the area of application for which this rheometer may be suitable for pure rheometry studies: it is clearly of value for control purposes.

- Ongoing in-process rheometry studies include examination of less well-behaved (non-polyolefin) materials.

- A six sensor in line rheometer is currently installed on an extrusion compounding line in the process technology facilities at Raychem Ltd, providing valuable data for use in compounding development operations.

References

1. **Dealy, J M,** (1982) *Rheometers for Molten Plastics: A Practical Guide to Testing and Measurement*, Van Nostrand Reinhold, New York

2. **Cogswell, F N,** (1981) *Polymer Melt Rheology: A Guide for Industrial Practice*, Godwin, London

3. **Braun, P,** (1992) *Quality Control: Capillary Rheometry and MFI Measurement*, Kunstoffe, V 82 Pg 11

4. **Dealy, J. M. and Broadhead, T. O.,** (1993) *Process Rheometers for Molten Plastics: A Survey of Existing Technology,* Polymer Engineering and Science, Vol 33 (23) p1513-1523

5. **Coates, P.D.,** (1995) *In-Line Rheological Measurements for Extrusion Process Control*, Measurement and Control, Vol. 28 p10-16

6. **Kelly, A L, Coates, P D, Dobbie, T and Fleming, D J,** (1996) *On-Line Rheometry: Shear and Extensional Flows*, Plastics, Rubber and Composites; Processing and Applications, Vol.25,No.7, Pg 313-318

7. **Dreiblatt, A,** (1987) *On-Line Quality Control for Improved Compounding*, Plastics Engineering, Oct Pg 31-34

8. Rose, R M, Coates, P D and Wilkinson, B, (1995) *In-process Measurements for Monitoring and Control of Gelation Levels in UPVC Compounding*, Plastics, Rubber and Composites Processing and Applications, Vol. 23 No. 5 Pg 295-303

9. Padmanabhan, M and Bhattacharya, M, (1994) *In-Line Measurement of Rheological Properties of Polymer Melts*, Rheologica Acta, V33 Pg71-87

10. Bagley, E. B., (1957). *End Corrections in Capillary Flow of Polyethylene.* Journal of Applied Physics, Vol 28, p624 - 627.

11. Cogswell F.N., (1972) *Measuring the Extensional Rheology of Polymer Melts*, Trans. Soc. Rheol., Vol 16, p383

Acknowledgements

The support of the EPSRC, the IRC in Polymer Science and Technology, Rosand Precision Ltd, Raychem, Dynisco and the DTI MMP13 programme is gratefully acknowledged.

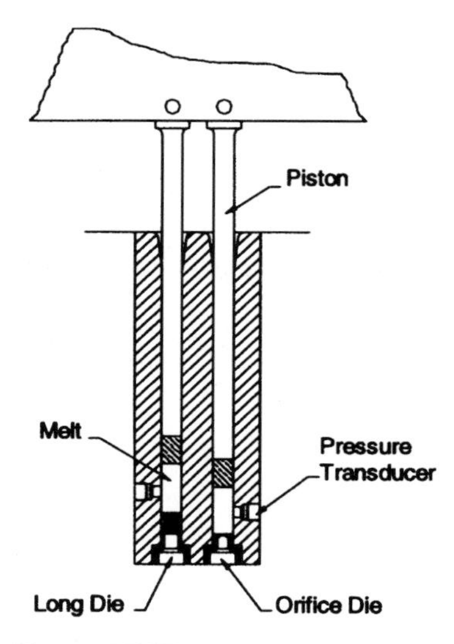

Figure 1 Rosand RH7 twinbore capillary rheometer

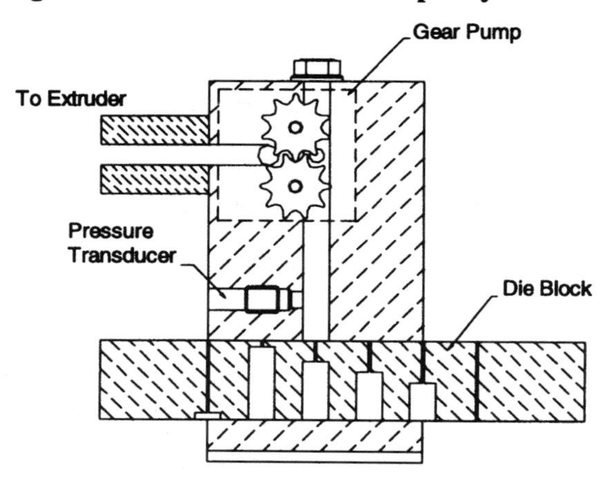

Figure 2 Rosand capillary on-line rheometer

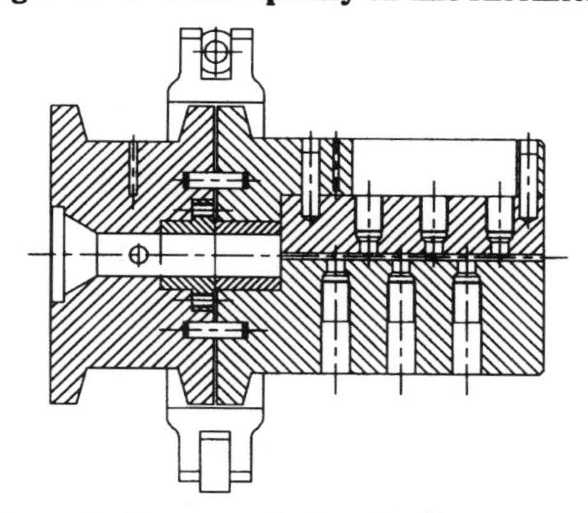

Figure 3 Six sensor in-line slit die rheometer

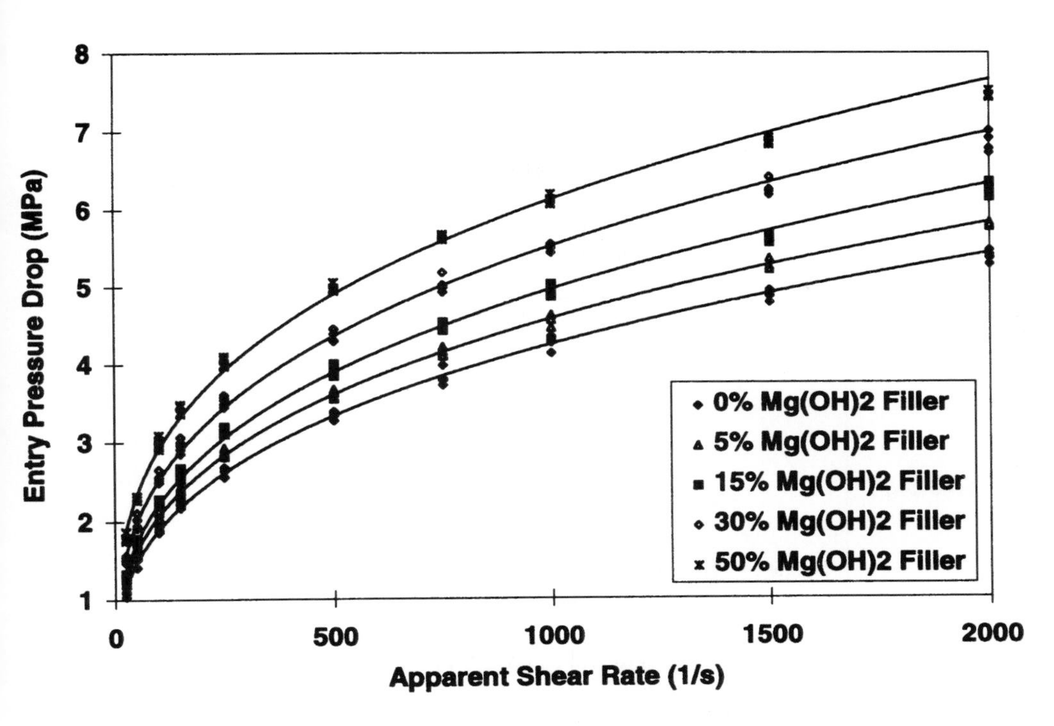

Figure 4 Off-line rheometry power law curve fits of LDPE capillary die entry pressure drop trends at 170C

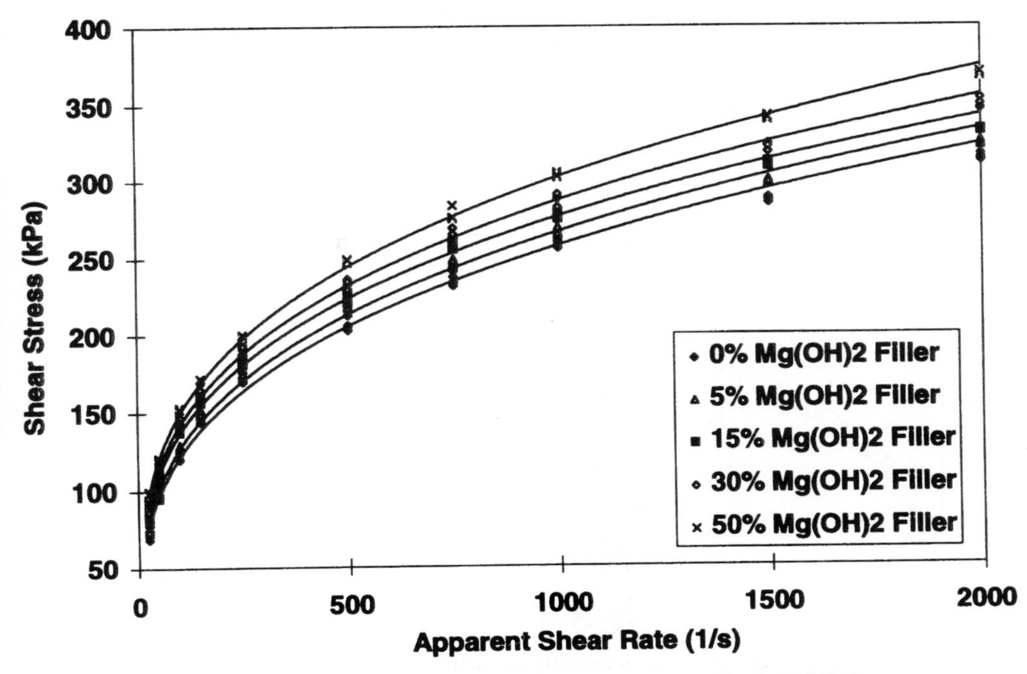

Figure 5 Off-line rheometry power law curve fits of LDPE wall shear stress trends at 170C

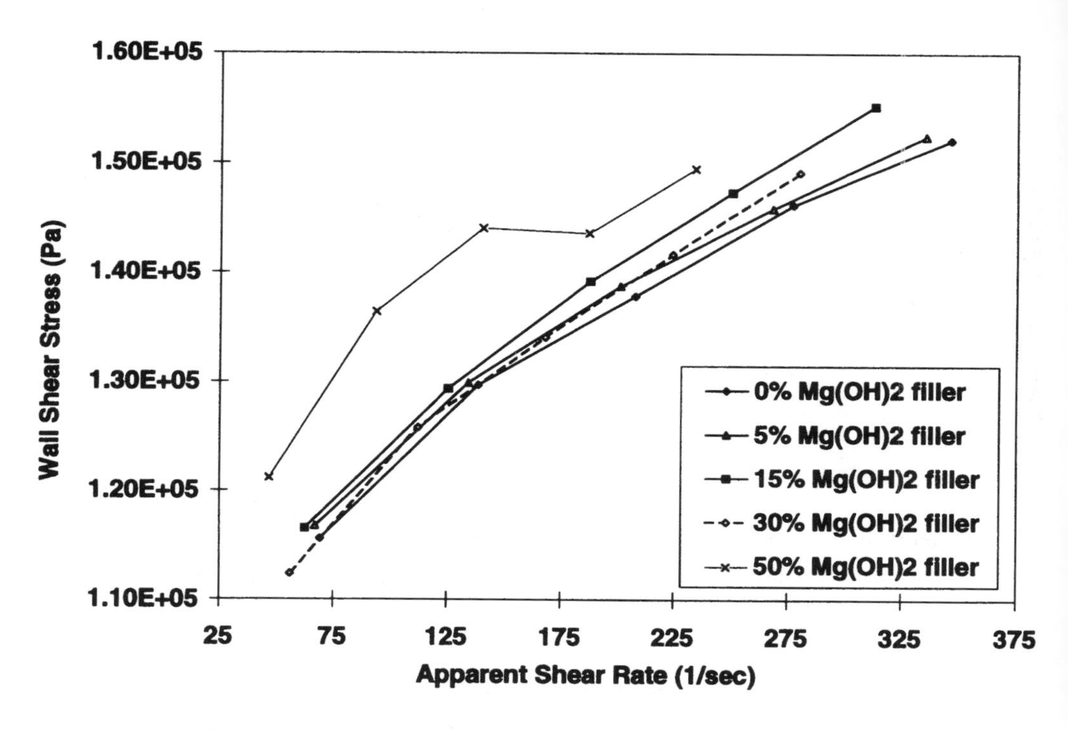

Figure 6 ILR wall shear stress trends for LDPE at 170C (ILR bulk melt temperature range 172C to 214C)

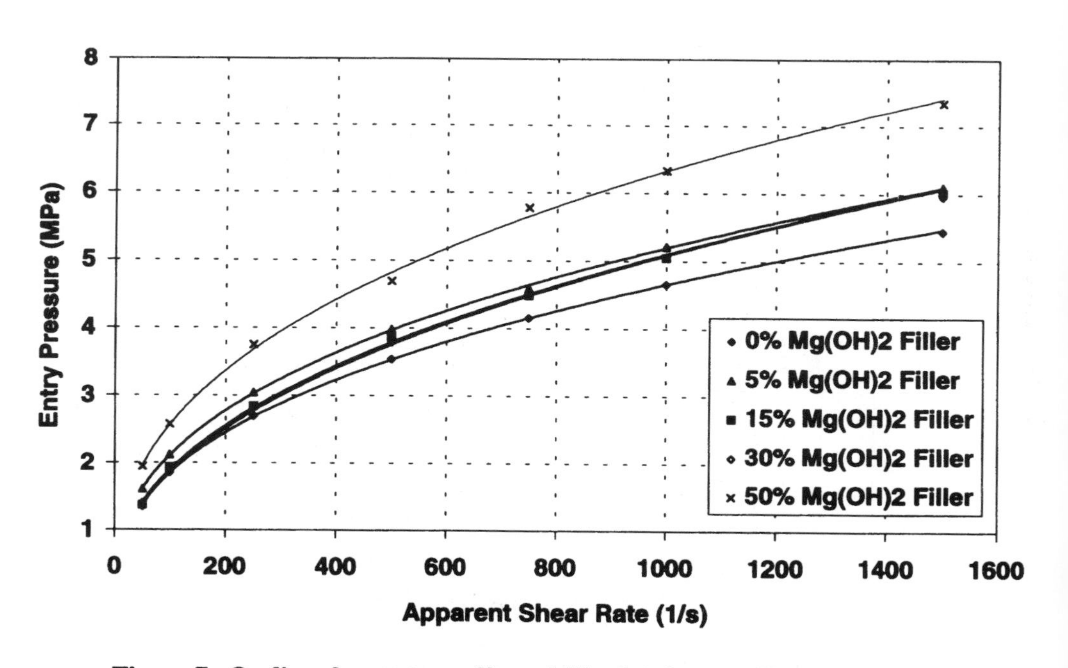

Figure 7 On-line rheometer: effect of filler level on capillary orifice die entry pressure drops for LDPE at 190C

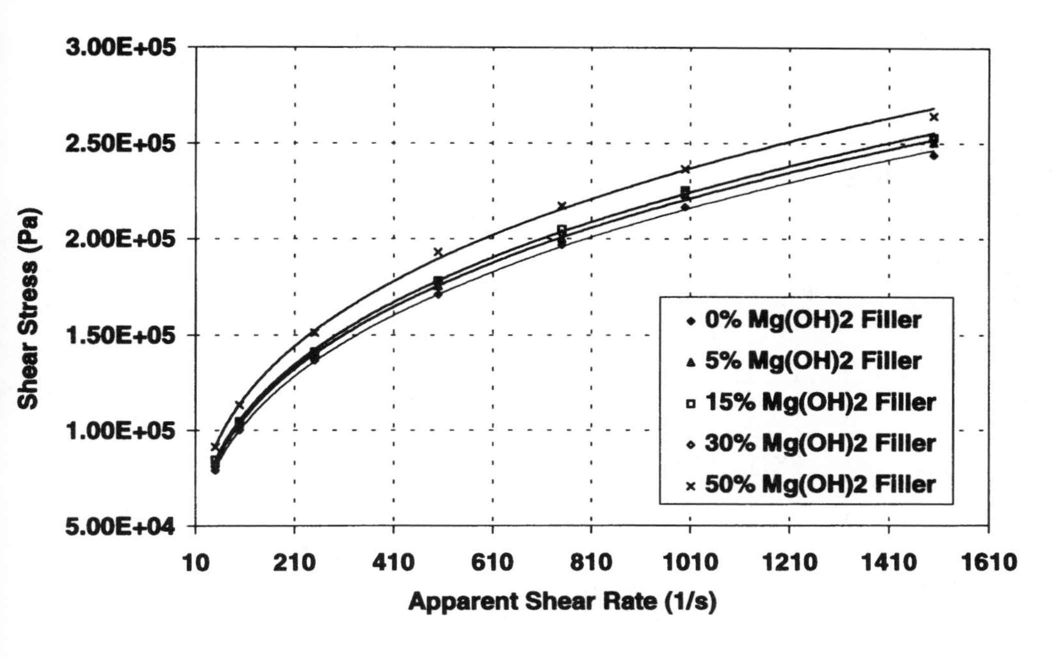

Figure 8 On-line rheometer: effect of filler level on shear stress power law functions for LDPE at 190C

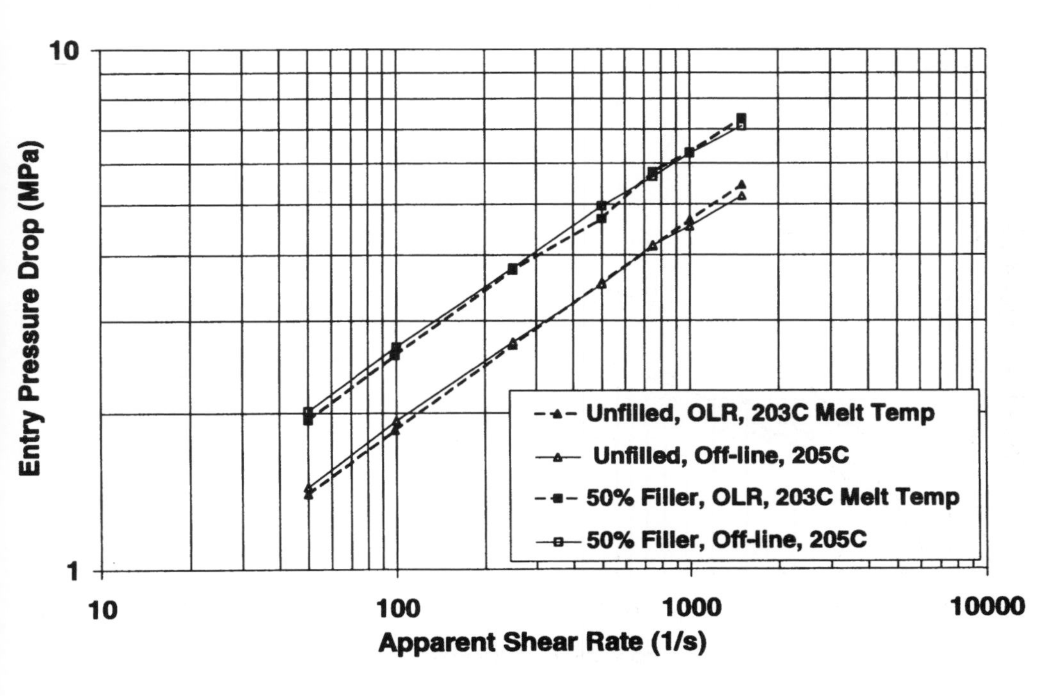

Figure 9 On-line and off-line rheometer comparisons of capillary entry pressure drop trends for LDPE

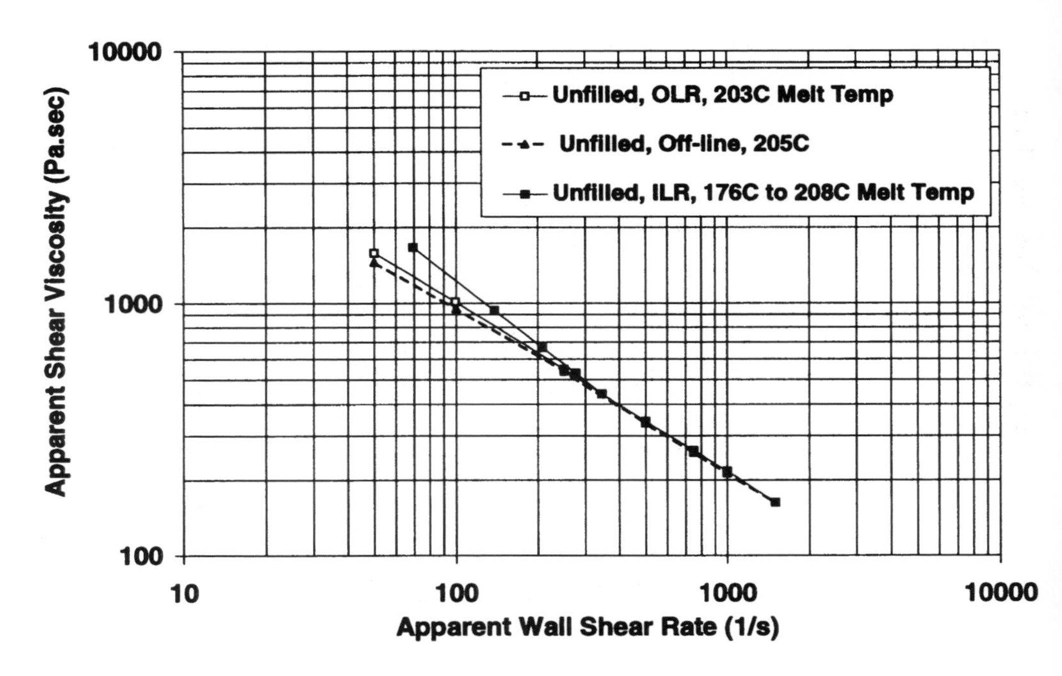

Figure 10 In-line, off-line and on-line rheometer comparison of apparent shear viscosity functions for unfilled LDPE

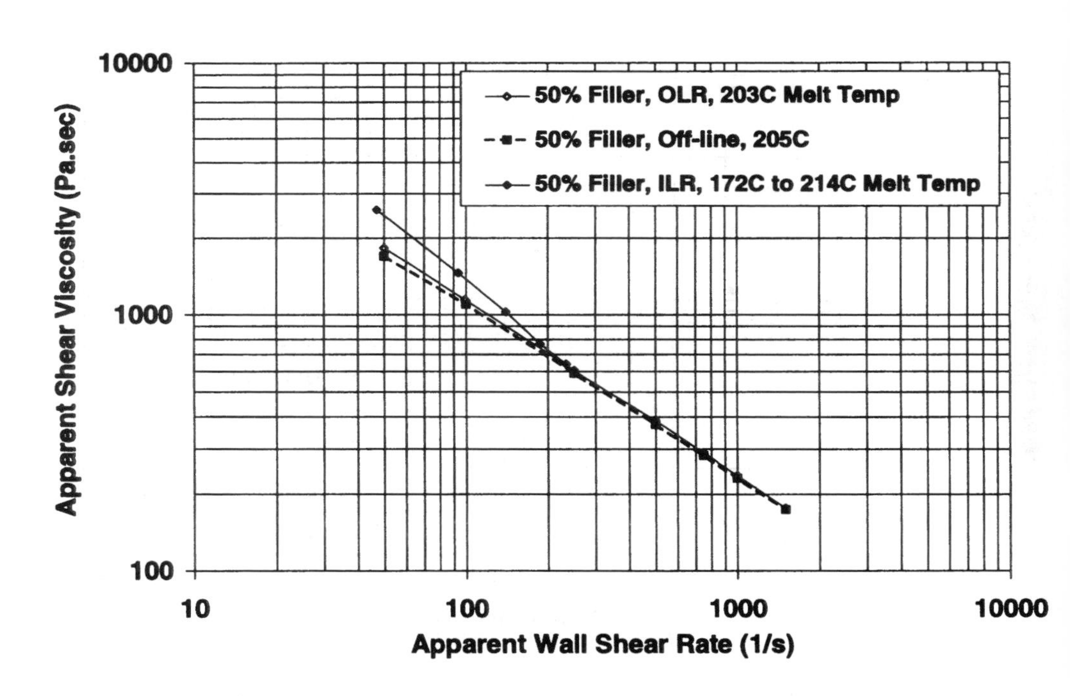

Figure 11 In-line, on-line and off-line rheometer comparison of apparent shear viscosity functions for filled LDPE

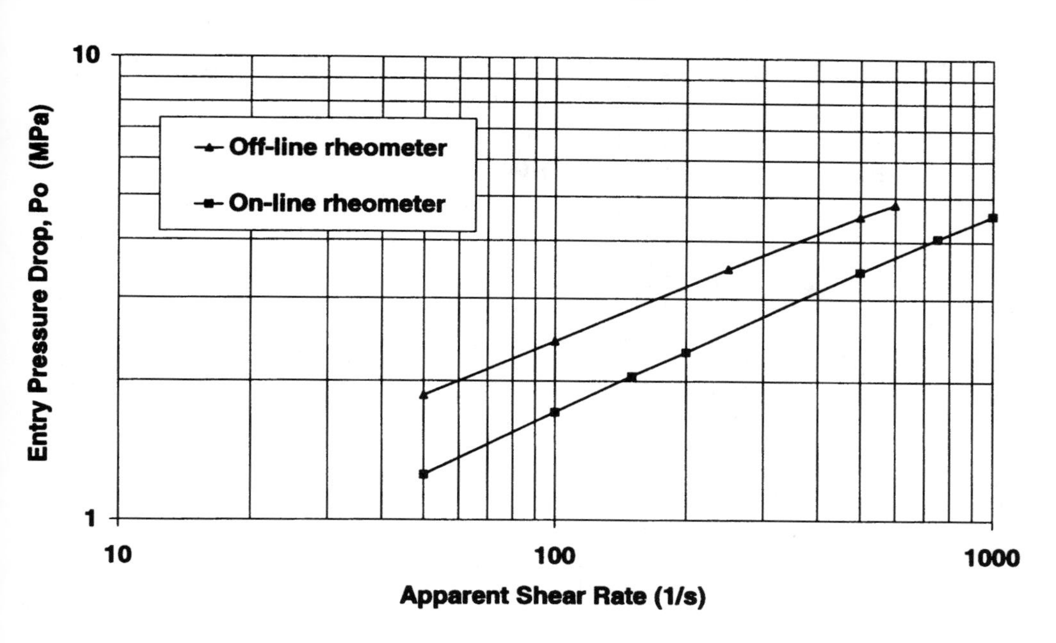

Figure 12 Off-line and on-line rheometer comparison of capillary orifice die entry pressure drops for 30wt% carbon black filled HDPE at 200C

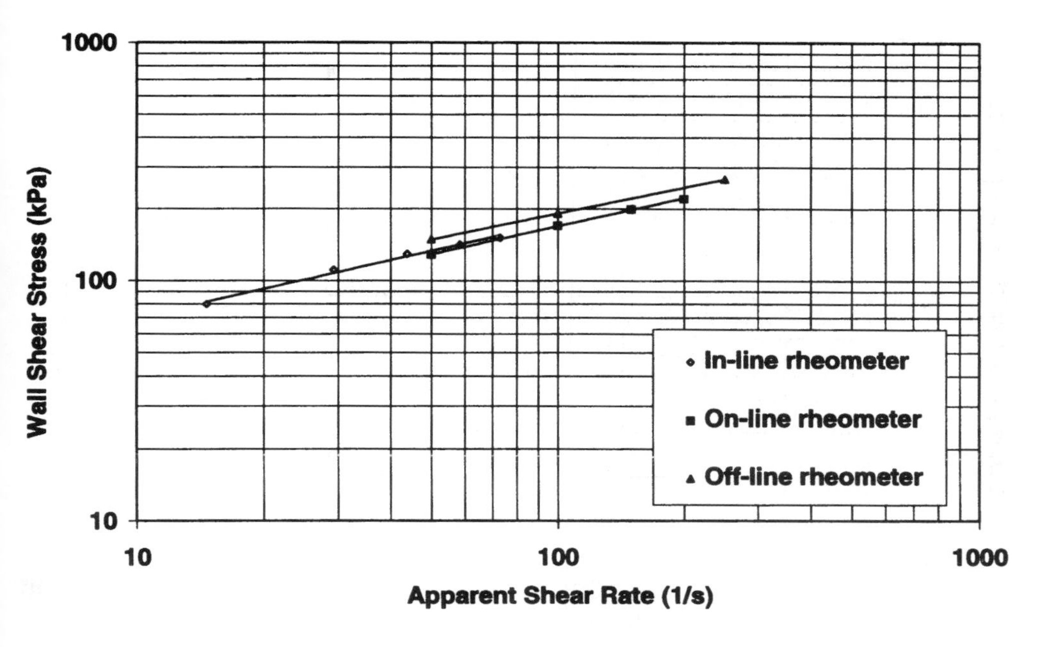

Figure 13 Off-line, on-line and in-line rheometer comparison of wall shear stress functions for 30wt% carbon black filled HDPE at 200C